Edward Kirk

The Cupola Furnance

A practical Treatise on the Construction and Management of Foundry Cupolas

Edward Kirk

The Cupola Furnance
A practical Treatise on the Construction and Management of Foundry Cupolas

ISBN/EAN: 9783337106300

Printed in Europe, USA, Canada, Australia, Japan

Cover: Foto ©ninafisch / pixelio.de

More available books at **www.hansebooks.com**

THE CUPOLA FURNACE:

A PRACTICAL TREATISE ON THE

CONSTRUCTION AND MANAGEMENT

OF

FOUNDRY CUPOLAS.

COMPRISING

THE BEST METHODS OF CONSTRUCTION AND MANAGEMENT OF CUPOLAS; DIFFERENT SHAPED CUPOLAS; HEIGHT OF CUPOLAS; PLACING TUYERES; SHAPES OF TUYERES; LINING; SPARK CATCHING DEVICES; BLOWERS; BLAST PIPES; AIR GAUGES; CHARGING; DIRECTIONS FOR THE MELTING OF IRON, TIN-PLATE SCRAP, AND OTHER METALS IN CUPOLAS; EXPERIMENTS IN MELTING; WHAT A CUPOLA WILL MELT; ETC.

BY

EDWARD KIRK,

PRACTICAL MOULDER AND MELTER, CONSULTING EXPERT IN MELTING.
Author of The Founding of Metals, and of Numerous Papers on Cupola Practice.

ILLUSTRATED BY SEVENTY-EIGHT ENGRAVINGS.

PHILADELPHIA:
HENRY CAREY BAIRD & CO.,
INDUSTRIAL PUBLISHERS, BOOKSELLERS AND IMPORTER
810 WALNUT STREET.
LONDON:
E. & F. N. SPON, LTD.,
125 STRAND.
1899.

PREFACE.

ALTHOUGH it is now more than twenty years since the publication of the author's volume, "The Founding of Metals," that book is still in demand. The reception which has been tendered to it, together with the urgent requests of many foundrymen for a more modern work on cupolas, has encouraged him to prepare the treatise now offered to the public.

This volume is designed to supply a want long felt, for a work on melting that would give practical details regarding the construction and management of cupolas, and the melting of iron for foundry work. Several valuable books have been written on the moulding and founding of iron and steel, but in these books, as in the foundry, but comparatively little attention is given to the cupola; and foundrymen and melters have been left to grope very much in the dark, and to rely solely on their own experience, in the construction and management of their cupolas.

This condition of things, and the great importance of the subject, have combined to induce the author in the present work to endeavor to develop the most vital principles connected with the cupola, its construction and its use, together with the best practice of this country. In order to accomplish these ends, he has supplemented his almost lifelong experience by consulting the latest works on foundry practice, and by visiting leading and thoroughly up-to-date foundries in different parts of the United States. He therefore trusts that this book will prove to be a useful aid to foundrymen, whether owners or workers, both here and abroad.

As is the general custom of the publishers, they have caused the work to be supplied with a copious table of contents, as well as a very full index, which will render reference to any subject in it both prompt and easy.

EDWARD KIRK.

PHILADELPHIA, *February 22, 1899.*

CONTENTS.

CHAPTER I.

THE CUPOLA FURNACE.

PAGE

Advantages of the cupola furnace for foundry work; Quantity of fuel required for melting iron in various kinds of furnaces; Attempts to decrease the amount of fuel consumed in the cupola by utilizing the waste heat 1
Description of the cupola furnace; Forms of cupolas; Sizes of cupolas; Foundation of a large cupola 2
Advantage of iron supports over brick-work; Height of the bottom of the cupola; Pit beneath the cupola 3
Bottom plate; Bottom doors; Support of the doors; Various devices for holding the doors in place; Construction of the casings . . . 4
Stack casing; Construction of the stack; Tuyere holes 5
Location of the charging door and its construction; Lining of the casing and materials used for it 6
The scaffold and its location; Construction of the scaffold; Size of the scaffold 7

CHAPTER II.

CONSTRUCTING A CUPOLA.

Proper location of a cupola; The scaffold 8
Conveyance of coal or coke to the scaffold; Cupola foundation and its construction 9
Prevention of uneven settling; Brick walls for the support of a cupola; Best supports for a cupola 10
Height of cupola bottom; Provision for the removal of the dump; Bottom doors. 11
Casing; Material for the casing or shell of the modern cupola and stack; Strain upon the casing due to expansion and shrinkage and its prevention; Contraction of the stack; Prevention of sparks . . . 12
What constitutes the height of cupola; Utilization of the waste heat; Table giving the approximate height and size of door for cupolas of different diameters 13
Melting capacity of a cupola; Charging door; Air chamber . . . 14

(v)

vi   CONTENTS.

PAGE

Construction of the air chamber when placed inside the casing and when placed upon the outside of the shell; Air capacity of the air chamber; Admittance of the blast to the air chamber 15
Location and arrangement of the air chamber when the tuyeres are placed high; Tap-hole 16
Arrangement when two tap-holes are required; The spout and its construction; Tapping of slag 17
Location of the slag-holes; Tuyeres; Number of tuyeres for small cupolas 18
Best shape of tuyere for a small cupola; Number of tuyeres for large cupolas; Size of combined tuyere area; Tuyere boxes or casings; Height at which tuyeres are placed in cupolas 19
Objection to high tuyeres; Two or more rows of tuyeres; Arrangement of a large number of rows 20
Area of the rows; Increase in the melting capacity with two or three rows of tuyeres; Lining; Material for lining the casing . . . 21
Grouting or mortar for laying up a lining; Manner of laying the brick; Thickness of cupola linings; Stack lining 22
Arrangement of brackets 23
Preference by many of angle iron to brackets; Mode of putting in angle iron; Reduction of the lining by burning out 24
Settling of the lining; Mode of reducing the size and weight of the bottom doors and preventing the casing from rusting off at the bottom; Prevention of the absorption of moisture into the lining . . . 25
Illustration of the triangular-shaped tuyere in position in the lining; Form of bottom plates. Fire-proof scaffolds; Exposure of the scaffold and its supports to fire 26
Devices to make scaffolds fire-proof; Novel plan of construction of a scaffold and cupola house in Detroit, Mich 27
Best and safest scaffolds; Cupola scaffold in the foundry of Gould & Eberhardt, Newark, N. J., and of the Straight Line Engine Co., Syracuse, N. Y. 28

CHAPTER III.

CUPOLA TUYERES.

Modes of supplying the cupola furnace with air; Admission of the air through tuyeres or tuyere holes; Former and present melting capacity of a cupola; Epidemics of tuyere invention 30
The round tuyere; Arrangement of round tuyeres in the old-fashioned cast-iron stave cupolas 31
Oval tuyere; Expanded tuyere 32
Doherty tuyere 33
Sheet blast tuyere; Horizontal slot tuyere; Mackenzie tuyere. . . 34
Blakeney tuyere 35

CONTENTS.

	PAGE
Horizontal and vertical slot tuyere	36
Reversed T tuyere or vertical and horizontal slot tuyere; Vertical slot tuyere; Truesdale reducing tuyere.	37
Lawrence reducing tuyere	38
Triangular tuyere; Results in melting with this tuyere obtained by the Magee Furnace Co., Boston, Mass.	39
Water tuyere	40
Colliau tuyere; Whiting tuyere	41
Chenney tuyeres; The double tuyere; Mode of placing the tuyeres in Ireland's cupola; Claims for the double tuyere	42
Consumption of fuel in a double tuyere cupola; Three rows of tuyeres; Cupola constructed by Abendroth Bros., Port Chester, N. Y.	43
Object in placing tuyeres in a cupola; Production of heat by consuming the escaping gases from the combustion of fuel	44
Greiner tuyere; Adjustable tuyeres	45
Cupola of the Pennsylvania Diamond Drill and Manufacturing Co., Birdsboro, Pa.; Bottom tuyere	46
Mode of covering the mouth of a bottom tuyere	47
Bottom tuyere patented in this country by B. H. Hibler; Thomas D. West on the bottom tuyere	48
Size of tuyeres; Size of the combined tuyere area of a cupola.	49
Height of tuyere; Great difference of opinion on this subject; Experiments with tuyeres at various distances above the sand bottom	50
Experiment to soften hard iron by bringing the molten metal in contact with charcoal in the bottom of a cupola; Reason given in favor of high tuyeres	51
Tuyeres in stove foundry cupolas; Location of tuyeres in smaller cupolas.	52
Tuyeres in machine and jobbing foundry cupolas, and in cupolas for heavy work	53
Number of tuyeres; Objection to the use of only one tuyere; Two tuyeres sufficient for the largest cupola in use	54
Arrangement of a double row of tuyeres; Shape of tuyeres; Tuyeres to improve the quality of iron	55
Tuyere boxes	56

CHAPTER IV.

CUPOLA MANAGEMENT.

Necessity of learning the peculiarities in the working of a cupola; A cupola cannot be run by any given rule or set of rules; Drying the lining	58
Drying a backing or filling between the casing and lining; Putting up the doors; Devices for raising the doors into place	59
Support of double doors; Sizes of props to support the bottom	60

CONTENTS.

	PAGE
Ring attachment to the prop; Superstition of older melters regarding the prop; Dropping the doors; Modes of releasing the prop	61
Sand bottom; Sand employed for this purpose; Objection to clay sands and other sands; Sand which makes the very best kind of bottom	62
Wetting the bottom sand; Bringing the sand into the cupola	63
Cause of leakage in the sand bottom	64
Boiling of iron due to a wet bottom; Pitch or slope of the bottom	65
Effect of a high pitch; Change in the action of the iron at the spout by the pitch of the bottom; How the bottom should be made	66
Slope of the bottom in cupolas with two tap holes; Spout; Spout lining material	67
Effect of the use of too much clay or of too much sand in the lining; Mode of making up the spout lining	68
Building up the sides of the lining; Place of the greatest strain upon the spout lining	69
Proper shape of the spout lining; Cause of pools of iron forming in the spout; Removal of slag from the spout	70
Front; Material used for putting in the front; Mode of putting in the front	71
Effect of working the front material too wet; Troubles due to poor front material	72
Sizes of tap hole; Locating the holes	73
Slag hole	74
Slag hole front; Chilling of slag in the tap hole	75
Lighting up; Mode of placing the wood and shavings in the cupola; Putting in the bed fuel	76
Effect of carelessness in arranging the wood and lighting up	77
The bed; The melting point in a cupola; The melting zone and determination of its exact location; Necessity of discovering the melting point in order to do good melting	77
To find the melting point	78
Cause of trouble in melting after a cupola has been newly lined; Fuel required for a bed in cupolas of different diameters	79
Charging; Old way of loading or putting the fuel and iron into a cupola; Modern way of stocking a cupola; Correct theory of melting iron in a cupola	80
Practical working of a cupola upon this theory; Effect of too heavy charges of iron, and of too heavy charges of fuel; Variation in the weight of the first charge of iron	81
Variations in the per cent. of iron to fuel; Placing the charges	82
Mode of placing the pieces of pig or other iron; Distribution of the charge of fuel; Charging additional iron	83
Poor melting may be due to bad charging; Improper and proper charging of a cupola; Charging flux; On what the quantity of flux depends	84
Blast; The old and the modern ways of giving blast to the cupola; Blast phenomena	85

CONTENTS.

	PAGE
Melting; When melting begins in a cupola; Difference in opinion as to the time for charging the iron before the blast is put on	86
Best time for putting on the blast; Chilling and hardening of the first iron; Running of a heat without stopping in; Mode of reducing the size of the tap-hole	87
No advantage in holding molten iron in a cupola to keep it hot; Proper management of hand-ladle work; Indication of how the cupola is melting by the flow of iron from the tap-hole	88
Poking the tuyeres	89
Fuel; Amount of fuel required in theory and in practice; Necessity of keeping an accurate account of the amount of iron melted	90
Chief object of melting iron in a cupola; The old story of "not enough blast;" Necessity of an even volume of blast.	91
Tapping bars; Shapes and sizes of tapping bars.	92
Steel bar for cutting away the bod before tapping; Bod sticks; Combination stick	93
Objection to the combination stick; Number of bod sticks for each cupola; Bod material; Importance of the material of which the bod is composed.	94
Mixture for a good bod; Bod for small cupolas; Qualities of a good bod.	95
Tapping and stopping in; Mode of making the bod; How to make the tap	96
Mode of stopping in; Difficulties in stopping in; Devices to assist in stopping in	97
The skill of the melter seen at the tap-hole; Uneven melting is the fault of the melter; Dumping.	98
Removal of the small props; Bridging over of small cupolas above the tuyeres and mode of removing the bridge	99
Various methods of handling the dump; Removing the dump and various devices for this purpose	100
Breaking up the dump and picking it over; Different ways of recovering the iron from the dump; Chipping out	101
Theory of some melters to prevent iron from running into the tuyeres; Objection to this theory; Cupola picks	102
Daubing; Materials used	103
Soaking fire clay; Amount of sand required for mixing with the clay; No advantage in using a poor cheap daubing	104
Putting on the daubing; Shaping the lining; Object of applying daubing to a lining; Mode of making new linings	105
Chipping off cinder and slag that adhere to the lining over the tuyeres; Not necessary or advisable to fill in the lining at the melting zone; Objections to sudden offsets or projections	106
Thickness of the daubing; Sectional view of a cupola, illustrating effect of excessive daubing	107
Shaping the lining of the boshed cupola; Special directions required for	

	PAGE
shaping and keeping up the lining of the patent and odd-shaped cupolas	109
Relining and repairing; Thickness of the lining; Location of the greatest wear on the lining; Destruction of the lining at and below the tuyeres; Length of time a cupola lining will last; Burning away of the lining	110
Thickness of lining required to protect the casing; Repairing the lining at the melting zone	111
Repairing a lining with split brick; Mode of making a split brick	112

CHAPTER V.

EXPERIMENTS IN MELTING.

Various opinions formerly held by foundrymen as to the point in a cupola at which the melting of iron actually took place; Different ways of charging or loading a cupola	113
Experiments to learn definitely at what point iron is really melted in a cupola; Construction of an experimental cupola	114
Results of the first experiment	115
Arrangement of the bars of iron for the next experiment	116
What was learned from this experiment; Arrangement of the bars and cupola for the next heat	117
High pressure of blast may be almost wholly due to the size of the tuyeres; Arrangement of the bars for the next heat and the result of this experiment	118
Arrangement of the cupola for the next heat and result of the experiment	119
Fuel used in the experiment; Reasons why iron is not melted in a cupola by the blast and flame of the fuel; Melting zone in a cupola	120
Fire under the tuyeres	121
Low tuyeres; Results of an experiment with low tuyeres.	122
Melting zone; What determines the location of a melting zone in a cupola; Lowering and raising the melting zone	123
Change in the location and depth of the melting zone	124
Experiments to learn the depth of the melting zone in practical melting.	125
Charges used in the experiments	126
Charges with which the most melting was done in these experiments; Necessity of passing the blast through a certain amount of heated fuel before a melting zone was formed in a cupola	127
Cause of iron melted high in a cupola being made dull	128
Development of the melting zone above the tuyeres; Experiments with a cupola with the tuyeres placed near the top; Failure of this plan	129
Melting with coal; Softening hard iron; Experiments in softening iron by passing it in molten state through charcoal in its descent from the melting zone to the bottom of the cupola	130

CONTENTS. xi

PAGE

Time for charging; Difference of opinion among foundrymen as to the proper time for charging; Experiments to ascertain the proper time for charging and putting on the blast after charging 132
Devices for raising the bottom doors 133
Device for raising heavy doors 134

CHAPTER VI.

FLUXING OF IRON IN CUPOLAS.

Definition of a flux; Use of fluxes; Materials used as fluxes; Purpose of the use of limestone in the production of pig-iron 135
On what the making of a brittle cinder in a cupola by the use of limestone depends; Limestone in large quantities 136
Variation in the quantity of limestone required to produce a fluid slag; Weight of slag drawn from a cupola 137
Constituents of the slag; Effect of flux upon iron 138
The action of fluxes on lining 139
How to slag a cupola; Cause of trouble in slagging; General method of charging the limestone; The slag hole 140
Slag in the bottom of a cupola; Importance of the time for drawing the slag; Does it pay to slag a cupola? Estimate of the cost of slagging . 141
Shells; Use of oyster, clam and other shells; Cause of the crackling noise of shells when the heat first strikes them; Marble spalls . . 142
Experiments with mineral and chemical materials with the view of making a cheap malleable iron; Reasons why iron is often ruined as a foundry iron by improper melting and fluxing; Increase in the per cent. of iron lost in melting by improper melting and fluxing . . 143
Effect of silicon on iron; Per cent. of silicon an iron may contain; Use at the present time of high silicon cheap southern iron . . . 144
Heavy breakage due to the use of high silicon irons; Effect of carbon upon cast iron; Removal of free carbon from iron 145
Fluor spar, and its use as a flux 146
Cleaning iron by boiling; Poling molten iron 147

CHAPTER VII.

DIFFERENT STYLES OF CUPOLAS.

Old Style Cupolas.

Old style cupola in general use throughout this country many years ago, described and illustrated 149
Practice of casting with the use of the old-style cupola . . . 151
The reservoir cupola, described and illustrated 152
Stationary bottom cupola; Old style English cupola, described and illustrated 154
Expanding cupola, described and illustrated 155

	PAGE
Ireland's cupola, described and illustrated	157
Ireland's center blast cupola, described and illustrated	159
Voisin's cupola, described and illustrated	161
Woodward's steam-jet cupola, described and illustrated	163
Objection to this style of cupola; Tank or reservoir cupola, described and illustrated	167
Production of soft iron by putting a quantity of charcoal on the sand bottom; Use of tanks in England	169
Mackenzie cupola, described and illustrated	170
Management of the Mackenzie cupola	172
The Herbertz cupola, described and illustrated; Movable hearth of this cupola	173
Advantages of the application of a steam jet to create draft in the cupola	175
Test heats with the Herbertz cupola	176
Composition of the escaping gases from the Herbertz cupola	177
Explanation why less carbon and silica are eliminated from the iron in the Herbertz cupola than in the ordinary blown cupola	178
Working of the Herbertz cupola at Elizabethport, N. J.	179
The hearth in the cupola used at Elizabethport.	180
Process of melting; Results of test-heats at Elizabethport	181
Herbertz's cupola used for melting steel, described and illustrated; Melting bronze.	182
Pevie cupola, described and illustrated	184
Object of Mr. Pevie in constructing a cupola; Stewart's cupola, described and illustrated.	186
Rapid melting of this cupola; The Greiner patent economical cupola, described and illustrated.	188
What the novelty of this invention consists of.	189
Principle of the workings of the Greiner cupola illustrated	190
Mr. Greiner's conclusions.	191
Points in favor of the Greiner cupola; Colliau patent hot-blast cupola, described and illustrated.	192
History and description of the cupola and results obtained in melting	193
Claims made for the Colliau furnace.	194
The Whiting cupola, described and illustrated.	196
Jumbo cupola, described and illustrated	198
Charge table for the Jumbo cupola	200
The Crandall improved cupola with Johnson patent center blast tuyere, described and illustrated.	202
Mode of applying the air to this cupola	203
Claims made for the Crandall cupola; Blakeney cupola, described and illustrated	204
Advantages of this cupola.	205

CHAPTER VIII.

ART IN MELTING.

The art of melting iron in a cupola but little understood by many foundrymen and foundry foremen; Troubles experienced in melting. 206
No chance work in nature or in art; Necessity of understading the construction and mode of operation of a cupola to do good melting . . 207
Location and arrangement of the tuyeres; Preparation of the cupola for a heat; Lighting up 208
Melting iron in a cupola a simple process; Things to be learned and practiced; Necessity of a close study of all the materials used in melting 209
What should be the aim of every moulder; Advisability of the foreman of a foundry being the melter; Duties of the melter 210

CHAPTER IX.

SCALES AND THEIR USE.

Necessity of an accurate scale upon the scaffold; Size of scale required; What the melting of iron in a cupola, when reduced to an art, consists in 211
Division of the fuel and iron into charges; The theory of melting not understood by many foundrymen; Incorrect methods of calculating the charges of iron and fuel 212
Use of old, worn-out scales condemned 213

CHAPTER X.

THE CUPOLA ACCOUNTS.

Value of cupola records; Manner of keeping the accounts . . . 214
Cupola report of Abendroth Bros., Port Chester, N. Y. . . . 215
Cupola report of Byram & Co., Iron Works, Detroit, Mich. . . 216
Daily report of Foundry Department, Lebanon Stove Works . . 217
Melting sheet of Syracuse Stove Works 218
Report of castings in ―――― Shop 219
Cupola slate for charging and cupola report 220
Blanks for reports and records and mode of making them out; Report on a slate; Correctness essential to the value of a cupola account . 221

CHAPTER XI.

PIG MOULD FOR OVER IRON.

Saving in iron and labor by the use of cast iron pig moulds for collecting over iron 222

CHAPTER XII.

WHAT A CUPOLA WILL MELT.

Chief use of the cupola furnace; Employment of the cupola furnace for other purposes than melting iron 223
Quantity of iron that can be melted in a cupola; Number of hours a cupola may be run; Size and weight of a piece of cast-iron that can be melted in a cupola; Charging large pieces of iron at the foundry of Pratt & Whitney Co., Hartford, Conn.; Melting of cannon and other heavy government scrap at the Lobdell Car Wheel Co., Wilmington, Del. 224

CHAPTER XIII.

MELTING TIN PLATE SCRAP IN A CUPOLA.

Various ways of preparing the scrap for charging; Attempts to recover the tin deposited upon the iron; Quality of the molten metal from the scrap 225
Susceptibility of the molten metal to the effect of moisture; Uses of the metal; Production of a gray metal from the scrap; Tests to learn the amount of metal lost in melting the scrap 226
Action of tin as a flux when melted with iron; Unsuitability of galvanized sheet-iron scrap for melting in a cupola; Doctoring the metal from tin-plate-scrap; Process of melting tin-plate-scrap . . . 227
Fluxing tin-plate-scrap; Construction of a cupola expressly for melting tin-plate-scrap 228
Cost of melting tin-plate-scrap and profit in the business. . . . 229

CHAPTER XIV.

COST OF MELTING.

Unreliability of melting accounts as generally kept; Objection to measuring fuel in baskets 230
Results of an accurate account of the melting in a foundry in New Jersey; Cupola book; Proper method of figuring the cost of melting per ton 231

CHAPTER XV.

EXAMPLES OF BAD MELTING.

Necessity of giving causes of poor melting; Trouble with the cupolas at the stove foundry of Perry & Co., Sing Sing, N. Y. 233
Cause of the trouble; Sectional views of lining out of shape . . . 236
Remedy of the troubles 240
Bad melting at a West Troy Stove Works; Visit at the foundry of Daniel E. Paris & Co., West Troy, N. Y.; Inspection of the foundry with a view of locating the trouble 242

CONTENTS. xv

PAGE

Trouble due to the use of too much fuel 243
Experiment of running the cupola with less fuel; Objection of the
  melter to the experiment 244
Result of the experiment; Heats with a still further reduction of fuel . 246
Cause of bad melting in this foundry 247
Warming up a cupola; Visit to the plant of the Providence Locomotive
  Works; Trouble with the cupola 248
Cause of the poor melting due to the bed being burned too much . . 249
Remedy of the trouble; Bad melting, caused by wood and coal; Cause
  of poor melting in one of the leading novelty foundries in Philadelphia 250
Poor melting in a Cincinnati cupola; Sectional elevation showing the
  condition of the cupola 251
Uneven burning of the bed; Reason for the necessity of dumping a
  cupola at the foundry of Perry & Co. . . , ... 253

CHAPTER XVI.

Melters.

Respect due to the practical and scientific melter; Unfortunate position
  of a poor melter; Interference with a good melter frequently the cause
  of poor melting 254
Necessity of furnishing proper tools for chipping out, and making up
  the cupola; What should be the aim of every melter 255
Interest of every foundryman to keep his melter posted 256

CHAPTER XVII.

Explosion of Molten Iron.

Conditions under which molten iron is explosive; Explosions caused by
  a wet spout or a wet bod 257
Cause of sparks; Various causes of the explosion of molten iron; Explosion due to thrusting a piece of cold, wet or rusted iron into molten
  iron 258
Explosion of molten iron when poured into a damp or rusted chill-
  mould or a wet sand-mould; Accident in the foundry of Wm. McGilvery & Co., Sharon, Pa. 259
Explosion of molten iron when poured into mud or brought into contact
  with wet rusted scrap; Accident in the foundry of James Marsh, Lewisburg, Pa.; Accident at the foundry of North Bros., Philadelphia, Pa. 260
Explosion at the foundry of the Skinner Engine Co., Erie, Pa., and at
  the foundry of the Buffalo School Furniture Co., Buffalo, N. Y.. . 261
Prevention of explosions 262

CHAPTER XVIII.

SPARK-CATCHING DEVICES FOR CUPOLAS.

	PAGE
Spark-catcher in old-style cupolas	263
Spark-catching device for modern cupolas.	264
Return flue cupola spark-catcher, designed by John O. Keefe.	266
Other spark-catching devices	268
The best spark-catching device; Cause of sparks being thrown from a cupola; Prevention of sparks being carried out of the stack; Enlarged stack	269

CHAPTER XIX.

HOT BLAST CUPOLAS.

Hot blast cupolas constructed by Jagger, Treadwell & Perry . . . 271
Cupola at the foundry of Ransom & Co., Albany, N. Y.; Arrangement with exhaust pipes 273
Heating the blast for a cupola; Waste heat from a cupola; Plans for utilizing the heat escaping from a cupola 274
Cupolas at the Carnegie Steel Works, Homestead, Pa.; Prevention of the escape of heat in low cupolas 275

CHAPTER XX.

TAKING OFF THE BLAST DURING A HEAT—BANKING A CUPOLA—BLAST PIPES, BLAST GATES.

Explosions in blast pipes, blast gauges, blast in melting; Length of time the blast can be taken off a cupola; Management of a cupola from which the blast is taken off 276
Banking a cupola; Communication from Mr. Knoeppel, Foundry Superintendent, Buffalo Forge Co., Buffalo, N. Y., on banking a cupola . 277
Blast pipes; Importance of the construction and arrangement of blast pipes; Underground blast pipes 279
Objection to underground blast-pipes; Materials used in the construction of blast-pipes; Galvanized iron pipes 280
Table prepared by the Buffalo Forge Co , Buffalo, N. Y., as a guide for increasing the diameter of pipes in proportion to the length; Diameter of blast-pipes; Friction in pipes 281
Frequent cause of a blower being condemned as being insufficient. . 282
Table showing the necessary increase in diameter for the different lengths 283
Connection of blast pipes with cupola ; Combined area of the branch pipes 284
Table of diameter and area of pipes 285
Connecting blast pipes direct with tuyeres; Perfect connection of air chambers; Poor arrangement of pipes in a "perfect cupola" . . 286

CONTENTS. xvii

PAGE

Mode of connecting a belt-air chamber with the tuyeres; Best way of connecting blast pipes with cupola tuyeres 288
Blower placed near cupola 289
Poor melting often caused by long blast pipes; Perfect manner of connecting the main pipe with an air chamber; Blast gates; Advantage of the employment of the blast gate 290
Explosions in blast pipes; Prevention of such explosions; Blast gauges; Variety of gauges 292
What an air-gauge to be of any value in melting must indicate . . 293
Blast in melting; Means for supplying the required amount of air to the cupola; Machines for supplying the blast; Relative merits of a positive and non-positive blast 294
Amount of air required for combustion of the fuel in melting a ton of iron 295
Theory of melting in the old cupolas with small tuyeres; Necessity of discarding the small tuyeres 296
Points to be remembered in placing tuyeres in a cupola; Best tuyere for large cupolas; Size of the largest cupolas in which air can be forced to the center from side tuyeres 297
Cupolas of the Carnegie Steel Works, Homestead, Pa.; Experiments with a center blast tuyere 298
Claims for the center blast 299

CHAPTER XXI.

BLOWERS.

Placing a blower; Convenient way of placing a blower near a cupola . 300
Fan blowers; Buffalo steel pressure blower; Claims for this blower . 301
Blower on adjustable bed, and on bed combined with countershaft . 303
Blower on adjustable bed, combined with double upright engine . . 305
Buffalo electric blower built in "B" and steel pressure types . . 306
Buffalo blower for cupola furnaces in iron foundries . . . 308
Table of speeds and capacities as applied to cupolas; Smith's Dixie fan blower 309
Forced blast pressure blowers; The Mackenzie blower . . . 311
Description of and claim for this blower 312
Sizes of the Mackenzie blower; Construction and operation of the machine 313
Directions for setting up blower; The Green patented positive pressure blower; Claims for this blower 314
Complete impeller, illustrated and described; Directions for setting up blower 316
Efficiency of blower 317
Power; Rule for estimating the approximate amount of power required to displace a given amount of air at a given pressure . . . 318

CONTENTS.

	PAGE
Standard foundry blowers driven by steam, dimensions in inches	319
Speed of foundry blowers	320
Connersville cycloidal blower	321
Special value of combining the epi- and hypo-cycloids to form the contact surfaces of impellers.	322
Advantages claimed for the Connersville cycloidal blower	323
Table of numbers, capacities, etc., of the cycloidal blowers; What is meant by ordinary speed; Vertical blower and engine on same bed plate.	326
Blower and electric motor	327
Garden City positive blast blowers	328
Root's rotary positive pressure blower	329
Claims for this blower	330

CHAPTER XXII.

CUPOLAS AND CUPOLA PRACTICE UP TO DATE.

Kinds of furnaces employed in the melting of iron for foundry work; Coke the almost universal fuel for foundry work; Quantity of fuel required in the different kinds of furnaces 332
Rule for charging a cupola; Height or distance the tuyeres should be placed above the sand bottom. 333
Function of the fuel placed below the tuyeres; Fallacy of the claim that it is necessary to have tuyeres placed high to collect and keep iron hot for a large casting; Objection to low charging doors . . . 334
Highest cupolas in use in this country; What is required for a cupola to do economical melting; Determination of the top of the melting zone. 335
Tests to ascertain the amount of fuel required in the charges and the amount of iron that can be melted upon each charge; General consumption of too great an amount of fuel 336
On what the per cent. of fuel required in melting depends; Necessity of reducing the melting to a system; Advisability of keeping an accurate cupola record 337

CHAPTER XXIII.

CUPOLA SCRAPS.

Brief paragraphs illustrating important principles; Terms used in different sections of the country to indicate the melting of iron in a cupola. 339
Best practical results for melting for general foundry work . . 343
Remarks by Mr. C. A. Treat; Difficulty experienced by a foundryman in obtaining reliable cupola reports for publication . . . 344

NOTE.

Paxson-Colliau Cupola.	345
Index	347

CHAPTER I.

THE CUPOLA FURNACE.

THE cupola furnace has many advantages over any other kind of furnace for foundry work.

It melts iron with less fuel and more cheaply than any other furnace, can be run intermittently without any great damage from expansion and contraction in heating and cooling. Large or small quantities of iron may be melted in the same furnace with very little difference in the per cent. of fuel consumed, and the furnace can readily be put in and out of blast. Consequently in all cases where the strength of the metal is not of primary importance, the cupola is to be preferred for foundry work.

In the reverberatory furnace from ten to twenty cwt. of fuel is required to melt one ton of iron.

In the pot furnace one ton of coke is consumed in melting a ton of cast iron, and two and a half tons in melting a ton of steel.

In the blast furnace twenty to twenty-five cwt. of coke is consumed in the production of a ton of pig iron.

In the cupola furnace a ton of iron is melted with from 172 to 224 lbs. of coke.

It will thus be seen that in the cupola furnace we have the minimum consumption of fuel in melting a ton of iron, although the amount consumed is still three or four times that theoretically required to do the work.

Many attempts have been made to decrease even this small amount of fuel consumed in the cupola, by utilizing the waste heat passing off from the top for heating the blast. But the cupola being only intermittently at work has rendered all such attempts futile.

The cupola furnace is a vertical furnace consisting of a hollow casing or shell, lined with fire-brick or other refractory material, resting vertically upon a cast iron bottom plate, having an opening in the centre equal to the inside diameter of the lining and corresponding in shape to the shape of the furnace. This opening is closed with iron doors covered with sand when the furnace is in blast. Two or more openings are provided near the bottom of the furnace for the admission of air by draught or forced blast. A small opening, on a level with the bottom plate, is arranged for drawing off the molten metal from the furnace. An opening, known as the charging door, is made in the side of the casing at the top of the furnace for feeding it with fuel and iron, and a stack or chimney is constructed above the charging door for carrying off the escaping smoke, heat and gases.

Cupolas have been constructed cylindrical, elliptical, square and oblong in shape, and they have been encased in stone, brick, cast iron and wrought iron casings. From one to a hundred or more tuyeres have been placed in a cupola, and the stationary and drop bottoms have been used. At the present time cupolas are constructed almost entirely in a cylindrical or elliptical form, and the casing is made of wrought iron or steel boiler plate. The stack casing is made of the same material and is extended up to a sufficient height to give draught for lighting up, and to carry off the escaping heat and gases. The drop bottom has been almost universally adopted, at least in this country.

Cupolas are constructed of various sizes, to suit the requirements of the foundry they are to supply with molten metal. Those of large size are, when charged with iron and fuel, of immense weight, and require a very solid foundation to support them. The foundation is generally made of solid stone work up to the level of the foundry floor; upon this is placed brick work laid in cement, or cast iron columns or posts, for the support of the iron bottom and cupola. In all cases where the cupola is set at sufficient height from the floor to admit of the use of the iron supports they are to be preferred to brick-work, as they admit of more freedom in removing the dump and repairing the lining. The columns or posts are placed at a sufficient distance apart to permit the drop doors to swing free between them. This arrangement removes the liability to breaking the doors by striking the cupola supports in falling, and admits of their being put back out of the way when removing the dump.

The height the bottom of the cupola is placed above the moulding floor depends upon the size of the ladles to be filled, and varies from fourteen inches to five feet. If placed too high for the sized ladle used, considerable iron is lost by sparks and drops separating from the stream in falling a long distance, and the stream is more difficult to catch in the ladles. For hand ladle work it is better to place the cupola a little higher than fourteen inches, and rest the ladle upon a hollow oblong pedestal eight or ten inches high, and open at both ends, than to set it upon the floor. The ladle can then be moved back or forward to catch the stream, and iron spilled in changing ladles falls inside the pedestal, and is prevented from flying when it strikes the hard floor, and is collected in one mass inside the pedestal. This arrangement reduces the liability of burning the men about the feet and renders it easier to lift the full ladle.

If a cupola is set very low, it is then necessary to make an excavation or pit beneath it to permit of the removal of the dump, and repairing of the lining. This pit is made as wide as it conveniently can be, and of a length equal to two or three times the diameter of the cupola. The distance from the bottom plate to the bottom of the pit should not be less than three feet. The bottom of the pit is lined with a hard quality of fire-brick set on edge, and the floor sloped from the edges to the centre, and from the end under the cupola outward, so that any molten iron falling within the dump will flow from under the cupola, and thus facilitate its removal. In the centre of the pit under the cupola a block of stone or a heavy block of iron is securely placed, upon which to rest the prop for the support of the iron bottom doors.

The bottom plate is made of cast iron, and must be of sufficient thickness and properly flanged or ribbed to prevent breaking. If broken when in place, it can not be removed, and it is then almost impossible to securely bolt it so as to hold it in place. The plate must be firmly placed upon the iron supports or brick work, so that no uneven strain will be put upon it by the weight of the cupola and stack.

The bottom doors are made in one piece or in two or more sections. For large cupolas they are generally made in two or four sections to facilitate raising them into place. Bottom doors are made of cast or wrought iron. Those made of cast iron are, when in place, the stiffest and firmest. Those made of wrought iron are the lightest and easiest to handle, but are also more liable to be warped by heat in the dump, and to spring when in place. The door, or doors, whether made of cast or wrought iron, have wide flanges to overlap the bottom plate and each other when in place, to prevent the sand, when dry, running out through cracks and making holes in the sand bottom. The doors are supported in place by a stout iron or wooden prop; and when the doors are light, or sprung, one or more additional props are put in for safety. Numerous bolts and latches have been devised for holding the doors in place, but they have all been abandoned in favor of the prop, which is the safest. Sliding doors, or plates, have been arranged upon rollers to slide into place under the cupola from the sides, and be withdrawn by a ratchet or windlass to dump the cupola. They admit of easy manipulation; but in case of leakage of molten iron through the sand bottom, they are sometimes burnt fast to the bottom plate and cannot be withdrawn, and for this reason the sliding door is seldom used.

The casings are made of cast or wrought iron plate. When made of cast iron they are cast in staves, which are put in place on the iron bottom and bound together by wrought iron bands;

these bands being shrunk on. Or they are cast in cylindrical sections, which are placed one on top of another, and bolted together by the flanges. This kind of casing generally cracks from expansion and shrinkage in a short time, and is the poorest kind of casing. With the cast iron casing a brick stack, constructed upon a cast iron plate supported by four iron columns, is generally used. The wrought iron casing is more generally employed at the present time than that of cast iron. It is made of boiler plate, securely riveted together with one or two rows of rivets; but one row of rivets, and those three inches apart, is generally found to be sufficient, as the strain upon the casing, when properly lined, is not very great.

The stack casing is generally made of the same material as that of the cupola, and is a continuation of the cupola casing; the two generally being made in one piece.

The stack is made the same size as the cupola, or is contracted or enlarged according to the requirements or fancy of the foundryman. A contracted stack gives a good draught, but throws out a great many sparks at the top. An enlarged stack gives a poor draught, unless it is very high, but throws out very few sparks at the top. As sparks are very objectionable in some localities, and not in others, different sized stacks are used. When surrounded by high buildings or hills, the stack must be made of sufficient height to give the necessary draught for lighting up in all kinds of weather, and they vary in height from a few feet above the foundry roof to twenty or thirty feet. Bands of angle iron are sometimes riveted to the inside of the cupola and stack casing to support the lining, and admit of sections being taken out and replaced without removing the entire lining.

The casing and lining are perforated with two or more tuyere holes near the bottom, for the admission of air by draught or forced blast. These tuyeres, when supplied with a forced blast, are connected with the blower by branch pipes to each tuyere, or are supplied from an air chamber riveted to the cupola casing either on the outside or inside. The air chamber is made three or four times the area of the blast pipe, and is supplied from the blast pipe connecting it with the blower. An opening is made through the casing and lining, just above the bottom plate, for drawing the molten iron from the cupola, and a short spout is provided for running it into the ladles. Another small opening is sometimes made, just under the lower level of the tuyeres, for tapping or drawing off the slag from the cupola. This opening is never used except when a large amount of iron is melted, and the cupola is kept in blast for a number of hours.

An opening for feeding the furnace, known as the charging door, is placed in the cupola at a height varying from six to twenty feet above the bottom plate, according to the diameter of the cupola. This opening is sometimes provided with a cast iron frame or casing on the inside to protect the lining around the door when putting in the fuel and iron. A door frame is placed upon the outside, upon which are cast lugs for a swinging door, or grooves for a sliding door. The door for closing the charging aperture may consist of a cast or wrought iron frame filled with fire-brick, or be made of boiler plate with a deep flange all around for holding fire-brick or other refractory material. The sliding door consists of an iron frame filled in with fire-brick, and is hung by the top, and moved up and down with a lever or balance weights. This door is moved up and down in grooves cast upon the door frames, which grooves frequently get warped by the heat, and hold the door fast. The hinge or swing door, with plenty of room for expansion and shrinkage, is the door generally used.

The casing is lined from the bottom plate to the top of the stack with a refractory material. A soft refractory fire-brick, laid up with a grout composed of fire-clay and sand, is used for lining in localities where such material can be obtained. In localities where fire-brick can not be procured, soapstone from quarries or the bottoms of small creeks, is laid up with a refractory clay. Some grades of sandstone or other refractory substances are also employed for lining. Native refractory materials are seldom homogeneous, and those which have been ground and moulded, or pressed into blocks, make the best linings. The thickness of the lining varies in large and small cupolas. Those in the large cupolas are from six to nine inches, and in small cupolas from four to six inches.

The cupola charging aperture is placed at too great a height from the floor to admit of the cupola being charged or loaded rom the floor, and a scaffold or platform is erected from which to charge it. The scaffold is generally placed in the rear of the cupola, so as to be out of the way when removing the molten iron in crane ladles. But for hand ladle work it is placed at any point most convenient for getting up the stock, and the charging aperture placed in the cupola at any point most convenient for charging. For very large cupolas the scaffold is frequently constructed to extend all around the cupola, and a charging aperture is placed in the cupola on each side, so that it may be more rapidly charged. The scaffold is constructed of wood or iron frame work, or is supported by a brick wall. The floor is placed level with the bottom of the charging aperture, or is placed from one to two feet below it. The scaffold should be made large enough to place a weighing scale in front of the charging door, to hold iron and fuel for several heats, and have plenty of room for handling the stock when stocking the scaffold and charging the cupola. Nine-tenths of the scaffolds are too small for the work to be done on them, and the cupola men work to a great disadvantage when handling the stock. Much of the bad melting done in foundries can be traced directly to the lack of room on the scaffold for properly charging the cupola.

Having thus given a general outline description of the cupola furnace, we shall in the next chapter describe in detail where to locate a cupola and how to construct it.

CHAPTER II.

CONSTRUCTING A CUPOLA.

WHEN about to construct a cupola to melt iron for foundry work, the first thing to be decided on is the proper location. In deciding this a number of points are to be taken into consideration, the two most important of which are the getting of the stock to the cupola and the taking away of the molten iron. It should be borne in mind that there is more material to be taken to a cupola than is to be taken away from it. For this reason the cupola should be located as convenient to the stock as possible. It must also be borne in mind that the object in constructing a cupola is to obtain fluid molten iron for the work to be cast, and if the cupola is located at so great a distance from the moulding floors that the molten metal loses its fluidity before it can be poured into the mould, the cupola fails in the purpose for which it was constructed.

If the work to be cast is heavy and the greater part of the molten metal is handled by traveling or swinging cranes, the small work may be placed near the cupola and the cupola located at one side or end of the foundry near the yard. But if the work is all light hand-ladle or small bull-ladle work, the cupola should be located near the centre of the moulding-room so that the molten iron may be rapidly conveyed to the moulds in all parts of the room.

SCAFFOLD.

It is often found difficult, owing to the shape of the moulding-room and location of the yard, to place the cupola convenient for getting the stock to it and the molten iron away from it. When this is the case, means must be provided for getting the stock to the cupola and the cupola located at a point from which the molten metal can be rapidly conveyed to the moulds. At the present low price of wrought iron and steel, a fire-proof cupola scaffold can be constructed at a very moderate cost, and the difficulty of locating the cupola convenient to the yard may be overcome by constructing a scaffold of a sufficient size to take the place of a yard for iron and fuel. The scaffold may be constructed under the foundry roof and made of proper size to hold one or two cars of coal or coke, a hundred tons of pig and scrap iron and all the necessary material for a cupola. The space under the scaffold can be utilized as moulding floors for light work or for core benches, core oven, ladle oven, sand-bins, etc. The cupola and its supplies are then under roof, and there is no trouble from cupola men staying at home in bad weather, as is often the case when the cupola and stock are out of doors.

When this arrangement is adopted, an endless chain or bucket elevator should be constructed to convey the coal or coke to the scaffold as fast as it is shoveled from the truck or car. Another elevator should be provided for pig and scrap iron, and as the iron is thrown from the car it is broken and at once placed upon the scaffold convenient for melting. This arrangement saves considerable expense for labor in the rehandling of iron and fuel, and also prevents the loss of a large amount of iron and fuel annually tramped into the mud in the yard and lost. The saving in labor and stock in a short time will pay the extra expense incurred in constructing this kind of scaffold.

CUPOLA FOUNDATION.

Too much care cannot be taken in putting in a cupola foundation, for the weight of a cupola and stack, when lined with fire-brick to the top, amounts to many tons, and when loaded with fuel and iron for a heat to many tons more. If the foundation gives way and the cast iron cupola bottom is broken by uneven settling, the cupola is rendered practically worthless, for it is impossible to replace the bottom with a new one without taking out the entire lining, which entails much expense, and it is almost impossible to bolt or brace the plate so as to keep it in place.

The foundation should be built of solid stone work, and if a good foundation cannot be had, piles must be driven. Separate stone piers should never be built for each column or post, for they frequently settle unevenly and crack the bottom plate. Uneven settling and breaking of the bottom are, to a large extent, prevented by placing a heavy cast iron ring upon the stone work upon which to set the cupola supports. This ring should be placed several inches below the floor to prevent it being warped and broken by the heat in the dump.

When brick walls are constructed for the support of a cupola, the bottom plate is made square, from two to three inches thick and strongly ribbed or supported by railroad iron between the walls, to prevent breaking. The walls do not admit of sufficient freedom in removing the dump and for this reason are, at the present time, seldom used in the construction of cupolas. Even when the cupola is set so low that a pit is required for the removal of the dump, the iron supports are used and the pit walls built outside of them. When the round cast iron columns are employed, the plate must be made square or with a projection for each column, to admit of the columns being placed at a sufficient distance apart to let the bottom doors swing between them. The best supports for a cupola are the T-shaped posts. They take up less room under the cupola and are less in the way when removing the dump than the round columns, and when slightly curved at the top, can be placed at a sufficient distance apart to permit of the drop doors swinging between them. When these posts are used, the bottom plate is made round. and of only a slightly larger diameter than the cupola shell or air chamber, and when made of good iron and the foundation plate is used, the bottom plate does not require to be more than $1\frac{1}{2}$ or 2 inches thick for the largest sized cupola. The supports when curved at the top must be bolted to the plate to hold them in place.

HEIGHT OF CUPOLA BOTTOM.

The height the bottom of a cupola or spout should be placed above the moulding floor or gangway, depends upon the class of work to be cast. For small hand-ladle work the proper height is 18 to 20 inches; for small bull- and hand-ladle work 24 to 30 inches; and for large crane-ladle work three to five feet.

It is very difficult and dangerous to change ladles and catch a large stream from a high cupola in hand-ladles; and when pieces are only cast occasionally, requiring the use of a large crane-ladle, it is better to place the cupola low and dig a pit in front of it, in which to set the ladle when a large one is required for the work.

When the cupola is set low, room must be made for the removal of the dump. This may be done by constructing a wall in front of the cupola to keep up the floor under the spout, and lowering the floor under and around the back part of the cupola. When the cupola is so situated that this can not be done, a pit should be constructed for the removal of the dump.

BOTTOM DOORS.

For cupolas of small diameter, but one bottom drop door is used. But when the cupola is of large diameter the door, if made in one piece, would be so large that there would not be room for it to swing clear of the foundation without setting the cupola too high, and the door would be very heavy and difficult to raise into place. For large cupolas the door is cut in the middle and one-half hung to the bottom on each side. Four and six doors are sometimes used, but they are always in the way when taking out the dump, and require more care in putting in place and supporting.

The doors are generally made of cast iron, and vary in thickness from a half-inch to an inch and a half in thickness, and are frequently very heavy and difficult to raise into place. If the doors are large they are much lighter and easier to handle when made of wrought iron, and if properly braced answer the purpose equally as well as the stiffer cast iron one. If the lugs on the bottom plate are set well back from the opening, and the lugs on the doors made long, the doors drop further away from the heat of the dump, and may be swung back and propped up out of the way when removing the dump.

CASING.

The casing or shell of the modern cupola and stack is made of iron or steel boiler plate, riveted together with one or two rows of rivets at each seam. The thickness of the plate required depends upon the diameter and height of the cupola and stack. The lining in the stack is seldom renewed, while the lining in the cupola is often removed every few months and replaced with a new one, and the casing must be of a sufficient thickness to support the stack and lining when the cupola lining is removed. The strain upon the casing due to expansion and shrinkage is not very great when properly lined; but when improperly lined with a poor quality of fire-brick, the expansion may be so great as to tear apart the strongest kind of casing. The only way to prevent this is to take care in selecting the fire-brick, and in laying up the lining. The greatest wear and tendency to rust is in the bottom sheet, and it is also weakened by cutting in the front, tuyere and slag holes, and should be made of heavier iron than any other part of the casing. Plate of $\frac{1}{4}$ inch or $\frac{3}{8}$ inch thickness is heavy enough for almost any sized cupola. The cupola and stack casing are generally made in one piece, the cupola ending at the charging door and the stack beginning at the same point. The stack may be contracted above or below the charging door, and made of smaller diameter than the cupola. This gives a better draught and requires less material for casing and lining; but it also increases the number of sparks thrown from the cupola when in blast. Where sparks are very objectionable, as in closely built up neighborhoods, it is better to make the cupola and stack of the same diameter, or to enlarge the stack from the bottom of the charging door. This may be done by placing a cast iron ring upon the top of the cupola shell, and supporting it by brackets riveted to the shell, and placing the stack shell upon the ring. The sparks then fall back into the cupola if the stack is of a good height, and very few are thrown out at the top.

The height of a cupola is the distance from the top of the bottom plate to the bottom of the charging aperture. Many plans have been devised for utilizing the waste heat from a cupola, but the only practical means so far discovered is to construct a high cupola. The heat lost in a low cupola is then utilized in heating the stock in the cupola before it escapes from it. But all the heat is not utilized in this way, for a great deal of gas escapes unconsumed. This is shown by the increase in flame as the stock settles in the cupola to a point at which the oxygen from the charging aperture combines with the escaping gas in sufficient quantity to ignite it, when it burns with a fierce flame above the stock. Still a great deal more heat is utilized in a high cupola than in a low one.

It is well known among iron founders that a high cupola will melt more iron in a given time and with less fuel than a low one of the same diameter. Therefore the charging aperture should be placed at the highest practicable point. There is a limit to the height at which the aperture in a small cupola can be placed, for where the diameter is small the iron in settling frequently lodges against the lining and hangs up the stock. When this occurs the stock has to be dislodged by a long bar worked down through from the charging aperture. If the aperture is placed at too great a height and the lodgment takes place near the bottom, the trouble cannot be remedied with a bar, and melting stops. Cupolas of large diameter may be made of almost any height desired, but there seems to be a limit to the height at which heat is produced in a cupola by the escaping gases, and we have arranged the following table from practical observation, giving the approximate height and size of door for cupolas of different diameters:

THE CUPOLA FURNACE.

Diameter Inside Lining, Inches.	Height of Cupola, Feet.	Size of Charging Door, Inches.	Melting Capacity per Hour, Tons.	Melting Capacity per Heat, Tons.
18	6—7	15 x 18	¼—¾	1—2
20	7—8	18 x 20	½—1	2—3
24	8—9	20 x 24	1—2	3—5
30	9—12	24 x 24	2—5	4—10
40	12—15	30 x 36	4—8	8—20
50	15—18	30 x 40	6—14	15—40
60	16—20	30 x 45	8—16	25—60

The melting capacity of a cupola varies with the kind of fuel used. One-fourth more iron can be melted per hour with coke than with coal, and the melting capacity per heat is greatly increased by the tapping of slag and number of tuyeres.

CHARGING DOOR.

The charging door may be made in one or two sections and lined with fire-brick or daubed with fire-clay; or it may be made of wire gauze placed in an iron frame. The charging door is of but little importance in melting, as it is seldom closed during the greater part of the heat, and is only of service to give draught to the cupola when lighting up, and to prevent sparks being thrown upon the scaffold during the latter part of the heat.

AIR CHAMBER.

The air chamber for supplying the tuyeres with blast may be constructed either outside or inside the cupola shell. When placed inside, the cupola must be boshed and the lining contracted at the bottom to make room for the chamber without enlarging the diameter of the cupola casing. When the cupola is large this can readily be done, and the boshing of the cupola increases its melting capacity; but small cupolas cannot be contracted at the bottom to a sufficient extent to admit of an air chamber being placed inside without interfering with the dumping of the cupola. When placed inside, the chamber may be formed with cast iron staves made to rest upon the bottom plate at one end and against the casing at the other. The staves are flanged to overlap each other with a putty joint, and when new make a very nice air chamber. But when the lining becomes thin they become heated and frequently warp or break, and permit the blast to escape through the lining to so great an extent that the lining has to be removed and the staves replaced with new ones.

The air chamber, when constructed inside the casing, should be made of boiler plate, and securely riveted to the casing to hold it in place and prevent leakage of blast through the lining. It must be constructed of a form to correspond with the boshing of the cupola, and of a size to supply a sufficient quantity of blast to all the tuyeres. If these conditions cannot be met without reducing the cupola below 40 inches diameter at the tuyeres, then the air chamber should be placed on the outside, and any desired boshing of the cupola made by placing common red brick behind the fire-brick lining.

When the air chamber is placed upon the outside of the shell, it may be formed by a round cast iron or sheet metal pipe extending around the cupola, with branches extending down to each tuyere; or it may be made of boiler plate and riveted to the shell. The great objection to the round or overhead air chamber is the numerous joints required in connecting it with each tuyere. These joints require continual looking after to prevent leakage of blast, and in many cases they are not examined from one year's end to another, and a large per cent. of the blast is frequently lost through leaky joints. The best air chambers are those made of boiler plate and riveted to the cupola shell and securely corked. These air chambers are made of any shape that may suit the fancy of the constructor, and in many cases are very much in the way of the melter in making up the cupola and of the moulders in removing the molten iron. They should not be made to extend out from the shell more than six inches, and any air capacity desired given by extending the chamber up or down the shell. The air capacity should not be less than three or four times the area of the outlet of the blower, and may be much larger. The blast should be admitted to the chamber from the top on each side of the cupola. This arrangement places the pipes out of the way where they are least likely to be knocked and injured. When the tuyeres are placed low, the chamber may be made to extend down to the bottom plate. In this case, the bottom plate must be made larger and the chamber cut away front and back for the tap and slag holes.

When the tuyeres are placed high, the chamber should be placed up out of the way of the tap and slag holes, and riveted to the shell at both top and bottom. An opening should be made in the air chamber under each tuyere and covered with a piece of sheet lead, so that any molten iron or slag running into the chamber from the tuyeres will flow out and not injure or fill up the chamber. An opening should be placed in front of each tuyere for giving draught to the cupola when lighting up, and for the removal of any iron or slag that may run into the tuyere during a heat. These openings should not be made over three or four inches in diameter, and should each be provided with a tight-fitting door to prevent the escape of the blast.

TAP HOLE.

One or more orifices are placed in the casing at the bottom plate for the removal of the molten iron from the cupola. These openings are known as tap holes, and in the casing are from six to eight inches wide and seven to nine inches high, curved or rounded at the top. The opening through the cupola lining is generally formed by the brick and presents a very ragged appearance after the lining has been in use a short time. This opening should be lined with a cast iron casting bolted to the cupola casing, and made to extend almost through the lining. The casing should be made slightly tapering with the large end inside, or ribbed, to prevent the front being pushed out by the pressure of molten iron retained in the cupola. For small cupolas, or a large cupola from which the iron is removed in large ladles, but one tap hole is required. But large cupolas melting over eight tons of iron per hour,

CONSTRUCTING A CUPOLA. 17

from which the iron is taken in hand ladles, require two tap holes. Two tap holes are sometimes placed in a cupola on opposite sides to shorten the distance of carrying the iron to the moulds. And two tap holes are also sometimes placed side by side so that each may be kept in better order through the heat. This is bad practice, for if the front is properly put in, one tap hole will run off all the iron a cupola is capable of melting. When two tap holes are put in they should be placed one in front and the other in the back or side of the cupola, so that the moulders will not be in each other's way when catching-in.

THE SPOUT.

A short spout must be provided for conveying the molten iron from the tap hole to the ladles. This spout is generally made of cast iron, and is from six to eight inches wide with sides from three to six inches high, and for small ladle work is from one to two feet long. For large ladle work it is made much longer. In some foundries where a long spout is only occasionally required, the spout is made in two sections and put together with cleats, so that an additional section may be put up to fill a large ladle and taken down when it is filled. The spout should be long enough to throw the stream near the center of the ladle when filling. In a great many foundries the spout is laid upon the bottom plate, and only held in place by the making up of the front, and is removed after each heat. This entails the loss of a great deal of spout material each heat, and sometimes the spout is struck in the careless handling of ladles and knocked out of place, when much damage may be done. When not in the way of removing the dump, the spout should be securely bolted to the bottom plate.

When it is desired to run a very small cupola for a greater length of time than an hour and a half, or a large cupola for a longer time than two hours and a half, slag must be tapped to remove the ash of the fuel and dross of the iron from the cupola, to prevent bridging over and bunging up. The slag hole from which the slag is tapped is placed between the tuyeres, and below the lower level of the lower row of tuyeres. A hole is cut through the casing and lining from three to four inches in diameter, and a short spout or apron is provided to carry the slag out, so that it will fall clear of the bottom plate. The slag hole should be placed at the back of the cupola, or at the greatest possible distance from the tap hole, so that the slag will not be in the way of the moulders when catching the iron. The height at which a slag hole should be placed above the sand bottom depends upon how the iron is tapped. The slag in a cupola drops to the bottom and floats upon the surface of the molten metal, and rises and falls with it in the cupola. If the molten iron is held in the cupola until a large body accumulates, the slag hole must be placed high and the slag tapped when it has risen upon the surface of the molten iron to the slag hole. When the iron is withdrawn, the slag remaining in the cupola falls below the slag hole, and the hole must be closed with a bod to prevent the escape of blast. If the iron is drawn from the cupola as fast as melted, the slag hole is placed two or three inches above the sand bottom at the back of the cupola. The slag then lies upon the molten iron, or upon the sand bottom, and the slag hole may be opened as soon as slag has formed, and allowed to remain open throughout the heat.

TUYERES.

A number of openings are made through the casing and lining near the bottom of the cupola for admitting the blast into the cupola from the air chamber or blast pipe. These openings are known as tuyeres. Tuyeres have been designed of all shapes and sizes, and have been placed in cupolas in almost every conceivable position, so there is little to be learned by experimenting with them, and the only things to be considered are the number, shape, size and position of tuyeres for different sized cupolas. For a small cupola, two tuyeres are sufficient. A greater number promotes bridging. They should be placed in the cupola on opposite sides, so that the blast will meet in the center of the cupola, and not be thrown against the lining at any one point with great force. The best shape for a small cupola is a triangular or upright-slot tuyere. These cause less bridging than the flat-slot or oval tuyere, and in small cupolas make but little difference in the amount of fuel required for the bed. When only two tuyeres are provided, a belt air chamber around the cupola is not required, and the blast pipes are generally connected direct with each tuyere. In large cupolas, the shape of the tuyeres selected makes but little difference in the melting, so long as they are of sufficient size and number to admit the proper amount of blast to the cupola, and so arranged as to distribute it evenly to the stock. The flat-slot or oval tuyeres are generally selected for the reason that they require less bed than the upright-slot tuyere.

The number of tuyeres required varies from four to eight, according to the size of the cupola and tuyeres. They should be of the same size and placed at uniform distances apart. A tuyere should never be placed directly over the tap or slag hole. The combined tuyere area should be from two to three times greater than the area of the blower outlet. The tuyere boxes or casings are made of cast iron, and should be bolted to the cupola shell to prevent any escape of blast through the lining when it becomes old and shaky, or when lined with poor material and the grouting works out, as is sometimes the case.

The height at which tuyeres are placed in cupolas above the sand bottom varies from one or two inches to five feet, and there is a wide difference of opinion among founders as to the height at which they should be placed. When the tuyeres are placed low, the iron must be drawn from the cupola as fast as melted, to prevent it running into the tuyeres. In foundries where the iron is all handled in hand-ladles, this can readily be done, and the tuyeres are placed low to reduce the quantity of fuel in the bed and make hot iron. In foundries in which heavy work is cast, and the iron handled in large ladles, the tuyeres are placed high, so that a large amount of iron may be accumulated in the cupola to fill a large ladle for a heavy piece of work.

We do not believe in high tuyeres, and claim they should never be placed more than 10 or 12 inches above the sand bottom for any kind of work; and if slag is not to be tapped from the cupola, they should not be placed more than two or three inches above the sand bottom. In stove foundries, in which cupolas of large diameter are employed and hot iron required throughout the heat, the tuyeres are placed so low that the sand bottom is made up to within one inch of the bottom of the tuyeres on the back, and two or three inches at the front. This gives plenty of room below the tuyeres for holding iron without danger of it running into the tuyeres. In cupolas of small diameter, two inches is allowed at the back and three or four inches at the front. This insures a hot, even iron throughout the heat, if the cupola is properly charged, and a much less quantity of fuel is required for the bed than if the tuyeres were placed high. Molten iron is never retained in the cupola for this class of work, and the tap hole is made of a size to let the iron out as fast as melted and the stream kept running throughout the heat.

Cupolas with high tuyeres are not employed for this class of work, for they do not produce a hot fluid iron throughout a heat without the use of an extraordinarily large per cent. of fuel, and when the tuyeres are extremely high they do not make a hot iron with any amount of fuel. Nothing is gained by holding molten iron in a cupola, for iron can be kept hotter in a ladle than in a cupola, and melted hotter with low than high tuyeres, and a cupola is kept in better melting condition throughout a heat by tapping the iron as fast as melted.

TWO OR MORE ROWS OF TUYERES.

It is the common practice to place all the tuyeres in a cupola at the same level, or in one row extending around the cupola. But two or more rows are frequently placed one above the other. When a large number of rows are employed, they decrease in area gradually from the lower to the top tuyere, and the rows are generally placed very close together. When two rows are put in, the second row is made from one-half to one-tenth the area of the first row, and the two rows are placed from 8 to 18 inches apart. If the area of the second row is one-half that of the first, it is generally placed from 8 to 10 inches above the first row, and only when the tuyeres are very small are they placed at a greater height above the first row. When three rows are put in, the second row is made one-half the area of the first row, and the third row one-fourth the area of the second, and the rows are placed from 6 to 10 inches apart. When tuyeres are placed in a cupola all the way up to the charging door, those above the first or second rows are made one inch diameter, and are placed from 12 to 14 inches above each other.

The tuyere in the upper row may be placed directly over the tuyere in the row beneath it, or may be placed between two lower ones. Some cupola men claim that much better results are obtained by this latter plan, but we have never observed that it made any difference whether they were placed over or between those of the lower rows.

Faster melting is secured with two or three rows of tuyeres than with one row in cupola of the same diameter, and the melting capacity per hour is increased about one-fourth in melting large heats. When melting a small heat for the size of the cupola, nothing is gained by the additional rows of tuyeres, since a much larger quantity of fuel is required in the bed, for which there is no recompense by saving of fuel in the charges through the heat, and fast melting is seldom any great object in small heats.

LINING.

The casing may be lined with fire-brick, soapstone or other refractory substances. In localities where fire-brick cannot be obtained, native refractory materials are used; but fire-brick are to be preferred to native mineral substances. Cupola brick are now made of almost any shape or size required in cupola lining, and can be purchased at as reasonable a price as the common straight fire-brick. The curved brick, laid flat, make a more compact and durable lining than the wedge-shaped brick set on end, and are most generally used. When laying up a lining, the grouting or mortar used should be of the same refractory material as the brick, so that it will not burn out and leave crevices between the brick, into which the flame penetrates and burns away the edges of the brick. This material is made into a thin grout, and a thin layer is spread upon the bottom plate. The brick is then taken in the hand, one end dipped in the grout, and laid in the grout upon the plate. When a course or circle has been laid up, the top is slushed with grout to fill up all the cracks and joints, and the next course is laid up and grouted in the same way. The joints are broken at each course, and the brick are laid close together to make the crevice between them as small as possible, and prevent the flame burning away the corners in case the grouting material is not good and burns out.

Brick that do not expand when heated are laid close to the casing. Those that do expand are laid from a fourth of an inch to an inch from the casing, to give room for expansion, and the space is filled in with sand or grout. Brick of unknown properties should always be laid a short distance from the casing, to prevent it being burst by expansion of the lining.

The lining is made of one thickness of brick, and a brick is selected of a size to give the desired thickness of lining. In small cupolas, a four or five-inch lining is used, and in large cupolas a six or nine-inch lining. A heavier lining than nine inches is seldom put in, except to reduce the diameter of the cupola or prevent the heating of the shell. In these cases, a filling or false lining of common red brick is put in between the fire-brick and shell. The stack lining is seldom made heavier than four inches for any sized cupola, as the wear upon it is not very great, and a four-inch lining lasts for a number of years. The stack lining is laid up and grouted in the same way as the cupola lining.

ARRANGEMENT OF BRACKETS, ETC.

In Fig. 1 is shown the manner in which brackets or angle iron are put into a cupola for the support of the lining in sections upon the casing. The brackets are made of heavy boiler plate from five to six inches wide, circled to fit the casing and

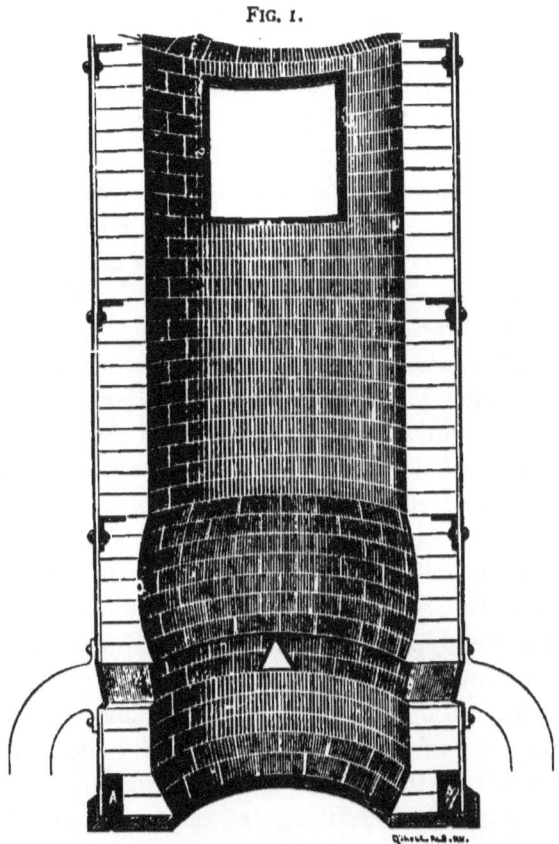

FIG. 1.

SECTIONAL VIEW OF CUPOLA.

bent at a square angle. The part riveted to the casing is made four inches long and secured to the casing with two or three rivets. The bracket or shelf for the support of the lining is made from one and a half to two inches long. The brackets are placed about two feet apart around the casing and in rows from two to three feet above each other. These brackets are but little in the way when laying up a lining, and support the latter so that a section may be taken out and replaced without disturbing the remainder of the lining.

Angle iron is by many preferred to brackets for the support of the lining. It is put in bands extending all the way around the casing and riveted to it. These bands not only support the lining but act as a brace to the casing, and in some respects are a better support for the lining than brackets. They catch and hold in place all the grouting or sand that may work out of the lining between the casing, and give a more even support to the lining, but with their use it is sometimes more difficult to fit the brick around when laying up a lining. Still, angle iron has generally taken the place of brackets and is put in all the modern cupolas. The brackets or angle iron should not be made to extend out from the casing more than one and a half or two inches, for if they do they are liable to be burned off when the lining becomes thin and let the iron or heat through to the casing. One and a half inches are sufficient to support the lining if the brick form a circle to fit the casing. No supports should be put in at the melting zone, for the lining frequently burns very thin at this point, even in a single heat. It is not necessary to put in any below the melting zone, and the first one should be placed at the upper edge of the zone, and from this up they should be put in at every two or three feet.

The weight of brick placed upon the lower courses in a cupola lining is sufficient to crush most of the soft cupola brick, and were it not for the support given to three sides of them in the lining they would, by the great weight placed upon them, be reduced to a powder. As a lining burns out it becomes thin more rapidly at the bottom, and it often happens that the lining at the melting zone is reduced to one-half its thickness, or even less, in a few heats, and this reduced lining often has to support a lining of almost full thickness for the entire cupola, and in some cases also the stack lining. The cohesive force of these bricks is reduced by the intense heat in the cupola, and when subjected to so great a pressure and heated they are crushed and the lining gradually settles and becomes shaky. This settling is so great with some qualities of brick that in cupolas having no frame riveted to the casing around the charging aperture, the arch over the door frequently settles so low that it becomes necessary to rebuild it to maintain the full size of the opening.

Brick do not give the best results when subjected to so great a pressure and heated to a high temperature. Therefore, in all cupolas, brackets or angle iron should be put in every two or three feet for the support of the lining on the casing, and the casing should be made heavy enough to support the entire lining when a section has been burned out or removed.

In the illustration (Fig. 1) is also shown a way for reducing the size and weight of the bottom doors and preventing the casing from rusting off at the bottom. In many of the large cupolas requiring heavy sand bottoms, the bottom plate can be made to extend into the cupola from three to six inches all round without in the least interfering with dumping, and the first few courses of brick sloped back from the edge of the plate to the regular thickness of lining to prevent sand lodging on the edges of the plate around the lining. By this arrangement in large cupolas the diameter of the doors may be reduced from six to ten inches and very much lightened, and less sand will be required, for the sand bottom and the dump falls as freely as when the doors are the full size of the cupola.

Cupolas that are not in constant use absorb a great deal of moisture into the lining and are constantly wet around the bottom plate, and light casings are eaten away by rust in a short time. To prevent this the first one or two courses of brick can be laid a few inches from the casing and a small air chamber formed around the cupola at this point. If this chamber is supplied with air from a few small holes through the iron bottom or casing, the latter is kept dry and rusting is prevented.

In the illustration (Fig. 1) is shown the triangular-shaped tuyere in position in the lining. This tuyere prevents bridging to a greater extent than any other, and is, for a small cupola, one of the very best shapes. It is formed with a cast iron frame set in the lining, and each tuyere may be connected with a separate pipe, as shown, or they may be connected with an air belt extending around the cupola.

Bottom plates may be cast with a light flange around the edge, as shown in the illustration (Fig. 1), or made perfectly flat on top; but it is better to cast them with a small flange or bead for holding the shell in place upon the plate, and thus cause the cupola to have a more finished look around the bottom.

FIRE PROOF SCAFFOLDS.

The charging door or opening through which fuel and iron are charged into a cupola is placed at so great a height from the floor that it is necessary to construct a platform or scaffold, upon which to place the stock, and from which to charge it into the cupola. For heavy work, this scaffold is generally placed on three sides of the cupola, leaving the front clear for the swinging of crane ladles to and from the spout; but for light work the scaffold frequently extends all the way around the cupola to give more room for placing stock upon it. The distance the floor of a scaffold is generally placed below the charging door is about two feet, but that distance varies, and floors are frequently placed on a level with the door or three or four feet below it to suit the kind of iron to be melted or the facilities for placing stock upon the scaffold from the yard. The scaffold and its supports are more exposed to fire than almost any other part of a foundry, for live sparks are thrown from the charging door upon the scaffold floor, and molten iron, slag, etc., are frequently thrown against its supports and the under side of the floor with considerable force when dumping the cupola. Numerous plans have been devised to make scaffolds fire-proof and prevent the foundry from being set on fire. In many of the wooden foundry buildings the scaffold is constructed entirely of wood, and to render it fire-proof the supports and under side of the floor are covered with light sheet iron to protect them from molten iron, slag, etc., when dumping. The covering of the woodwork of a scaffold in this way is very bad practice, for while it protects the wood from direct contact with the fire, it also prevents it from being wetted, and in a short time the wood becomes very dry and very combustible. The thin covering of sheet iron is soon eaten away with rust, leaving holes through which sparks may pass and come in contact with the dry wood and ignite it under the sheet iron where it cannot be seen, and the cupola men, after wetting down the dump very carefully, may go home leaving a smoldering fire concealed by the sheet iron covering which may break forth during the night and destroy the foundry. It is better to leave all the woodwork entirely uncovered and exposed to the fire and heat, and wet it in exposed places before and after each heat; the wood is then kept dampened and is not so readily combustible as when covered with sheet iron, and if ignited the fire may be seen and extinguished before the men leave for home after their day's work is done. At many of the wooden foundry buildings the cupola is placed outside the foundry building and a small brick house or room constructed for it and the molten iron run into the foundry by a cupola spout extending through the wall. In this way a scaffold may be made entirely fire-proof by putting in iron joist and an iron or brick floor, and putting on an iron roof. We saw a scaffold and cupola house at a small foundry in Detroit, Mich., about twenty years ago, that was constructed upon a novel plan and was perfectly fire-proof. The house was twelve feet square and constructed of brick, the scaffold floor was of iron and supported by iron joist, the walls were perpendicular to five feet above the scaffold floor, and from this point they were contracted and extended up to a sufficient height to form a stack three feet square at the top. The cupola was placed at one side of this room and the cupola-house, and the spout extended through the wall into the foundry; the open top of the cupola extended about two feet above the scaffold floor, and its stack was formed by the contracted walls of the cupola-house. There were no windows in the house, and only one opening above for placing stock upon the scaffold and one below for removing the dump and making up the cupola, both of which openings were fitted with iron door frames and doors, and could be tightly closed. When lighting up, the scaffold door was closed to give draught to the cupola, and when burned up the door was opened and the cupola charged from the scaffold. Sparks from the cupola when in blast fell upon the scaffold floor and were never thrown from the top of the stack or cupola-house upon the foundry roof or the roofs of adjoining buildings, and when the doors were closed the scaffold was as fireproof as a brick stack. The great objection to this scaffold was the gas from the cupola upon the scaffold when the blast was on, and the intense heat upon the scaffold in warm weather or when the stock got low in the cupola.

The best and safest scaffolds are those constructed entirely of iron, or with brick floors and supported by iron columns, or brick walls and made of a sufficient size to admit of wood or other readily combustible cupola material being placed at a safe distance from the cupola. The cupola scaffold in the foundry of Gould & Eberhardt, Newark, N. J., is constructed of iron supported by iron columns and brick walls, and is of sufficient size and strength to carry two car-loads of coke, one hundred tons of pig and scrap iron, and all the wood shavings and other material required for the cupola. In the new iron foundry building recently erected by The Straight Line Engine Company, Syracuse, N. Y., the scaffold is constructed entirely of iron and supported by the iron columns which support the foundry roof. It extends the entire length of the foundry, affording ample room for storing iron, coke, wood, and all cupola supplies, thus doing away with a yard for storing such material, and placing them under the foundry roof and convenient for use. Scaffolds of this kind greatly reduce the expense of handling cupola stock, and also reduce the rate of insurance of foundry buildings.

CHAPTER III.

CUPOLA TUYERES.

THE cupola furnace may be supplied with the air required for the combustion of the fuel by natural draft induced by a high stack, a vacuum created by a jet of steam, or by a forced blast from a fan or blower. In either case the air is generally admitted to the cupola through openings in the sides near the bottom. These openings are known as tuyeres or tuyere holes. The location, size, number and shape of these tuyeres are a matter of prime importance in constructing a cupola, and are a subject to which a great deal of attention has been given by eminent and practical foundrymen for years, and to these men is due the credit for the advancement made in the construction of cupolas.

It is only a few years since 10 to 15 tons was considered a large heat for a cupola, and when a large casting was to be poured two or more cupolas were run at the same time and the greater part of a day consumed in melting. Now 60 tons are melted in one cupola in four hours for light foundry work, and hundreds of tons are melted in one cupola in steel works without dropping the bottom. This improvement in melting is largely due to the improvement in the size, shape and arrangement of tuyeres.

There have been epidemics of tuyere inventing several times in this country in the past twenty-five years, and during these periods it has been almost impossible for an outsider to get a look into a cupola for fear the great secret of melting would be discovered in the shape of the tuyere and made public. During these epidemics tuyeres of almost every conceivable shape have been placed in cupolas, and great results in melting claimed for them. Many of these tuyeres were soon found to be complicated and impracticable, or the advantage gained by their use in melting was more than offset by extravagant use of fuel.

It would be useless for us to describe all the tuyeres we have seen employed, for many of them were never used out of the foundry in which they were invented, and only used there for a short time. We shall, therefore, describe only a few of those that have been most extensively used or are in use at the present time.

The *round* tuyere is probably the oldest or first tuyere ever placed in a cupola. It was used in cupolas and blast furnaces in Colonial days in this country, and long before that in France and other countries. In the old-fashioned cast iron stave cupolas three round tuyeres were generally placed in a row, one above another, on opposite sides of the cupola. The first or lower tuyere was placed from 18 to 24 inches above the sand bottom, and the others directly over it from 3 to 4 inches apart. The tuyere nozzle or elbow was attached to the blast-pipe by a flexible leather hose, and first placed in the lower tuyere and the two upper tuyeres temporarily closed with clay. When a small heat was melted the nozzle was permitted to remain in the lower tuyere through the heat. But when a large heat was melted and the cupola melted poorly at any part of the heat, or if molten iron was to be collected in the cupola for a large casting, the clay was removed from the upper tuyeres, and the nozzle removed from one to the other, as required, and the lower tuyeres closed with clay.

In these cupolas the tuyeres were generally too small to admit a proper volume of blast to do good melting. In one of 28 inches diameter we recently saw at Jamestown, N. Y., the original tuyeres were only 3 inches in diameter. Two tuyeres of this size could not possibly admit a sufficient volume of blast to do good melting in a cupola of the above diameter, and in this one they had been replaced by two of a much larger diameter placed at a lower level than the old ones. The round tuyere is still extensively used in small cupolas where the tuyeres can be made of a diameter not to exceed 5 or 6 inches, but in large cupolas it has generally been replaced by the flat or oval tuyere, which admits the same volume of blast and permits of a smaller amount of fuel being used in the bed than could be used with a round tuyere of large area.

OVAL TUYERE.

In Fig. 2 is shown the oval or oblong tuyere now extensively used. It is made of different sizes to suit the diameter of cupola, the most common sizes used being 2 x 6, 3 x 8, and 4 x 12 inches. They are laid flat in the lining and generally supplied from an outside belt air chamber. This tuyere is the one most commonly used by stove, bench and other foundries requiring very hot iron for their work. They are placed very low, generally not more than two or three inches above the sand bottom, and in large cupolas the slope of the bottom frequently brings it up to the bottom of the tuyeres on the back side of the cupola. This tuyere admits the blast to a cupola as freely as a rounded tuyere of the same area, and the tendency of the stock to chill over the tuyeres in settling and bridge the cupola is no greater than with a round tuyere of the same capacity. It admits of a lower bed than the round tuyere, and is to be preferred to the round form for cupolas requiring tuyeres of larger area.

EXPANDED TUYERE.

In Fig. 3 is seen the expanded tuyere, which is made larger

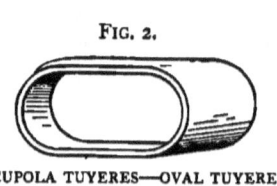

FIG. 2.

CUPOLA TUYERES—OVAL TUYERE.

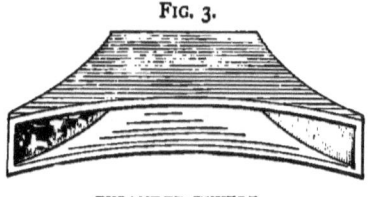

FIG. 3.

EXPANDED TUYERE.

at the outlet than at the inlet. It is reduced at the inlet so that the combined tuyere area may correspond with the outlet of the blower and equalize the volume of blast entering the cupola at each tuyere from the air belt. It is expanded at the outlet to permit the blast to escape freely from the tuyeres into the cupola, and in case the stock settles in the front of the tuyere in such a way as to close up part of it, there may still be sufficient opening for the full volume of blast entering the tuyere to pass into the cupola. The tuyere is made from two to four inches wide at the inlet and six to twelve inches long. The width of the outlet is the same as that of the inlet, and the length of the outlet is from one-fourth to one-half longer than the inlet. The tuyere is laid flat in the lining, the same as the oval tuyere, and the only advantage claimed for it over that tuyere is that it cannot be closed so readily by the settling of the stock and the chilling of the iron or cinder in front of it. The expanded tuyere is preferred by many to the oval tuyere on this account and is extensively used at the present time.

DOHERTY TUYERE.

In Fig. 4 is seen the Doherty arrangement of tuyeres, designed by Mr. Doherty of the late firm of Bement & Doherty, Philadelphia, Pa., and employed in the Doherty cupola, a cupola that was extensively used in Philadelphia about twenty-five years ago. The arrangement consists of two or more round tuyeres placed in the lining and at an angle to it, instead of passing straight through the lining as tuyeres generally do. The blast pipes connecting with each tuyere were placed at the same angle as the tuyere, the object being to give the blast a whirling or spiral motion in the cupola. The blast took the desired course, as could be plainly seen by its action at the charging door, and it had the appearance of making a more intense heat in the cupola than when delivered from the straight tuyere. But this appearance was deceptive, and after careful investigation it was found that no saving in fuel was effected or faster or hotter melting done on account of this motion of the blast. The cupolas and tuyeres were, however, constructed of proper proportions, and were a decided improvement on the small tuyere cupolas in use at that time. Many of them were placed in foundries and are still in use, but no importance is attached to the spiral motion of the blast.

SHEET BLAST TUYERE.

In Fig. 5 is seen the horizontal slot tuyere. This tuyere

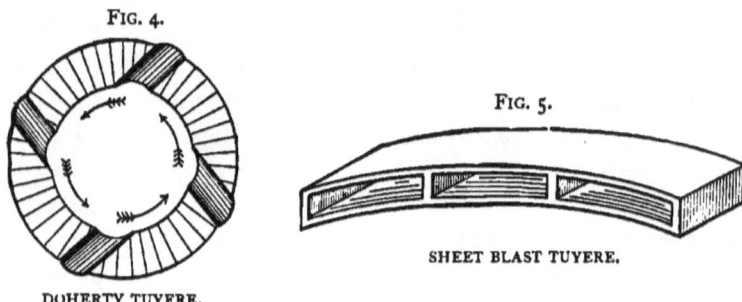

FIG. 4.

DOHERTY TUYERE.

FIG. 5.

SHEET BLAST TUYERE.

consists of a slot from one to two inches wide, extending one-third around the cupola on each side, or a continuous slot extending all the way around the cupola. The slot is formed by two cast iron plates, on one of which are cast separating bars to prevent the plates being pressed together by the weight of the lining or warped by the heat. This tuyere is known as the sheet blast tuyere. It admits of a smaller amount of fuel being used for a bed than any other tuyere placed in a cupola at the same height above the bottom. It distributes the blast equally to the stock, and does fast and economical melting in short heats. But the tendency of the cupola to bridge is greater than with almost any other tuyere, and a cupola with this tuyere cannot be run successfully for a greater length of time than two hours.

MACKENZIE TUYERE.

In Fig. 6 is seen the Mackenzie tuyere, designed by a Mr. Mackenzie of Newark, N. J., and used in the Mackenzie cupola. This is a continuous slot or sheet blast tuyere, but differs from the one just described in that the cupola is boshed and the bosh overhangs the slot from four to six inches. The slot is protected by the overhanging bosh and cannot be closed up by

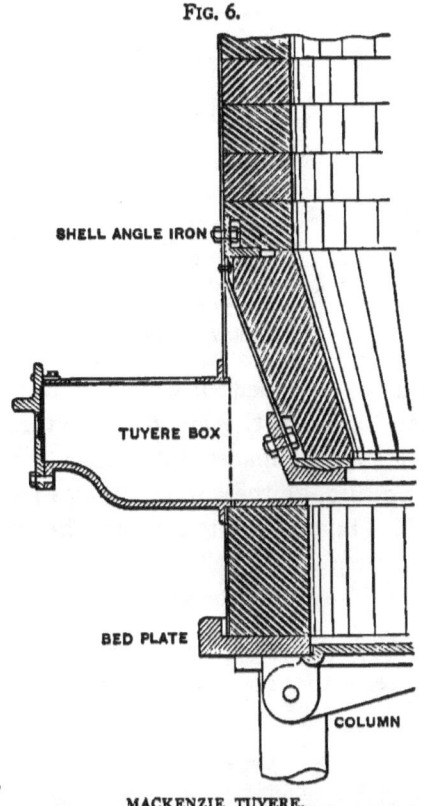

FIG. 6.

MACKENZIE TUYERE.

the settling of the stock. The Mackenzie cupolas with this tuyere are constructed of an oval or oblong shape, with an inside belt air chamber. The blast enters the air chamber from a tuyere box at each end of the cupola, and passes into the cupola through a two-inch slot extending all the way round the cupola.

BLAKENEY TUYERE.

In Fig. 7 is seen the Blakeney tuyere used in the Blakeney cupola constructed by The M. Steel Company, Springfield, Ohio.

This tuyere is a modification or an improvement on the sheet blast tuyere, and extends all the way around the cupola. It is

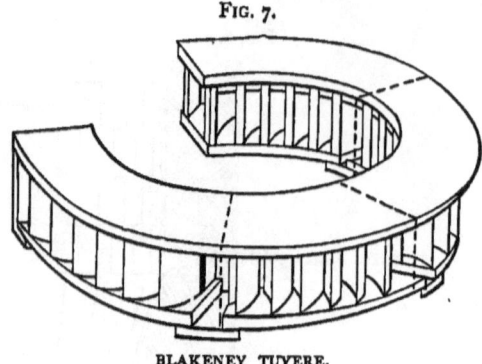

FIG. 7.

BLAKENEY TUYERE.

supplied from an outside belt air chamber riveted to the shell. The blast is conducted to the air chamber through one pipe, and, striking the blank spaces sidewise in rear of chamber, passes all around through the curved tuyeres into the centre of the furnace. This tuyere admits the blast freely and evenly to the cupola and very good melting is done with it. All the tuyeres described above may be used with either coal or coke.

HORIZONTAL AND VERTICAL SLOT TUYERE.

In Fig. 8 is seen the horizontal and vertical slot tuyere.

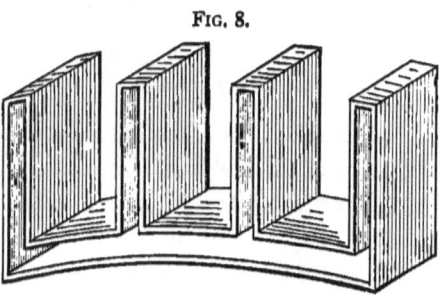

FIG. 8.

HORIZONTAL AND VERTICAL SLOT TUYERE.

This was designed for coke, and we have seen it used in but one cupola, a 40-inch one. One tuyere was placed on each side of the cupola. The horizontal slot of each tuyere, 1 inch wide, extended one-third way round the cupola, and the vertical slots, 1 inch wide and 12 inches long, were placed above it as shown. The tuyere did excellent melting, and the cupola could be run for a long time without bridging.

REVERSED T TUYERE.

In Fig. 9 is seen a vertical and horizontal slot or reversed T tuyere, also used for coke. The slots in this tuyere are from two to three inches wide and ten to twelve inches long. From two to eight of these tuyeres are placed in a cupola, according to the diameter. This tuyere has been extensively used, and is said to be an excellent tuyere for coke melting.

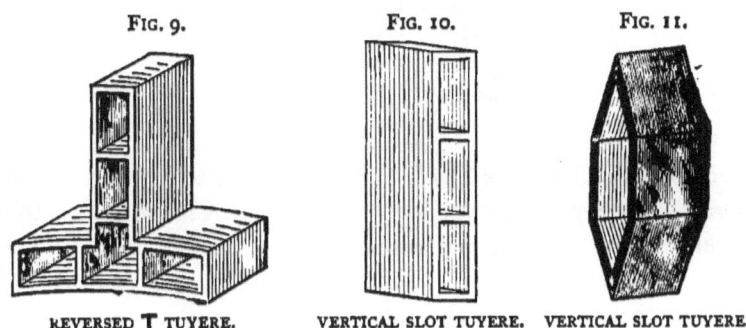

FIG. 9.   FIG. 10.   FIG. 11.

REVERSED T TUYERE.   VERTICAL SLOT TUYERE.   VERTICAL SLOT TUYERE.

In Figs. 10 and 11 are seen the vertical slot tuyeres used principally in cupolas of small diameter to prevent bridging. They are made from two to three inches wide and ten to twelve inches long, and two or more are placed in a cupola at equal distances apart.

TRUESDALE REDUCING TUYERE.

In Fig. 12 is seen the Truesdale reducing tuyere designed by a Mr. Truesdale of Cincinnati, Ohio, and extensively used in cupolas in that vicinity about 1874. The tuyere consisted of one opening or tuyere placed directly over another until six, eight or ten tuyeres were put in. The lower tuyere was made three or four inches in diameter, and tuyeres above it were placed one inch apart, and each one made of a smaller diam-

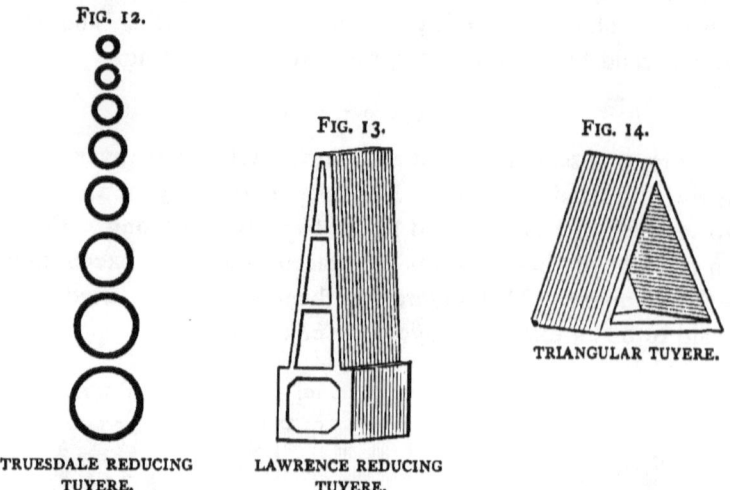

Fig. 12.

Fig. 13.

Fig. 14.

TRIANGULAR TUYERE.

TRUESDALE REDUCING TUYERE.

LAWRENCE REDUCING TUYERE.

eter until they were reduced to one inch. The bottom row of tuyeres were placed two, four and six inches apart, and the tuyeres in each succeeding row were placed further apart, were of a smaller diameter and admitted less blast to the cupola toward the top of the bed than at the bottom. The cupolas were generally boshed, and the tuyeres supplied from an inside belt air chamber, formed of cast iron staves, to which the tuyeres were attached by cleats or dovetails cast on the stays. Very fast melting was done in cupolas with this tuyere, but the tendency to bridge in cupolas of small diameter is so great that it could not be used. In large cupolas, however, it gave excellent results, and is still in use in numerous foundries.

LAWRANCE REDUCING TUYERE.

In Fig. 13 is seen the Lawrance reducing tuyere designed by Frank Lawrance of Philadelphia, Pa., and used in the Lawrance cupola, built by him. This tuyere was designed for either coal or coke melting, and works equally well with either. The opening at the bottom is from 3 to 4 inches square, and the slot from 10 to 12 inches long, from 1 to 1½ inches wide at the bottom, and tapers to a point at the top. The tuyeres are placed in the cupola from 6 to 12 inches apart, and supplied from a belt air chamber inside the casing. The air chamber in this cupola was first formed with cast iron staves, and the tuyeres held in place by cleats cast upon the staves. But the staves were found to break after repeated heating and cooling, and a boiler iron casing is now used for the air chamber. This tuyere and cupola do excellent melting, and a great many of them are now in use.

TRIANGULAR TUYERE.

In Fig. 14 is seen the triangular tuyere, designed by the writer over 25 years ago to prevent bridging in small cupolas and extensively used in both small and large cupolas, with either coal or coke. This tuyere may be made with the base and sides of the tuyere of an equal length, forming an equilateral triangle, or the sides may be made longer than the base, bringing the tuyere up to a sharp point at the top to prevent bridging; or the sides may be extended up to a sufficient height to form a reducing tuyere.

The Magee Furnace Company, Boston, Mass., placed this tuyere in their large cupola, constructed to melt iron for stove plate, about twelve years ago, and it has been in constant use ever since, giving excellent results in melting with coal and coke. In this cupola, which is 5 feet 4 inches diameter at the melting point, the tuyere is 9 inches wide at the base and 16 inches high. It was not thought best to extend the tuyere up to a point at so sharp an angle, and the top was cut off, leaving the opening 2 inches wide at the top. This tuyere has been arranged to take the place of the Truesdale reducing tuyere, and has been made from 6 to 8 inches wide at base and 24 to 30 inches high, running up to a point. It has also been used in imitation of the Lawrance reducing tuyere and made from 3 to 4 inches wide at base and 12 to 16 inches high.

40   THE CUPOLA FURNACE.

WATER TUYERE.

In Fig. 15 is seen the water tuyere. This tuyere is designed to be used in cupolas or furnaces where the whole or part of the tuyere is exposed to an intense heat and liable to be melted or injured, as is the case with tuyeres placed in the bottom of a cupola or in furnaces where a hot blast is used.

The tuyere or metal surrounding the tuyere opening is cast hollow and filled with water, or one end is left open and a spray thrown against the end exposed to the heat from a small pipe, as shown in illustration. The tuyere is also made with a coil

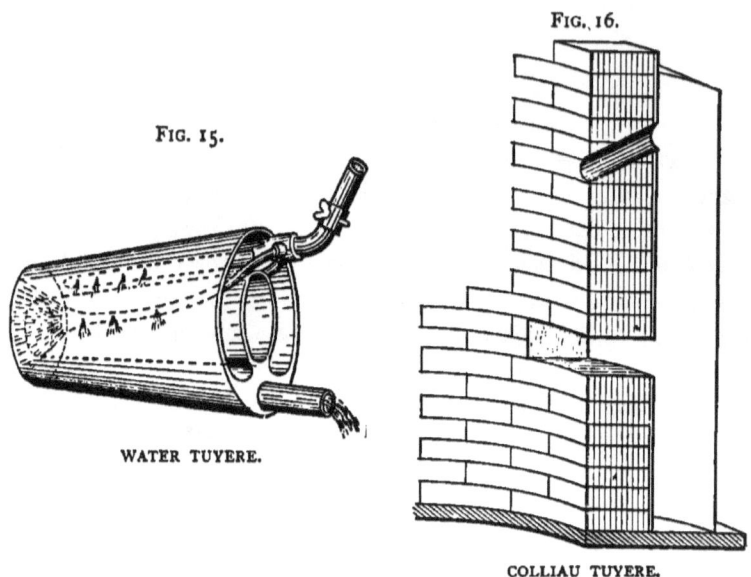

FIG. 15.

WATER TUYERE.

FIG. 16.

COLLIAU TUYERE.

of gas pipe cast inside the tuyere through which water constantly flows. The water tuyere is never used in cupolas when the tuyeres are placed in the sides of the cupola, but it has been used in cupolas in which the tuyere was placed in the bottom and exposed to the heat of molten iron, cinder and slag. When used in this way the tuyere is placed in the centre of the bottom and is made from 1 to 3 feet long, the mouth being placed at a sufficient height above the sand bottom to prevent molten iron or slag overflowing into it. The part of the tuyere extending up in the cupola and exposed to the heat is protected and prevented from melting by the stream of water. For this purpose the coil gas pipe tuyere is better than the hollow or spray tuyere just described.

COLLIAU TUYERE.

In Fig. 16 is seen the Colliau double tuyere designed by the late Victor Colliau of Detroit, Mich., and used in the Colliau cupola. In this cupola the tuyeres are placed in two rows one above the other in place of one row as in the ordinary cupola. The first row is placed at about the same height above the sand bottom as in the ordinary cupola and the second row from 12 to 18 inches above the first row. The first row are flat, slightly expanded tuyeres similar to that shown in Fig. 2, and are made from 2 to 4 inches wide and 6 to 14 inches long, according to the size of the cupola. The tuyeres in the second row are made round and from 2 to 4 inches diameter. The tuyeres in the first row pass straight into the cupola through the lining, and those in the second row are pointed downward at a sharp angle, as shown in the cut. The object of the second row is to furnish sufficient oxygen to consume the escaping gases and create a more intense heat at the melting point than is obtained with the single row of tuyeres from the same amount of fuel.

WHITING TUYERE.

The Whiting tuyere, used in the Whiting cupola, manufactured by the Whiting Foundry Equipment Company, Chicago, Ill., was designed by Mr. Whiting, a practical foundryman of Detroit, Mich., as an improvement on the Colliau tuyere. The Whiting tuyere is a double tuyere, but differs somewhat in arrangement from the Colliau tuyere. The first row are flat, slightly expanded tuyeres, and the second row are of the same shape and made larger in proportion to the lower row than the Colliau, and the two rows are not placed at so great a distance apart. Both the upper and lower rows pass straight into the cupola.

CHENNEY TUYERES.

The Chenney tuyere, designed by the late Mr. Chenney, a practical foundryman of Pittsburgh, Pa., is a double tuyere very similar in arrangement to the Colliau and Whiting tuyeres, the only difference being that both the upper and lower rows point downward at a sharp angle to the lining.

THE DOUBLE TUYERE.

The double or two rows of tuyeres appears to have first been designed and put into practical use about 1854 by Mr. Ireland, a practical English foundryman and cupola builder. In Ireland's cupolas, many of which were in use in England about that time, the tuyeres were placed in two rows about 18 inches apart. Those in the upper row were of only one-third the diameter of those in the lower, and twice the number of tuyeres were placed in the upper row as were in the lower. The slag hole was also used by Ireland in his cupola, which was run for a great many hours without dumping or raking out, as was the custom in those days. These cupola appear to have given very good results in long heats, but in short heats they were not so satisfactory, and in more recent patents obtained by Mr. Ireland the upper row of tuyeres was abandoned. The double tuyere was also used in Voisin's cupola, by another English cupola designer and constructor, and in Woodward's steam jet cupola, also an English cupola, many years before they were introduced into this country by Mr. Colliau about 1876.

It is claimed for the double tuyere that the second row consumes the gases which escape with the single tuyere, and, therefore, a great saving in fuel is effected in melting. That a more intense heat is created in the cupola at the melting zone by the double tuyere cannot be disputed, for the destruction of lining is much greater at this point than with the single tuyere; but on the other hand, that any saving in fuel is effected has not been proven by comparative tests made in melting with the double tuyere cupola and the single tuyere cupola, when properly constructed and managed. On the contrary it has been proven that the single tuyere cupola is the most economical in fuel and lining. That the double tuyere melts iron faster than the single in cupolas of the same diameter is undisputed, and as between the single and double it is only a question whether the time saved in melting more than compensates for the extra expense of lining. When a double tuyere cupola is run to its full capacity, the consumption of fuel per ton of iron is about the same as the single tuyere, but in small heats it is much greater. This is due to the large amount of fuel required for a bed, owing to the great height of the upper tuyeres above the sand bottom; for the bed must be made about the same height above the upper tuyeres as above the lower in a single tuyere cupola, and no greater amount of iron can be charged on the bed with the double tuyere than with the single. When constructing or ordering a double tuyere cupola, the smallest one that will do the work should be selected, so that the cupola may be run to its fullest capacity each heat and the best results obtained in melting.

THREE ROWS OF TUYERES.

A number of large cupolas have been constructed with three rows of tuyeres, for the purpose of doing faster melting than can be done with the single or double tuyere cupola. Probably one of the best melting cupolas of this kind in use at the present time is one constructed by Abendroth Bros., Port Chester, N. Y., to melt iron for stove plate, sinks, soil pipe and plumbers' fittings. This cupola is 60 inches diameter at the tuyeres and 72 inches at the charging door, and is supplied with blast from 36 tuyeres, placed in the cupola in three horizontal rows 10 inches apart, 12 tuyeres being placed in each row. The tuyeres in the first row are 6 inches square, those in the second row 4 inches square, and those in the third row 2 inches square. This cupola melts 60 tons of iron in four hours, which is probably the fastest melting done in this country for the same number of hours for light work requiring hot iron.

THE CUPOLA FURNACE.

In the double or triple tuyere cupola the upper tuyeres may be placed directly over a tuyere in the lower row, or they may be placed between the tuyeres of the lower row at a higher level. In Ireland's cupolas double the number of tuyeres were placed in the upper row as were in the lower row, so that one was placed directly over each tuyere in the lower row and one between. In the modern double tuyere cupola the same number of tuyeres are placed in each row, and the upper tuyeres are generally placed between those in the lower row. The object in placing these tuyeres in a cupola, as stated before, is to supply the oxygen to burn the unconsumed gases escaping from the combustion of fuel at the lower tuyeres. If a proper amount of blast is admitted at the lower tuyere the cupola is filled with gases at this point, and it does not make any difference whether the upper tuyeres are placed over or between the lower ones, so long as the tuyeres are only to supply oxygen to consume the gases with which the cupola is filled. If this theory of producing heat by consuming the escaping gases from the combustion of fuel is correct, they can be consumed at any point in the cupola, and the row of tuyeres for this purpose should be placed above the bed, and the gas burned in the first charge of iron to heat it and prepare it for melting before it settles into the melting zone. To consume these gases only the tuyeres should be small, and the number of tuyeres in the upper rows should be two or three times greater than in the lower row, so as to supply oxygen to all parts of the cupola, and not permit the gases to escape unconsumed between the tuyeres. If the tuyeres in the second or third rows are made too large in proportion to the lower row, the supply of oxygen is too great for the combustion of the gases, and the effect is to cool the iron. In the modern double tuyere cupola this theory is not carried out, for the tuyeres in the second row are made big, and admit such a large volume of oxygen at one point that if they were placed high their effect would be to cool the iron rather than heat it. But they are placed low so as to force the blast into the bed and give a deeper melting zone, and their effect is to cause a more rapid combustion of fuel and do faster melting than is done in the single tuyere cupola of the same diameter.

GREINER TUYERE.

In Fig. 17 is seen the Greiner tuyere. The novelty of this device consists in a judicious admission of blast into the upper zones of a cupola, whereby the combustible gases are consumed within the cupola and the heat utilized to pre-heat the descending charges, thereby effecting a saving in the fuel necessary to melt the iron when it reaches the melting zone. This device consists of a number of upright gas pipes attached to the top of the wind box around the cupola, with branch pipes of 1 inch diameter extending into the cupola through the lining and about 1 foot apart, from a short distance above the melting zone to near the charging door. It is claimed that these small pipes admit a sufficient amount of oxygen to the cupola to burn the carbonic oxide produced by the carbonic acid formed at the tuyeres absorbing carbon from the fuel in its ascent. A great saving in fuel is thus effected by consuming this gas and preparing the iron for melting before it reaches the melting zone. A large number of cupolas with this device are in use in Europe, and quite a number in this country.

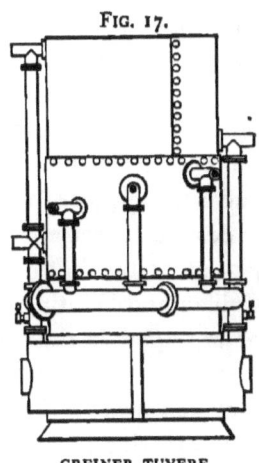

FIG. 17.

GREINER TUYERE.

ADJUSTABLE TUYERES.

Tuyeres are sometimes placed in a cupola so that they may be adjusted to conform with the size of the heat to be melted or the way the iron is to be drawn from the cupola, and thus save fuel in the bed. They are placed low when the heat is small or the iron is drawn from the cupola as fast as melted, and placed high when the heat is large or when iron is to be held in the cupola for a large piece of work. One of the best arranged cupolas of this kind we have seen is the cupola of the Pennsylvania Diamond Drill & Mfg. Company, Birdsboro, Pa. The air belt extending around the cupola is riveted to the shell about 4 feet from the bottom plate. From this belt a cast iron air box bolted to the shell extends down nearly to the bottom plate in front of each tuyere. The front of this box has a sliding door extending full length of the box. The cupola shell has a slot in front of each box the full length of the box. On each side of this slot a piece of angle iron is riveted to the shell to hold the lining in place. The slot is filled in with fire-brick, and a tuyere opening is left at any desired height from the bottom. When it is desired to lower the tuyere the brick are removed from the bottom of the tuyere and placed at the top, and held in place by a little stiff daubing or clay, and when it is desired to raise it the brick are removed from the top and placed at the bottom when making up the cupola. With the Colliau and Whiting style of air belt an adjustable tuyere can be arranged in this way at a very moderate cost, and foundrymen who think they must have their tuyeres placed high so they can make a large casting and only make such a casting once or twice a year, can save a great deal of fuel from the bed by having their tuyeres arranged in this way. The old plan of putting in two or three tuyere holes one above the other, and adjusting the tuyeres during the heat by raising the tuyere pipe from one to the other, is not practicable with the modern way of charging a cupola, and has long since been abandoned.

BOTTOM TUYERE.

In Fig. 18 is seen the bottom or center blast tuyere. This tuyere, as will be observed, passes up through the bottom of the cupola instead of through the sides, and admits the blast to the center of the cupola at the same level as the side tuyeres. It is not designed to change the nature of the iron by forcing the blast through the molten iron in the bottom of the cupola, and, in fact, the blast has no more effect upon the quality of iron when admitted in this way than when admitted through side tuyeres. A tuyere when placed in the bottom of a cupola, unlike a side tuyere, is brought in direct contact with heated fuel and molten iron, and it must be made of a refractory material, or protected by a refractory material if

FIG 18.

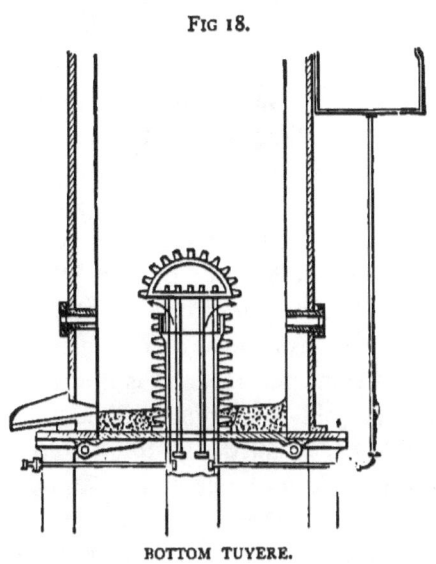

BOTTOM TUYERE.

made of metal. The tuyere shown in the cut is made of cast iron and is provided with a water space between the outside and the inside, through which a stream of water constantly flows, when the tuyere is in use, from a small pipe connected with a tank placed alongside the cupola or on the scaffold. But it has not been found necessary to keep the tuyere cool with water in short heats, for the heat in a cupola under the tuyeres is not sufficiently intense to melt cast iron, and the tuyere may be sufficiently protected against molten iron dropping upon it or coming in contact with it by a thick daubing of refractory material held in place by the prickers cast on the tuyere. The mouth of a bottom tuyere must be covered to prevent molten iron, slag and fuel dropping into it in their descent to the bottom of the cupola. This is done with a rounded cap placed on top of the tuyere to throw off the molten iron and slag, and the blast is admitted to the cupola through an opening around the tuyere under the cap, as indicated by the arrows. The tuyere must be carefully dried and daubed before it is put in place. It cannot be attached to the bottom doors and must be put in place through a hole in the doors after they are put up, and withdrawn in the same way and removed before the cupola is dumped, to prevent it being broken or injured in falling or by the heat in the dump. It must have an adjustable and removable support, and the sand bottom must be made up very carefully around it to prevent leakage of molten iron. The tuyere often gets fast in the bottom and the men are frequently burned in removing it, and it sometimes gets filled with iron or slag, and spoils a heat.

The bottom tuyere has been tried a great many times by foundrymen at different periods, and is nothing new. In conversing with several old foundrymen in Massachusetts about 20 years ago we learned that the bottom tuyere had been used in that State away back in the 40's, and at one time was quite popular with foundrymen there; and we have met a number of other old foundrymen in different sections of the country who had tried the tuyere years ago and given it up. A bottom tuyere was patented by B. H. Hibler in this country August 13, 1867. Ireland & Voisin used a bottom tuyere in their cupola many years ago, and had these practical men found any advantages in it over the side tuyere it would, no doubt, have been brought into general use in cupolas before this.

The bottom tuyere was brought prominently before the foundrymen of this country by an ably written article by Thomas D. West, read before the Western Foundrymen's Association at Chicago, Ill., October 18, 1893, in which he describes his experiments with the tuyere and claims for it a great saving in fuel and cupola lining. Since the publication of Mr. West's article a number of foundrymen have published their experience with the tuyere and all claim it effects a great saving in lining and fuel. But if these foundrymen have not discovered some new feature in the tuyere that was overlooked by experimenters with it years ago, it will never come into general use.

SIZE OF TUYERES.

Foundrymen make a great mistake in placing small tuyeres in their cupolas, with a view of putting the blast into the cupola with greater force and driving it to the center of the cupola with the blower. Air may be driven from a small opening by a blower with greater velocity than the same volume of air from a large opening, but the air from a small opening loses its velocity when it strikes a solid body, just the same as the air from a large opening. When the blast from a small tuyere strikes the solid fuel in front of it, its velocity is gone and it will not penetrate any further into the stock than the same volume of blast from a large tuyere. It is not the velocity at which the blast passes into a cupola that drives it to the center, but the force behind the blast. Neither is it the velocity of the blast that does the melting. It is the volume of blast. It therefore follows that nothing is gained in melting by forcing the blast through a small tuyere into a cupola with great velocity, and much is lost by increasing the power required to run the blower to force the blast through a small tuyere.

The small tuyere was one of the greatest mistakes made in the old-fashioned stave cupola. In these cupolas, many of which we have seen, only two tuyeres of 3 or 4 inches diameter were placed in a 30-inch cupola, and the improvement made in melting in the modern cupola is largely due to the enlargement of the tuyeres and the free admission of blast to the cupola.

The combined tuyere area of a cupola should be equal to three times the area of the outlet of the blower when the blower is of a proper size for the cupola. These dimensions may seem large at first sight, but it must be remembered that the size or area of a tuyere when a cupola is not in blast does not represent the area of the tuyere when a cupola is in blast or the volume of blast that may be admitted to the cupola by the tuyere. When a cupola is in blast the space in front of the tuyere is filled with fuel weighted down by tons of iron. This fuel closes the mouth of the tuyere, and the outlet is represented by the number of crevices between the pieces of fuel through which the blast may escape. Should a large piece of fuel fall in front of a tuyere the blast cannot remove it and the tuyere may be closed and rendered useless. Small tuyeres are more liable to be closed in this way than large ones, and for this reason they should never be placed in a cupola. Small tuyeres, furthermore, are not only more liable to be stopped off by the fuel but also tend to promote bridging by admitting an insufficient amount of blast at certain points.

HEIGHT OF TUYERE.

There is a wide difference of opinion among foundrymen as to the height or distance tuyeres should be placed in a cupola above the sand bottom. So great is this difference of opinion at the present time that tuyeres are placed in cupolas at from 2 inches to 5 feet above the sand bottom. This wide variation in the height of tuyeres is due to some extent to the different classes of work done in different foundries, it being claimed by foundrymen making heavy work that it is necessary to have the tuyeres high to hold molten iron in the cupola and keep it hot for a large casting. Foundrymen making light castings requiring very hot iron draw the iron as fast as melted, and do not think it necessary to have high tuyeres to hold iron in the cupola. In the many experiments we have made in melting iron in a cupola, we have placed the tuyeres at various distances above the sand bottom, and closely observed the effect of tuyeres at different heights. We learned by these experiments that the fuel under the tuyeres is not consumed in melting, nor is it wasted away to any extent by the heat or molten iron coming in contact with it. Charcoal may be placed in the bottom of a cupola, and if care is taken to prevent it being consumed by admission of air through the front before the blast is put on, the charcoal will not be consumed during the heat and may be found in the dump. We have tried this in our experiments to soften hard iron by bringing the molten metal in contact with charcoal in the bottom of a cupola, and found it correct. Pieces of charred wood used in lighting up are often found in the dump after having remained in the cupola through a heat. If these soft combustible substances are not consumed under the tuyeres, then it is not at all likely that the less combustible hard coal and coke are consumed. No iron can be melted in a cupola under the tuyeres, and the only function of the fuel below the tuyeres is to support the stock in a cupola above the tuyeres. If there is not sufficient heat in the bottom of a cupola to consume wood or charcoal, then there is not sufficient heat to keep molten iron hot for any length of time; and it is a well-known fact among practical foundrymen that large bodies of molten iron can be kept hot and fluid for a greater length of time in a ladle when covered with charcoal to exclude the air than it can be in a cupola.

Another reason given in favor of high tuyeres is that it is necessary to have them high to tap slag in long heats. The only slag in a cupola that can be drawn through a slag hole is a light fluid slag that floats on top of the molten iron or rests on the bottom of the cupola when there is no molten iron in it, and this slag may be drawn at any point between the sand bottom and tuyeres. When a slag hole is placed high, slag only can be drawn when the cupola is permitted to fill up with molten iron and raise the slag upon its surface to the slag hole. Slag may then be drawn for a few minutes while the cupola is filling up with iron to the slag hole. As soon as the iron reaches the slag hole, however, it flows out and must be tapped from the front. The slag then falls in the cupola with the surface of the iron as it is drawn off and the slag hole must be closed to prevent the escape of blast through it. Iron tapped after permitting a cupola to fill up to a high slag hole is always dull.

When a slag hole is placed low it is not necessary to have the cupola fill up with iron before slag can be tapped, for the slag may be drawn off the bottom of the cupola, and, furthermore, the slag hole may be opened and permitted to remain open throughout a heat without waste of blast. The flow of slag regulates itself when the hole is of proper size. It is, therefore, not necessary to place tuyeres high that slag may be drawn from a cupola, nor is it necesssary to hold iron in a cupola for a large casting or to keep it hot. Molten iron should be handled in a ladle and not in a cupola.

Hot iron for light work cannot be made in cupolas with high tuyeres, and for this reason the tuyeres in stove foundry cupolas are always placed low. In cupolas of large diameter, having a large bottom surface for molten iron, the tuyeres are placed so low that those at the back of the cupola are not more than 1 inch above the sand bottom, and those in front not more than 2 or 2½ inches above the sand bottom. Tuyeres placed in this way give ample space below them to hold molten iron for this kind of work, for the iron must be very hot and is drawn from the cupola as fast as melted, and the cupola is large enough to melt iron as fast as it can be handled, and it is only when the cupola is not working free that it is stopped up to accummulate iron. The tuyeres in any cupola may be placed as low as in these large ones, if provision is made for handling the iron as fast as melted.

In smaller cupolas not capable of melting iron sufficiently fast to fill a 40 pound hand-ladle, every 8 or 10 seconds the tuyeres are placed from 2 to 4 inches above the sand bottom, so that a sufficient quantity of iron may be collected before tapping to give each man in the section catching a hand-ladle full, and fill the ladle in about 6 seconds.

In cupolas of very small diameter the tuyeres should be placed from 6 to 10 inches above the sand bottom. These very small cupolas melt so slow that if the iron is drawn as fast as melted the stream is so small that the iron is chilled in flowing from the cupola to the ladle more than it is by holding it in the cupola until a body of iron is collected sufficient to supply a large stream.

In machine and jobbing foundry cupolas tuyeres are generally placed from 18 to 24 inches above the sand bottom. The object in placing the tuyeres so high is to hold iron in the cupola for a large casting. But, as before explained, this is not neecssary or advisable. Another reason for these high tuyeres is that they are necessary for tapping slag. The slag from many cupolas is drawn off at the tap hole with the iron, and a number of spouts have been invented for separating the slag from the iron and preventing it running into the ladle. Slag may be drawn from the back of a cupola on a level with the sand bottom at that point, if the iron is drawn as fast as melted, or it may be drawn 1, 2 or more inches above the sand bottom at that point. It is, therefore, not necessary to place tuyeres at so great a height to tap slag.

The tuyeres in cupolas for heavy work should be placed from 6 to 8 inches above the sand bottom when slag is not to be tapped. This gives an abundance of room in a cupola for holding iron while removing or placing a large ladle, and that is all that is necessary. The tuyeres in many of the cupolas used in Bessemer steel works are placed 5 feet above the bottom. They are probably placed at so great a height because the tuyeres in the first cupola constructed for this work were placed at that height. Tuyeres in all cupolas should be placed as low as they can be for the size of the cupola and facilities for handling the iron, for the fuel placed in a cupola under the tuyeres is not consumed in melting and is wasted by being heated in the cupola and crushed and burned in the dump. The value of fuel wasted every year in the United States by the use of high tuyeres in cupolas is sufficient to make a man rich.

THE CUPOLA FURNACE.

NUMBER OF TUYERES.

A cupola may be supplied with blast from one tuyere placed on one side of the cupola, but the objection to one tuyere arranged in this way is that the heat is driven by the blast against the opposite side of the cupola, and the destruction of lining at this point is very great. For this reason, at least two tuyeres are always placed in a cupola, and they are located on opposite sides so that the blast will meet in the center and be diffused through the stock. When a greater number of tuyeres than two are placed in a cupola they are located opposite each other and at equal distances apart, to admit an equal amount of blast on all sides and prevent an uneven destruction of lining from the heat being forced unevenly against it by the blast. Any number of tuyeres desired may be placed in a cupola, and as high as 100 have been used in a 40-inch cupola, and a greater number in larger cupolas. But these large numbers have given no better results in melting than two or four tuyeres in the same cupolas. It is not necessary to place a large number of small tuyeres in a cupola to distribute the blast evenly to the bed, and it is not advisable to put in small tuyeres, which are easily closed by the fuel, cinder and iron, and are oftener rendered useless than large ones. Better results are obtained from large tuyeres and fewer of them.

The largest cupola in use may be supplied with blast by two tuyeres if they are big enough. The large cupola of the Buffalo School Furniture Company, Buffalo, N. Y., is supplied with blast by two tuyeres 12x18 inches, placed on opposite sides. This cupola, which is 60 inches in diameter inside, does excellent melting with only these two tuyeres, and the destruction of lining in melting is very light. We saw a large cupola with two tuyeres of about the above dimensions in use in a stove foundry in St. Louis, Mo., about 20 years ago, and it did excellent melting. The results obtained from these two cupolas would go to show that there is nothing gained in distributing the blast to the bed evenly by a large number of small tuyeres. When a number of tuyeres are placed in one row, every other tuyere is sometimes placed about the width of the tuyeres higher than the tuyeres on either side of it. We have, however, never observed that anything was gained in melting by placing tuyeres in this way. When a double row of tuyeres is used the upper row should be made very small in comparison with the lower row, for if they are made of the same size as the lower one, or even half the size, and the two rows are placed at any great distance apart, the heat is so concentrated upon the lining between them that it may be burned out to the casing in one or two heats. Foundrymen using the double tuyeres, who find the destruction of lining very great, may prevent it to some extent by reducing the size of the upper tuyeres.

SHAPE OF TUYERES.

The shape of a tuyere has nothing to do with the melting, except as it may tend to prevent bridging or increase the depth of the melting zone by supplying blast to the fuel at different heights in a cupola. A small horizontal slot tuyere extending around a cupola, or the greater part of the way around it, tends to promote bridging, and it is generally conceded that a cupola with a tuyere of this kind cannot be run for a greater length of time than two hours without bridging and clogging up. Vertical slot and reducing tuyeres supply blast to the bed at different levels and increase the depth of the melting zone the same as the double tuyere. For this purpose the Truesdale, Lawrence and triangular tuyere, with elongated sides, are excellent when made of a proper size and placed a proper distance apart. When it is not desired to admit the blast to the bed at different levels, the flat or oval tuyeres are generally considered the best shapes, for they admit the blast freely, and a less amount of fuel is required for a bed with these shapes than with a round or square tuyere of the same area.

TUYERES TO IMPROVE THE QUALITY OF IRON.

All kinds of fancy-shaped tuyeres have been placed in cupolas to improve or change the quality of iron in melting.

They have been placed to point up, point down, point across each other at certain angles, and to point to the center of the cupola. There is nothing more absurd than to attempt to improve the quality of iron in a cupola by the shape or angle of the tuyeres. The instant the blast leaves the mouth of a tuyere it strikes the fuel in front of it. The shape or angle given to it by the tuyere is then instantly changed, and it passes through the crevices in the fuel until its oxygen enters into combination with the carbon of the fuel and produces combustion. It then escapes at the top of the melting zone, where it comes in contact with the iron as carbonic acid gas. This is the result, no matter what the shape or angle of the tuyeres, if a proper amount of blast is supplied. It may be claimed that the blast acts upon the iron as it drops through the fuel in the bed after being melted; but as before stated, the shape or angle given to the blast by the tuyeres is changed by the fuel, and the effect on the iron of the blast from one tuyere would be the same as from another.

TUYERE BOXES.

The tuyeres may be and are often formed in the lining of a cupola when laying the brick, but this is a very poor way of making tuyeres, for there is nothing to support the brick and maintain the shape of the tuyeres, and they are often broken or burned away until there is no regular shape to the aperture, and it is difficult to put the blast into the cupola at the point desired or to prevent iron or slag getting into the tuyere. Tuyeres are more generally formed with a cast iron lining or tuyere box, having the shape and size of tuyere desired. This box may be cast with a flange on one end and be bolted to the casing, or it may be cast without a flange and placed in the lining at the desired point as it is laid up. The boxes are made in both ways, but it is better to cast it with a flange and bolt them to the casing, making an air-tight joint, as it then insures the blast going directly into the cupola at the point desired, Tuyere boxes laid in a lining answer the purpose very well when the lining is new, but when it becomes old and shaky, or a section is removed and replaced, the lining often settles and the grouting or filling falls out, leaving crevices through which the blast escapes between the casing and lining, and from there enters the cupola at points where it does no good.

The cold blast supplied to a cupola keeps the tuyere box cool, and it is not necessary to cast it hollow and fill it with water to prevent it being melted or injured by the heat. The only part of the box that is exposed and liable to be injured is the end next the fire, and to protect it the box at this point is generally cast about ½ inch shorter than the thickness of the lining and the end covered with a little clay or daubing.

CHAPTER IV.

CUPOLA MANAGEMENT.

THE peculiarities in the working of every cupola must be learned before it can be run successfully, and this can only be done by working it in different ways. It is a question very much disputed whether a cupola constructed upon the latest improved or patented design is superior to one of the old style. This question can only be decided by the intelligent working of each cupola, and the advantage will always be found in favor of the one that is properly worked, no matter what its construction. It is the duty of every foundryman to give his personal attention to the working of his cupola if he has time. If he is not a practical founder or has not the time to devote to this branch of the business that it requires, then he should have his foundry foreman give it his personal attention for a sufficient length of time each day to see that everything is right in and about the cupola.

No cupola can be run successfully by any given rule or set of rules, for conditions arise to which the rules do not apply. We shall therefore not only give directions for the proper working of a cupola at every point, but shall also give the results or effect of bad working at every point, so that the founder when he finds his cupola is not operating well may have some data from which to draw conclusions and be able to overcome the difficulty.

DRYING THE LINING.

The cupola having been newly lined, nothing is to be done to the lining for the first heat but to dry it. A very high or prolonged heat is not required for this when only one thickness of brick is put in and laid up in thin grout. The lining may be dried by making a wood fire after the sand bottom is put in, or by starting the fire for the heat a little sooner than usual. But the fire must not be started too early or the bed will be burned too much and the cupola filled with ashes, which will retard the melting.

When a backing or filling of wet clay or sand several inches thick is put in between the casing and lining, more time and care are required in drying. It must then be dried slowly and evenly, or the filling will crack, and when jarred in chipping out will crumble and work out through cracks in the lining or holes in the casing and leave cavities behind the lining. When a lining is put in in this way, the doors are put up and covered with sand and a good coal or coke fire is made in the cupola and allowed to remain in over night. In the morning the bottom is dropped to remove the ashes and cool off the lining before making up the sand bottom for a heat.

PUTTING UP THE DOORS.

The first thing to be done when making up the cupola for a heat is to put up the bottom doors. When the cupola is of small diameter and the door light it may be raised into place and supported by one man. But when the door is heavy two men are required, and if the cupola is a large one and the door made in two parts, three men are required to lift and support them. Two men get inside the cupola and raise one-half into place while the third man supports it with a temporary prop; they then raise the other half as far as it can be raised with their bodies between the two doors, where it is supported by a temporary prop. The men then get under the door on their hands and knees and raise it into place on their backs, and it is then supported by a prop.

Numerous devices have been arranged for raising the doors into place, but they soon get out of order from the heat of the dump or carelessness in manipulation, and they have almost all been abandoned. When the cupola is very small and the door light, it is sometimes supported by an iron bolt attached to the under side of the bottom plate at the front, where it can be readily withdrawn with an iron hook to drop the bottom. But the doors are generally supported by a stout iron prop or post placed under the door near the edge opposite the hinges. Double doors are supported by a stout iron prop in the center and generally a light one at each end of the doors to prevent them springing when charging the fuel and iron, or by a sudden settling of the stock, as may occur when melting large chunks. A great many melters have no permanent foundation under the cupola upon which to place the main prop, but make one every heat by laying down a small plate upon the sand and setting the prop upon it. The plate is often placed too high or too low, making the prop too long or too short, and the plate must be raised by putting a little more sand under it or lowered by scraping away a little sand. While this is being done the heavy iron prop, which frequently requires two men to handle in the cramped position in which they are placed under the cupola, has often to be put up and taken down two or three times before it is gotten into the right position to support the doors.

All this extra labor can be avoided and time saved by imbedding a heavy cast iron block in the floor or foundation under the cupola for the prop to rest upon. It must extend down a sufficient distance to insure its not being disturbed when shoveling out the dump. A block 6 inches square and 10 inches long, placed with the end level with the floor, will seldom be displaced, and makes a sure foundation for the prop. The size of prop required to support a bottom depends upon the size of cupola. In small cupolas the stock is supported to a large extent by pressure against the lining, while in large cupolas the stock is supported almost entirely by the prop. For small cupolas the props are made from $1\frac{1}{2}$ to 2 inches diameter, and for large cupolas from 3 to $3\frac{1}{2}$ inches diameter.

The props for large cupolas not only have a greater weight to support, but they are seldom pulled out of the dump and are therefore, if light, liable to be bent and twisted to such an extent as to render them useless. For this reason they are often made heavier than is actually necessary for the support of the bottom. Quite a number of foundrymen have adopted the plan of attaching a ring to the prop near the top or bottom with which to draw it from the dump and avoid heating it. The ring is made large and hangs loosely, or as a long loop which stands out from the prop. When the prop is to be removed a hook is placed in the ring or loop and a quick jerk given, which releases it, and it is at once drawn from under the cupola.

Some of the older melters never use the iron prop, but measure and cut a new wood prop for their cupola every heat. Many of them are so superstitious that they think the cupola would not melt without the new prop, and they would rather give up their job than try it. Such melters are not so plentiful now as they were 20 years ago, when we first began traveling as a melter through this country and Canada, but we find when visiting foundries there are still a few of them left.

DROPPING THE DOORS.

When it is desired to drop the doors it is done by removing the props or drawing the bolt. The small props are first taken out, being released by a stroke of the hammer, and are carefully laid away so that they will not be bent by the heat of the dump. A long bar with a handle on one end and a large hook on the other is then placed under the cupola with the hook behind the main prop and about 10 or 12 inches from it. By a sudden jerk of the bar the hook is made to strike the bottom of the prop a sufficiently hard blow to knock it out of place and permit the door or doors to drop. Two or more blows of the bar are sometimes necessary to release the prop, but it can always be released in this way. The prop can also be released by striking it at the top with a straight bar, but it is oftener missed than hit, and many thrusts are sometimes required to bring it down. Bolts are only used on small cupolas from which the dump falls slowly, and the bolt can generally be withdrawn by a blow of the hammer without danger to the melter. If it cannot be withdrawn in this way without danger of burning the melter, a hook is made on the end of the bolt or a ring placed in it so that it may be drawn with a hooked bar or struck with a long straight bar.

SAND BOTTOM.

When the door or doors are in place and properly supported, any openings or holes that may have been burned through them are carefully covered with a thin plate of iron, and all cracks through which the bottom sand might escape when dry are closed with clay. The doors are then covered with a bed of sand several inches in thickness, which is known as the sand bottom. The sand employed for this purpose must not be of a quality that will burn away and permit the molten iron to get down to the doors, or melt and form a hard mass that will not fall from the cupola when the doors are dropped, neither must it be so friable as to permit the molten iron to run through it when dry.

The clay sands when used for a bottom burn into a hard, tough mass that adheres to the lining all around the cupola, and in a small cupola frequently remains in place after the door is dropped and has to be dug out with a bar before the cupola can be dumped. Parting sand, sharp and fire sands are very friable and difficult to keep in place. They do not resist the action of the molten iron well, but melt and form a slag. Mixtures of clay and sharp sand burn too hard and do not drop well. The loam sands are the only ones suitable for a sand bottom, and sand that has been burned to a limited extent makes a better bottom than new sand.

In stove and other foundries with large gangway floors the scrapings from the gangways are collected in front of the cupola, passed through a No. 2 riddle to recover the scrap iron, and the sand used for the cupola bottom. This sand makes the very best kind of bottom. It is clean and free from cinder, soft and pliable, packs close, resists the action of the molten iron and drops free. In foundries where the daily gangway cleanings are not sufficient to make the bottom, part of the old bottom is used over and the gangway cleanings are mixed with it or placed on top. In foundries where there are no regular gangways to clean every day, the heavy part of the dump is thrown out and the sand bottom passed through a No. 2 riddle and used over again. When the bottom sand is used over day after day it must not be riddled out too close, and a little fresh material must be added to it each day to prevent it becoming rotten from repeated burnings and containing too many small particles of cinder, which render it fusible and easily cut away by the molten iron. The cleanings from the molding floors are generally added or a few shovels from the sand heaps, and in case it becomes too rotten a few shovels of new molding sand are mixed with it.

When the material contains so much cinder that it does not make a smooth bottom, a few shovels of burned sand from the heaps are put on top to give an even surface and prevent the molten iron coming in contact with the cinder and cutting the bottom. The bottom sand is generally wet with water, but some melters wet it with clay wash, to make it more adhesive and give it more strength to resist the action of the molten iron. A thick clay wash gives strength to a rotten sand when mixed with it, but it also increases the tendency of the bottom to cake and hang up, and it is better to improve the bottom material in the way above described and wet it with water only. The sand when wet is cut over and evenly tempered, and should be no wetter than molding sand when tempered for a mold.

The sand may be thrown into the cupola through the front opening, or may be thrown in at the charging door, but it is generally thrown in at the front, for it is more convenient to the material, and is also convenient for spreading it in the cupola. When the cupola is small the melter stands by the side of it and makes up the bottom by passing his arm in through the front opening, but when the cupola is large he goes inside, and his helper shovels the sand in as he wants it. The first sand thrown in is carefully packed around the edges with the hands to insure a tight joint. As the balance of the sand is thrown in it is spread evenly over the bottom in layers from 1 to 2 inches thick, and each layer is evenly rammed or trampled down until the required thickness of bottom is obtained, which is from 3 to 6 inches, according to the rise of the cupola. The desired pitch or slope for throwing the iron to the front is then given, and the bottom butted evenly and smoothly all over. The melter next goes carefully around the edges with his hands and feels for any soft spots there may be near the lining, and slightly raises the edges of the bottom around the lining to throw the iron off and prevent it working its way down between the lining and sand bottom. The bottom is then carefully brushed and smoothed off, and in small cupolas a bucket of thin clay wash is sometimes thrown in at the front and caught in the bucket as it runs out. This is called slushing the bottom, and is done to give a smooth, hard surface.

The sand bottom does not always remain impervious to the molten metal, but is sometimes penetrated or cut up and destroyed by it, in which case a leakage of molten iron takes place from the bottom of the cupola that is difficult to stop. Leakage of this kind may be due to springing of the bottom doors when charging and the cracking or loosening of the sand bottom around the lining. This can be prevented by placing more props under the doors to support them. Sand that has been used over and over in a bottom until it has become worn out and filled with cinder is readily cut up and converted into a slag by the molten iron, and it is only a question of the time occupied in running off the heat whether the bottom gives way or stands. When the bottom sand gets into this condition, it must be renewed by the addition of new sand, or the bottom covered with a layer of sand from the molders' sand heaps.

Molten iron will not lie upon a wet, hard substance, but will explode or boil and cut up the material upon which it is placed. If the bottom sand is made too wet, or rammed too hard, or rammed unevenly, the iron will not lie upon it, but will boil and cut up the sand until it gets down to the doors, which it will melt and run through. When a bottom cuts through, melters frequently attribute it to the bottom being too soft; and we have seen them take a heavy pounder and ram a bottom as hard as a stone. In these cases, if the sand was worked very dry, or the bottom was well dried out before any molten iron came in contact with it, it did not cut up or leak; but if the sand was wet when the molten iron came down, boiling at once took place and the bottom soon cut through—and in such cases they generally cut through about every other day. In the sand bottom of a cupola we have the same elements to contend with, so far as molten iron is concerned, as we have in a mold; and the sand should be worked no wetter, rammed no harder, and rammed as evenly as the sand for a mold. The sand should not be worked wet for a bottom, under the impression that it is dried out before the iron comes down, for the ashes of the shavings, wood, coal or coke cover the bottom soon after the fire is started, and protect it from the heat to such an extent that it is only dried to a very limited degree before the iron comes down upon it. Water may be seen dripping from a very wet bottom long after the blast is on. Even if it were dried out, wet sand cracks when dried rapidly and should not be used. We shall not attempt to give any directions for stopping a leak after it occurs, for the time and place to stop a leak is when putting in the sand bottom; and if all the remedies we have given for preventing leaks fail, then it is time to change the melter.

    The pitch or slope given to the bottom to cause the molten iron to flow to the tap hole from all parts of the bottom has a great deal to do with the temperature of the iron and nice working of a cupola. When the bottom is made too low and flat, molten iron lies in the bottom of the cupola and becomes dull. As the melted iron falls into this iron drop by drop, it is instantly chilled and the iron when drawn from the cupola is dull. This effect is more marked in a cupola melting very slowly, and a low bottom may be the cause of very dull iron when a sufficient quantity of fuel is consumed to make very hot iron. A high pitch throws the iron from the tap hole with great force and spouting velocity, and it is almost impossible to run a continuous stream from a cupola with such a bottom. It is more difficult to keep the tap hole and spout in order, and the stream must be closely watched to prevent it shooting over the ladle and burning the men. Slag flows freely from the tap hole with the stream of iron when the bottom has a high pitch, even when there is very little slag in the cupola. But the flow of slag from the tap hole with the iron may be entirely stopped by changing the pitch of the bottom, no matter how great the quantity of slag in the cupola. The action of the iron at the spout is entirely changed by the pitch of the bottom. A hard iron may be made to run smooth from the spout, while a soft iron may be made to sparkle and fly, giving all the indications of a hard iron. The best expert on the quality of iron at the spout may be deceived in the iron by the pitch of the bottom, and it is only in the extremely hard and extremely soft iron they cannot be deceived. The bottom should never be made hollow in the center and high all around the outside with an outlet or trough to the spout. This concentrates the iron in the center in such a way that a few hundred weight places as great a pressure upon the front as a ton would do if the bottom were flat, and the front may therefore be forced out by a comparatively small body of iron. The instant the tap hole is open the iron rushes out with great force, and it is almost impossible to stop it as long as there is any molten iron in the cupola.

The bottom should be made flat and level from side to side with only a slight rise around the lining, which should not extend out more than 1 or 2 inches from the lining. The pitch from back to front should not be more than $\frac{1}{2}$ to $\frac{3}{4}$ inch to the foot. This has been found to be a sufficient slope to throw all the iron to the front in an ordinary cupola. But in cupolas that melt very slowly a little more slope may be given, so as to concentrate the iron more rapidly and prevent it chilling on the bottom.

In cupolas with two tap holes the bottom must be sloped so that all the melted iron in the cupola can be drawn from either tap hole. It is very difficult for a melter to see what slope he is giving a bottom when inside the cupola, and for this reason many of them seldom get the slope alike. The melter should be provided with a notched stick or some other gauge, for measuring down from the top or bottom of each tuyere, to serve as a guide in sloping the bottom, so that it may be given the proper pitch and put in alike every heat.

SPOUT.

The old way of making a cupola spout is to place a short piece of pig iron on the bottom plate on each side of the front, and build up a spout between them with clay or loam. The modern spouts are made of cast iron with a flat or eight-square bottom, and are from 4 to 6 inches deep, 7 to 10 inches wide and 1 to 10 feet long. They are given a fall from the cupola of about 1 inch to the lineal foot, and are lined with a refractory material to protect them from the molten iron. The spout lining is made of a different material from the sand bottom, and generally consists of molding sand, loam or a mixture of fire clay and sharp sand. Some of the molding sands make an excellent spout lining that is not cut or fused by the stream of molten iron, while others crumble and break up too readily when cleaning the spout of dross and dirt, and cannot be used for this purpose. When a molding sand can be used it makes a nice clean spout that is easily and quickly made up. It is readily dried, and when making up the spout the crust of the old lining can be removed with a bar, and the sand wet up and used over with a coating of sand without removing it from the spout. For long spouts, requiring a good deal of material to line them, molding sand is the most economical lining that can be used.

Some of the loam and blue clays make excellent spout linings alone or when mixed with sand, and are the only materials used for this purpose in some sections of the country where they can be procured at a moderate cost. They make a stronger lining than molding sand—that is, not so liable to be broken up when cleaning the spout of dross and slag—and, furthermore, they dry quickly. The lining material probably more extensively used than any other is a mixture of fire clay and sharp sand. These two refractory substances when combined in right proportions and thoroughly mixed make one of the very best spout linings. But when not properly mixed they make one of the poorest linings.

When too much clay is used the lining does not give up the water of combination until heated to a very high heat, and it is almost impossible to get the lining dry so that the iron will not boil in the spout the first few taps when the spout is long, or sputter and fly when it is short. It cracks when dried rapidly, and is melted into a tough slag that bungs up the spout and cannot be removed without destroying the lining. When too much sand is used the lining crumbles when touched with the bar and is cut and melted by the stream. When the clay and sand are not thoroughly mixed the lining crumbles and cuts or melts in spots. A spout lining made of these two materials in right proportions, properly mixed and dried, becomes as refractory as a fire-brick, and 50 or 100 tons of iron may be run from a spout lined with them without a break in the lining. There are a number of other materials used for spout linings that are only found in certain localities, and their use is restricted to the districts where they can be procured at a moderate cost. But those above described are the materials most commonly used for this purpose.

The spout lining is made up new every heat, and when putting it in the spout is wet to make it adhere to it. The sand bottom is cut away from the front and the spout lining made to extend into the cupola past the tap hole. A perfect joint is made between the sand bottom and spout lining, and a little clay wash is generally brushed over the joint to make it more perfect and prevent cutting. Care must be taken to not get the bottom of the spout at the tap hole higher than the sand bottom, and also to give it the same pitch as the sand bottom. The bottom is put in first and is made about 1 inch thick when the spout has been given the proper pitch. If the spout has not been given a proper pitch, the lining is made heavier at the end next the cupola and light at the outer end and the pitch given in the lining. This is the common practice in short spouts.

The sides of the lining are built up full at the bottom, so as to leave only a narrow groove in the middle and keep the stream always in one place, but are sloped back from the middle to the top of the spout to give a broad spout surface for carrying the stream of iron. A half round groove 1 inch deep and 2 inches wide at the top is sufficient to carry off the stream of iron from almost any cupola. But the spout is liable to be choked up by dirt from the tap hole or slag, and it is made large for safety. A rammer is seldom used in making up a spout and it is generally made up with the hands and one of the bod sticks, or the small round stick used to make the tap hole.

When molding sand is used it is worked a little wetter than for molding and is beaten down with the bod stick and shaped up with the hands and bod stick. When clay or a mixture of clay and sand is used, it is worked wet and placed in the spout in balls and beaten or pressed into shape with the hands, and the bod stick is used to true it up and form the groove in the middle. Short spouts are made up with but little difficulty, but great care must be taken in making up a long spout to have it perfectly true and properly pitched, so that it will clean itself of molten iron the moment the cupola is stopped in.

The greatest strain upon the spout lining is under and around the tap hole, where it is liable to be cut away by the pressure and current of the stream or to be melted if the material is not very refractory, and it may be broken up by the tap bar if not very tenacious when heated to a high temperature. When molding sand or other materials that do not stand a high temperature well or are not very tenacious when heated are used, a layer of fire clay and sharp sand is placed over the lining material under the tap hole. When the heat is very heavy and a large amount of iron is drawn from one tap hole, a split fire brick is embedded under the tap hole to prevent cutting and insure a good tap hole throughout the heat. The spout is seldom coated or painted with blacking after it is made up or dried, but when a friable material is used for lining it is sometimes coated with clay wash.

If the spout is made with a broad, flat bottom the stream takes a new course every time the cupola is tapped, and before the heat is over the spout is so bunged up that the iron collects in pools. A continuous stream cannot, therefore, be maintained the length of the spout, and two or more streams may fall from the end of the spout at the same time. To prevent this, shape the lining to form a small groove for the stream in the center and keep it there every tap. The quality of the lining material has a great deal to do with the condition of a spout during the running out of a heat. The spout may be cut out in holes by the stream and pools of iron form in the spout at every tap. This is due to the lining material crumbling and being washed away by the stream. When this does not occur every heat with the same material, it is due to the material not being properly mixed; but if it does occur every heat, it is due to poor material. The spout may become choked or bunged up with slag when no slag flows from the tap hole with the iron. This is due to the lining melting and forming a slag. It is very difficult to keep a spout in order through a long heat when this occurs, and the lining material should at once be changed. Slag should be removed from the spout when very hot by lifting it up with a bar, or chipped away with a sharp bar when quite cold. All attempts to remove a tough semi-fluid slag break up and destroy the lining.

FRONT.

The front opening of the modern drop botton cupola is made so small that it is not necessary to place an apron or breast plate over it to hold the front or breast in place, as is done with the draw front cupola. The material used for putting in the front is generally the same as is used in making up the spout. The front is generally put in after the fire has burned up, but some melters put in the front before lighting, and light from the tuyeres. Others make up the tap hole and half the front with a stiff mixture of fire clay and sharp sand before lighting up, and fill in the other half after the fire has burned up. But as a general rule the entire front is left open to give draft for lighting, and the front is put in after the fire is burned up and about ready for charging. This gives sufficient time for drying it before the blast is put on.

When about to put in the front the ashes and dust are carefully brushed from the spout where the front is to be made, and the spout and front opening are wet all around with water or clay wash to make the front material adhere and insure a good joint. A breast of small pieces of coke is built in front of the fire, or a small board cut to fit the front, with a notch in the bottom for the tap hole, is placed in front of the fire to prevent the front material from being rammed or pressed too far back into the cupola. A small iron bar or a round wooden stick is then laid in the bottom of the spout to form the tap hole.

If the front is made of molding sand or other material that is likely to crumble at the tap hole and be cut away by the stream of iron or be broken away by the tap bar, a little fire clay and sharp sand, or other refractory material, is placed around the bar or stick to form the tap hole. The front material of molding sand or loam is then thrown in and rammed solid against the board, sides, top and bottom of the opening. If the front is made of clay or sand, and worked wet, it is made into balls and pressed into place with the hands. When the opening has been filled the front is cut away downward and inward from the top and sides of the opening to the bar forming the tap hole, until the tap hole is not more than 1½ inches long. The surplus material from the front is then removed from the spout, the bar drawn from tap hole and the front and spout carefully trimmed up.

If the spout lining and front have been made up with clay and sand, or other wet material, a wood fire is built on the spout to dry it and the front. When the spout and front are made up with molding sand or loam they are generally dried by the flame from the tap hole before stopping in, and an iron plate is sometimes laid on top of the spout to concentrate the heat upon it.

The front is generally made the full thickness of the lining and cannot be forced out by the pressure of molten iron if properly put in. When the front material is worked too wet, it falls away from the opening at the top when drying, and the opening must be closed to prevent the escape of the blast. If the tap hole is made too long the iron may chill in it, and the cupola cannot be tapped without cutting a new hole. This makes very bad work, for the iron is generally melted from the old tap hole by the stream passing through the new one, and the two holes become one. It is then very difficult to stop in or control the flow of iron.

When the front material is poor it melts into a semi-fluid slag that settles down and closes up the tap hole with a tough adherent slag that is difficult to remove. When this occurs the tap hole can only be kept open by continually opening it up with a tap bar. The only way to overcome this difficulty is to use a more refractory front material. Mineral fluxes sometimes make a front material fusible that is not otherwise fusible. When trouble is experienced in keeping the tap hole open when using a flux, or after one flux has been substituted for another, the composition of the front material must be changed or another material used.

When no board is used and the front material is rammed back into the fire until it becomes solid in the front, the front is ragged and soft on the inside and melts and makes a bad

CUPOLA MANAGEMENT.

tap hole even when the material is good. A good front or spout lining can always be made from fire clay and sharp sand by mixing them in right proportions for the purpose for which they are to be used.

SIZES OF TAP HOLE.

The sizes the tap hole is made depends upon how the iron is to be drawn from the cupola. If it is desired to run a continuous stream from the cupola, the tap hole is made small to suit the melting capacity of the cupola. If it is desired to accumulate a large body of iron in the cupola and fill a large ladle rapidly when the cupola is tapped, the hole is made large. The tap holes are made of various sizes from $5/8$ inch to $1\frac{1}{4}$ inches diameter, to suit the different kinds of work. When it is desired to run a continuous stream it is very desirable that the tap hole should not be cut and enlarged by the stream. This is generally prevented by placing a very refractory material around the rod forming the tap hole. But some melters have a form in which they mold a tap hole from a carefully prepared material that will not cut and dry it in an oven or on a stove. This tap hole form, when thoroughly dried, is placed in position on a split fire-brick and the front made up around it, which always insures a regular sized hole throughout the heat.

LOCATING THE TAP HOLES.

We have already described the manner of putting in the front and forming the tap hole, and shall here only consider the location and number of tap holes. The tap hole is placed in the side of the cupola from which it is most convenient to convey the iron to the work to be poured, and it makes no difference in the working of the cupola upon which side it is placed if the bottom is sloped to throw the iron to the hole. One tap hole is sufficient to run the iron from any ordinary cupola, but two are frequently put in. In some cupolas two fronts and tap holes are put in side by side only a few inches apart, and two spouts are made up so that the tap hole can be kept in better order for drawing off the iron. They are tapped turn about, and in case too great a quantity of melted iron accumulates in the cupola they are both opened at one time. Two tap holes placed in this way can only be worked for hand ladle work at the same time, and they cannot be worked to advantage even for that, for they are so close together the men are in each other's way. One tap hole if properly made and managed will run off all the iron a cupola will melt, and it is poor cupola practice to put in two fronts and tap holes in this way.

Two tap holes are frequently placed in a cupola for convenience in carrying the iron from the cupola to the molding floors. They are generally placed on opposite sides of the cupola, to save carrying the iron around the cupola or from one molding room to another. Two tap holes are also placed in cupolas to facilitate the removal of the iron in hand ladles. Six 40-pound hand ladles are all that can be safely taken from a spout per minute. When more than this number of ladles are filled and removed per minute, the men have to move so rapidly there is danger of a clashing of ladles and spilling of iron, and when a heavy stream once gets away from the men and falls to the floor, it spatters and flies so that it is difficult to stop in or again catch it. When more than 8 tons are melted per hour in a cupola for hand ladle work, two tap holes are always put in. They are placed in the side of the cupola that is nearest the work to be poured, but always at a sufficient distance apart to admit of the men catching at one spout being out of the way of those catching at the other.

SLAG HOLE.

A slag hole for drawing off slag is sometimes placed in a cupola, but it is not used except when the cupola is run beyond the capacity to which it can be run successfully without slagging. The hole is placed below the level of the tuyeres, and when it is desired to accumulate a large body of molten iron in a cupola the slag hole is placed high. When the iron is drawn from a cupola as fast as melted the hole is placed low.

The opening through the casing and lining is generally made oval and about 3 x 4 x 5 inches.

The slag hole front when the hole is placed high consists of a plug of the same material used for the tap hole front. The plug is placed in the outer end of the opening and is from 2 to 3 inches thick. A hole 1 inch diameter is made through it for a tap hole, and the plug or front is cut away from the edges of the casing to the hole until the hole is not more than $1\frac{1}{2}$ inches long. When the hole is placed low and the slag permitted to flow throughout the heat after it is opened the plug is made of loam or molding sand mixed with a little blacking to make it porous when heated, and the plug is placed in the hole on the inside flush with the lining. No tap hole is made through the plug when placed in this way, and when it is desired to tap slag a hole is cut through it with a sharp pointed tap bar. This material does not bake hard, and the entire plug may be cut out when necessary.

Slag chills more rapidly in a tap hole than iron, and is more difficult to tap or draw from a cupola, and when the slag hole is not properly arranged it cannot be drawn at all. If the tap hole is made small and long the slag chills in the hole and it is difficult to open the hole or keep it open. When the lining is very thick it must be cut away and the hole made large inside of the front, or the slag will chill in the lining the same as it might in the hole in the front. The hole in the lining can be made 6 or 8 inches diameter without injuring the lining, and a hole of this size will admit a sufficient quantity of slag to the tap hole to prevent it chilling. There is never any difficulty from the slag chilling when the front and tap hole are placed flush with the inside of the lining, for the slag is kept hot and fluid in the cupola, and may be drawn off whenever there is a sufficient quantity in the cupola to flow from it. It is therefore better to cut away the casing and lining, and place the front flush with the inside of the lining.

LIGHTING UP.

When the cupola is small the shavings are thrown in from the charging door and evenly distributed over the bottom. The wood is cut short and split fine and dropped down, a few pieces at a time, and so placed that the fire will burn up evenly and quickly. When the cupola is large the melter goes down into it and his helper passes him down the shavings and wood from the charging door. The shavings are evenly spread over the bottom, care being taken to get plenty around the outside to insure a good light. A layer of fine, light dry wood is then laid over the shavings, and on this a layer of heavier wood, and so on until the required quantity of wood for lighting the bed is placed in the cupola. Care is taken to arrange the wood so that it will burn up evenly and quickly. A light dry wood should be used, and the pieces must not be very large or too much time will be consumed in burning them, and the bed will settle unevenly.

When the wood has been arranged the melter gets out and a thin layer of small coal or coke is placed over the wood. The bed fuel is then thrown in evenly over the wood. All the bed is put in but a few shovelfuls, which are kept to fill up any holes that may be formed by an uneven settling. The charging door is then closed and the shavings lighted at the front opening. The tuyere doors are opened to give draft and the fire left to burn up. When the wood is nearly burned out and there is a good fire of hot coals at the front and tuyeres, the melter generally puts in the frontspout and builds a wood fire on the spout to dry them. He then looks in at the charging door, and if the smoke is burned off and the fire beginning to show through the top of the bed, he puts in the remaining few shovels of fuel and makes the top of the bed as level as possible. He then closes all the tuyere doors but one and begins charging the iron into the cupola.

Straw may be used in place of shavings for lighting up when shavings cannot be procured. The wood should be dry pine or other light wood, and it must not be used in too large sticks or the bed will be burned too much before the wood is burned out; and if the iron is charged before the wood is burned out, it smokes and the melter cannot see how to place the iron or fuel. For the same reason, hard or green wood should not be used in lighting up.

When the bed burns up on one side and not on the other in a small cupola, the bed may be burned up on the other side after the blast is put on and the heat run off successfully. But when the bed burns up on one side and not on the other in a large cupola the bottom had better be dropped at once. We once had to drop the bottom of a 60-inch cupola before the heat was half off, for the reason that the melter was careless in arranging the wood and lighting up, and charged the iron with the bed only burned up on one side. He thought the blast would make it burn up on the other side, but it did not, and the heat was a failure. Never burn the bed up to warm or heat up the cupola, for a cupola does not require to be heated before it is charged, and the lining burns out fast enough without wasting fuel to burn it out.

THE BED.

Iron is melted in a cupola within a limited space, known as the melting point or melting zone. The melting point is the highest point in a cupola at which iron is melted properly, and the melting zone is the space between the highest and lowest point at which iron melts properly. Iron may be melted to a limited extent above or below these two points, but it is burned, hardened and generally dull. The melting zone extends across the cupola above the tuyeres, and is from 6 to 8 inches in depth. Its exact location is determined by the volume of blast and the nature of the fuel employed in melting. A large volume of blast gives a high melting point, and a small volume a low melting point. A soft, combustible fuel gives a high melting point, and a hard fuel a low melting point, the blast being equal in volume with both fuels.

To do good melting the melting point must be discovered,

and only a sufficient quantity of fuel placed in the bed to bring the top of the bed up to the melting point. When the fuel is hard anthracite coal, the rule is to use a sufficient quantity of coal in the bed to bring the top of the bed 14 inches above the top of the tuyeres when the wood is burned out; with hard Connellsville coke 18 inches, and with soft coke 20 to 25 inches. But the melting point is varied by the volume of blast and these rules do not always hold good. So the melting point in each cupola must be learned to get the best results from the cupola.

To find the melting point a bed is put in according to the rule and iron charged upon it. If the iron is a long time in coming down after the blast is put on, or the iron melts very slowly during the melting of the first charge, but melts faster at the latter end of the charge and is hot, the bed is too high and the iron is being melted upon the upper edge of the melting zone. Fuel and time are then being wasted, and the fuel should be reduced so as to place the iron at the melting point when melting begins. If the iron comes down quick but is dull, or if it comes slow and dull and does not grow hotter at the latter end of the charge, the melting is being done on the lower edge of the melting zone and the quantity of fuel should be increased to bring the top of the bed up to the melting point. When the top of the bed is placed only half way up the melting zone the iron comes down hot and fast, but the bed does not melt the quantity of iron it should and the latter part of the charge on the bed is dull. The latter part of the charge on the bed when the bed is the proper height is also dull if the charge is too heavy for the bed, and care must be taken in noting this point.

If by comparison with the charges of iron in various sized cupolas the charge on the bed is found to be light, the bed should be raised uutil the melting indicates that it is at a proper height; then the weight of iron on the bed may be increased, if the charge is too light. When raising or lowering a bed, it should be done gradually by increasing or decreasing the fuel from 50 to 100 pounds each heat until the exact amount of fuel required in the bed is found. If the changes in the bed are made gradually in this way, the effect of the changes upon the melting may be observed more accurately and better results obtained than when a radical change is made by increasing or decreasing the fuel in large amounts at one heat. When the amount of fuel is found that brings the top of the bed to a height that gives the best results in melting, the top of the bed is maintained at that point each heat.

When a cupola is newly lined the diameter is decreased from what it was with the old lining, and the weight of fuel in the bed must be decreased to bring the top of the bed down to the melting point, and as the lining burns out and the cupola gets larger the fuel must be increased to keep the bed up to the melting point. Trouble is often experienced in melting after a cupola has been newly lined. This is because the diameter of the cupola is reduced from 6 to 10 inches, and the bed and charges are not changed to correspond with the reduced size of the cupola. There is never any trouble of this kind in foundries where a cupola book is provided and a record kept of the melting from one year's end to another, for the melter or foreman can look back and see the weight of the bed and charges when the cupola was newly lined, and the increase made in the weight as the lining burned out and the diameter increased.

No definite or even approximate weight can be given of the amount of fuel required for a bed in cupolas of different diameters, for the tuyeres are placed at such a variety of heights above the sand bottom that for two cupolas of exactly the same diameter twice the quantity of fuel may be required for a bed in one as is required for a bed in the other. Cupolas with two or three rows of tuyeres require a larger amount of fuel for a bed than cupolas with but one row, but the same general directions for burning and managing the bed apply to all cupolas.

CHARGING.

The old way and the way still in vogue in some localities of stocking, loading or putting the fuel and iron into a cupola is to place a sufficient quantity of fuel in a cupola to fill it above the tuyeres. On this fuel or bed are placed from 50 to 500 pounds of iron, according to the size of the cupola, then from one to four shovels of fuel are put in and from 50 to 200 pounds of iron, and so on until all the iron to be melted is placed in the cupola.

This way of stocking a cupola mixes the fuel and iron in the cupola and they come down to the melting point together. The fuel fills a space that should be filled with iron, and a great deal of the melting surface of the cupola is lost, and the cupola's melting capacity reduced in proportion.

The modern way of stocking a cupola is to put in the fuel and iron in layers or charges. Each layer or charge of fuel is separated from the layer or charge above and below it by a layer or charge of iron, and each layer of iron is separated by a layer of fuel. This way of stocking a cupola is known as charging the cupola. When a cupola is charged in this way the iron comes down to the melting point in a body extending over the melting surface of the cupola, and the entire melting surface is utilized. The melting capacity of a cupola is about one-half greater when charged in this way than when the fuel and iron are mixed, and the consumption of fuel is also less.

The first charge of iron is placed on the bed at the melting point. In melting this charge of iron a certain amount of fuel is consumed and the top of the bed settles down from the top of the melting zone to the bottom of the melting zone. The charge of fuel on top of the charge of iron that has just been melted settles with the iron until it unites with the bed and places the top of the bed again at the top of the melting zone, ready to melt the next charge of iron, and so on with each succeeding charge of fuel and iron throughout the heat. This is the correct theory of melting iron in a cupola, and the practice that must be followed to obtain the best results from a cupola.

Now, having described the theory of charging and melting, let us consider the practical working of a cupola upon the theory. The amount of iron placed upon the bed in the first charge and in each charge through the heat must be the exact amount of iron the fuel will melt while settling from the melting point to the bottom of the melting zone. The amount of fuel in each charge must be the exact amount required to raise the bed from the bottom of the melting zone to the melting point. If the charges of iron are made too heavy the iron comes dull at the latter end of the charge and hot at the first of the charge until a few charges have been melted, when it comes dull all through to the end of the heat. When the charges of iron are too light the iron comes hot, but there is a stoppage in melting at the end of each charge, changing to continuous but very slow melting as the heat progresses.

When the charges of fuel are too heavy the iron melts slowly and unevenly, and if the heat is a long one it comes dull and is hardened in melting. When the charges of fuel are too light and the charges of iron heavy the result is dull iron. When the charges of fuel and iron are both too light the iron generally comes hot but slowly throughout the heat, and the full melting capacity of the cupola cannot be realized.

There is no rule for making the weight of the first charge of iron of any definite proportion to the weight of the bed of either anthracite coal or coke that holds good in all cupolas. Manufacturers of some of the patent cupolas have such a rule for their cupolas that is approximately correct, but the tuyeres in different sizes of these cupolas are always placed at the same height and about the same amount of fuel is required for a bed. The bed will melt a heavier charge of iron in settling than the other charges of fuel, and the first charge is generally made from one-third to one-half heavier than the subsequent charges. The weight of the first charge of iron varies from two and one-half to four and one-half times the weight of the bed with anthracite coal; with coke the weight of the first charge varies from one and one-half to four and one-half times the weight of the bed.

These wide variations in the weight of the first charge of iron in proportion to the weight of the bed are largely due to the difference in the height of tuyeres and the large amount of fuel required for a bed in a cupola with very high tuyeres. But variation is also due in many cases to bad judgment in estimating the weight the first charge should be. The greater the weight of the first charge in proportion to the weight of the bed, the better the average will be in melting, and careful experiments should be made with every cupola to learn the largest amount of iron it will melt on the bed with safety, and that amount should always be placed in the first charge.

There is no rule for making the weight of the charges of fuel or iron of any definite proportion to the weight of the bed or first charge of iron, and the weight of the charges of both fuel and iron is frequently changed in different parts of the heat, to give a hotter iron for some special work or to make the iron run of an even temperature through the heat. In practice, the weight of the charges of iron to the charges of anthracite coal varies from 6 to 14 pounds of iron to the pound of coal. With coke they vary from 6 to 15 pounds of iron to the pound of coke. These variations in the per cent. of iron to fuel are due in many cases to the quality of fuel and in many other cases to poor judgment in working the cupola. In all cases the charge of iron should be made as heavy as the charge of fuel will melt and produce good hot iron for the work, for this is the only way a good per cent. of iron to the pound of fuel can be obtained.

PLACING THE CHARGES.

The top of the bed is made as level as it can be before charging the iron, and the smoke must all have disappeared so the melter can see how to place the charges. When the cupola is very high a few hundred of stove plate or other light scrap is placed upon the bed to prevent the heavy pieces of pig or other iron breaking up the fuel and settling down into the bed when thrown in. The pig should be broken into short pieces and placed in the cupola with the end toward the lining.

The pieces of pig or other iron are placed close together so as to utilize all the heat and prevent its escape up the stack, and each charge is made as level as it can be on top. The gates and cupola scrap are placed on top of the pig and are used to fill up holes and level up the charge. Old scrap is generally charged with the pig when heavy, and on top of the gates when light. Rattle barrel iron and gangway scrap or riddlings go in with the gates, a few few shovels to each charge.

The charge of fuel is distributed evenly over the charge of iron, and the second charge of iron is put in the same as the first, and the second charge of fuel the same as the first, and so on until the cupola is filled to the charging door. Charging is then stopped and the door closed until the blast goes on. When melting begins the stock begins to settle, and the door is opened and charging continued as before until all the iron to be melted is placed in the cupola. While charging is going on the cupola is kept filled to the charging door to prevent the gas igniting and making a hot flame at the charging door, which makes it hot for the men and difficult to place the charges of fuel and iron properly. When charging is finished the charging door is closed to prevent sparks or pieces of burning fuel being thrown upon the scaffold.

When the iron is all or nearly all melted that has been charged, and it is discovered there is not sufficient iron in the cupola to pour off the work, more iron is sometimes charged. At this stage of the heat the stock is so low in the cupola and the heat is so intense that the cupola is in a very bad condition for resuming charging to melt more iron. It is only a waste of fuel to charge it into the cupola at this stage of the heat, and the only iron that can be melted on the fuel already in the cupola is light scrap, and but a limited quantity of it. When the charging door is opened the heat at the opening is so intense that the men cannot go near it, and the scrap must be thrown in from a distance or by standing alongside of the cupola out of the heat and throwing the iron around into the door on a shovel.

Poor melting may be due to bad charging. Iron or fuel should never be dumped into a cupola from a barrow, for it all falls on one side of the cupola. The iron generally lays where it falls in a pile, and the fuel rolls to the other side of the cupola, and good melting cannot be done with the fuel on one side and the iron on the other. This way of charging is about equal to the old way of mixing the fuel and iron, and only about one-half of the melting capacity of the cupola can be realized. Fuel should never be emptied into a cupola from a basket or box, for it all falls in one place and cannot be spread evenly over the charge of iron. To charge a cupola properly the iron must all be thrown in with the hands, and the fuel with a shovel or fork.

CHARGING FLUX.

When it is desired to tap slag, the slag-producing material or flux is charged in the cupola on top of the iron and evenly distributed. The flux is sometimes put on each charge of iron, but generally about one-sixth of the heat is charged without flux. After that, flux is put in on every charge of iron except the last one or two charges, where it is not required if the proper amount has been used through the heat. The quantity of flux required depends upon the slag-producing propensity of the material used and the condition of the iron charged, and is from 30 to 100 pounds to the ton of iron melted. If the iron to be melted is all clean iron, the amount of flux required is less than when the iron is dirty scrap or a large per cent. of the heat is sprues and gates that have not been milled and are melted with a heavy coating of sand on.

If it is not desired to tap slag and the flux is used only to make a brittle slag in the cupola, it is charged in small quantities of from 5 to 10 pounds to the ton of iron, and is placed around the outside of the charge near the lining. Flux is sometimes charged in a cupola in a sufficient quantity to produce a large body of slag through which to filter the molten iron and cleanse it of impurities, but not in a sufficient quantity to admit of slag being drawn from the cupola. This way of fluxing works very well in a short heat, but in a long heat the slag sometimes absorbs a large amount of impurities, becomes overheated and boils up in the cupola and fills the tuyeres, and when boiling the slag cannot be drawn from the cupola at the slag hole and the bottom generally has to be dropped.

BLAST.

Before the blast is put on the tuyere doors are all tightly closed and luted to prevent the escape of any of the blast during the heat, and they should be examined from time to time to see that the luting has not blown out and the blast is not escaping. The blower is speeded up to the full speed at once, and the full volume of blast given the cupola from the start. The old way of putting on the blast light and increasing or decreasing it at different stages of the heat has been abandoned by practical foundrymen, and the cupola is given the same blast from the begining to the end of the heat. This is the only way good melting can be done in a cupola charged in the manner before described. If the cupola does not work properly, remedy the evil by changing the charges, but never vary the blast in different parts of a heat to improve the melting.

When the blast is first put on, it is indicated by a rush of blast from the tap hole and at the charging door by a volume of dust passing up the stack. Then follows a bluish colored gas which bursts into a bluish flame as the stock settles, changing to a yellowish hot flame as the stock sinks still lower. If the stock is kept up level with the charging door in a cupola of good height the gas does not ignite, and it is the aim of the chargers to keep the stock up to this point until they are through charging. When the blast is shut off from a cupola for any cause or at the end of the heat before the bottom is dropped, one or more of the tuyere doors are at once opened to prevent gas from the cupola passing into the blast pipe, where it is liable to explode and destroy the pipe.

The full consideration of the blast for a cupola would take up more space than we care here to give to it and would lead our readers too far from the subject of working a cupola. We shall therefore leave it for fuller consideration under another heading.

MELTING.

Melting begins in a cupola soon after the blast is put on, and the exact length of time is indicated by the appearance of molten iron at the tap hole. When the iron is charged two or three hours before the blast goes on, and the bed is not too high, iron flows from the tap hole in from three to six minutes after the blast is on. When the iron is charged and the blast is put on, immediately iron appears at the tap hole in from 15 to 20 minutes, after the cupola is filled if the bed is not too high. When the bed is too high, iron melts when the surplus fuel is burned up and permits it to come down to the melting point, and it is very uncertain when melting will begin; and it is generally from half an hour to an hour before any molten iron appears at the tap hole. If iron does appear at the tap hole within 15 or 20 minutes after the blast is on, the bed is either too high or the fire has not been properly lit, and the bed is not doing its work efficiently.

Foundrymen differ as to the time for charging the iron before the blast is put on. Some claim that fuel is wasted by lighting the fire early and charging the iron two or three hours before the blast is put on, while others claim fuel and power are only wasted by putting on the blast as soon as the iron is charged. We have melted iron in both ways, and we prefer to charge the iron from two to three hours before the blast is on, except when the cupola has a very strong draft that cannot be shut off, as is sometimes the case when there is no slide in the blast pipe for shutting off the blast, and as air is supplied to the cupola through the pipe. When iron is charged and the cupola filled to the charging door with fuel and iron, and the draft shut off from the cupola, the cumbustion of fuel in the bed is very light and the heat that rises from it is utilized in heating the first charge of iron. When the blast is put on, this charge of iron is ready to melt and iron comes down in a few minutes. When the blast is put on immediately after the cupola is charged the iron is cold, and time is required to heat it before it will melt, and the blower must be run 15 or 20 minutes before iron appears at the tap hole, and the first charge melts more slowly than when the iron has been heated before the blast is put on.

We think the best way is to put on the blast about two hours after the charging begins. When the blast goes on, the tap hole is open and is left open until the iron melts and runs hot and fluid from it. From 10 to 20 pounds are generally permitted to run from the spout to the floor to warm the spout and insure the iron being sufficiently hot not to chill in the tap hole after stopping in. The first iron melted is always chilled and hardened to some extent by the dampness of the sand bottom and spout, and when the work is light and poured with hand-ladles, a small tap of a few ladles is made in a few minutes after stopping in, and the iron used for warming the ladles and it is then poured into the pig bed or some chunks. In some foundries the cupola is not stopped in at all after the iron comes down. The first iron is used to warm the ladles, and as soon as the iron is hot enough for the work the molders begin pouring it. We recently ran off a heat of 31 tons in this way from a cupola for hand-ladle work without using a single bod for stopping in.

When the iron is handled in large ladles a tap is not made until a sufficient quantity is melted to fill a ladle or there is a sufficient body of iron in the cupola to insure it not chilling in the bottom of a large ladle before another tap is made. When the blast blows out at the tap hole after a tap is made, it indicates that the melted iron is all out of the cupola or the tap hole is too large, and the cupola should be stopped in until iron collects in the bottom, or the size of the tap hole should be reduced to prevent the escape of the blast. The size of the tap hole is reduced when it becomes too large to run a continuous stream without blowing out, by stopping in with a bod of stiff clay and sand that will not cut, and as soon as the bod is set, cutting a new tap hole through the bod with the tap bar. Iron should be melted hot and fast, and it should never be drawn from a cupola for any kind of foundry work if it is not hot and fluid enough to run stove plate or other light castings. Iron is not burned in a cupola by melting it hot and fast, but it is burned and hardened by melting it too high in the cupola and melting it slow and dull.

Nothing is gained by holding molten iron in a cupola to keep it hot, for it can be kept as hot in a ladle as in a cupola, and iron should be drawn from a cupola as fast as melted or as fast as it can be handled in pouring the work. If we were running a foundry we should never stop in the cupola except to get enough iron to give a gang of men a hand-ladle full all round, or to remove a large-ladle from the spout. When the cupola is very small and melts slowly it is sometimes necessary to stop in and collect iron in the cupola, but it is not necessary to stop in a large cupola for this purpose. If the iron is all poured with hand-ladles the men should be divided into gangs, with only enough men in each gang to take away the iron as fast as melted. If this is not done and there is a large number of men, the ladles get so cold between catches that they chill the iron before it can be poured, and the melter is blamed for not making hot iron.

The flow of iron from the tap hole indicates how the cupola is melting. If it has been properly charged the flow will be even in quantity and temperature throughout the heat, except in a very long heat, when the stream will get smaller and the cupola not melt so fast toward the end of the heat. When too much fuel is used the iron melts slowly and grows dull as the heat progresses. When the charges of iron are too heavy the iron is not of an even temperature throughout the heat, but is dull at the latter end of every charge and hot at the beginning of the next charge. When the charges of fuel are too heavy the iron melts very slowly at the beginning of each charge and fast at the latter end, and if the charges of iron are also too heavy the iron is dull at the latter end of the charge. If the cupola melts unevenly it is not being properly worked, and the mode of charging should be changed until it does melt evenly from the beginning to the end of the heat.

POKING THE TUYERES.

When the blast is first put on, the fuel in front of each tuyere is bright and hot, but it is soon chilled and blackened by the large volume of cold blast passing in, and the tuyere presents the appearance of being closed up and admitting no air to the cupola. The blast when first put on does not remove the fuel and make a large opening in front of each tuyere to get into the cupola, but works its way between the pieces of fuel in front of the tuyeres, and these openings remain open for the passage of blast after the fuel becomes cold and black. The blast, therefore, passes into the cupola just the same as when the fuel was hot, and it is not necessary to poke the tuyere with a bar or break away the cold fuel in front of each tuyere to let the blast into the cupola.

Toward the end of a long heat, slag and cinder settle and chill at the bottom of a cupola, and often not only close off the blast at the tuyeres, but prevent it passing freely through the stock and out at the top. If the tuyeres are poked at this stage of the heat an opening may be made well into the stock. But in working the bar around in forming this opening most of the natural passages the blast has made for entering the stock are closed up. The new opening is only a hole bored into a tough slag or cinder from which there is no way for the blast to escape into the stock, and less blast enters a cupola after the tuyeres have been poked and opened up than entered before. The only time a tuyere should be poked with a bar is when cinder or slag has lodged or formed in front of it in such a way as to run a stream of molten iron into the tuyere. The tuyere door should then be opened and the slag or cinder broken away with a bar to prevent the iron running into the tuyere.

THE CUPOLA FURNACE.

FUEL.

Theoretically ten pounds of iron are melted with 1 pound of anthracite coal, and 15 pounds with a pound of Connellsville coke. But this melting is done in the foundry office or in the mind of the foreman, and it takes a little more fuel to melt iron in a cupola for foundry work. Six pounds of iron to 1 of anthracite coal and 8 pounds of iron to 1 of Connellsville coke is by practical foundrymen considered good melting. A little better than this can be done in a full heat for the size of the cupola and under favorable circumstances, but in the majority of foundries fewer pounds of iron are melted to one of fuel than the above amount.

It is sometimes necessary for the melter to put in a few extra shovelfuls of fuel when the bed has been burned too much before charging, or to level up the charges when two or three men are shoveling in fuel at the same time and get it uneven. The melter is generally blamed if the iron from any cause comes dull, and he will generally put in a few extra shovelfuls of fuel the next heat to make it hot, and if the iron does come hot the next heat the extra shovelfuls are put in every heat, but are not put on the cupola report. In this way foundrymen are often misled by the cupola report and suppose they are melting more pounds of iron to the pound of fuel than they really are. The only way the foundryman can know exactly how many pounds of iron he is melting to the pound of fuel is to have an accurate account kept of the amount of iron melted and compare it with the amount of fuel bought and delivered for the cupola, after deducting from it any amount that may have been consumed in stoves or core ovens.

We recently met a foundryman who thought he was melting 14 pounds of iron to the pound of fuel, but when he came to compare the iron melted with the fuel bought and delivered for the cupola he found he was only melting about 7 pounds of iron to the pound of fuel; and about the same results would be found in every foundry that is claimed to be melting a very large per cent. of iron to fuel. There is nothing gained by saving a few cents' worth of fuel in the cupola and losing a dollar's worth of work on the floor by dull iron, and there is nothing gained by using too great a quantity of fuel, for too much fuel in a cupola makes dull iron as well as too little fuel.

Iron is not melted in a cupola for the fun of melting it or to learn how many pounds of iron can be melted with a pound of fuel, but is melted to make castings. What the foundryman wants from the cupola at the tap hole is an iron hot and fluid enough to make a sound casting, regardless of the amount of fuel required to produce it. As before stated, iron cannot be melted hot and fast in a cupola with either too much or too little fuel, and foundrymen have only to melt their iron as hot and fast as it can be melted in a cupola of the size they are using, to know that they are not using either too much or too little fuel in melting.

If the foundryman will ask his neighbor what is the size of his cupola, how many tons does he melt per hour, how long does it take him to run off a heat, he will get a better guide to run his cupola by than if he asks him how many pounds of iron he melts to the pound of fuel. As soon as the founder undertakes to imitate his neighbor and do faster melting or get better results from his cupola, he will hear the old, old story from both melter and foreman: "We haven't enough blast." More cupolas have too much blast than too little, and the apparent deficiency of blast is due in the majority of cases to too much fuel in the cupola and the iron being melted only on the upper edge of the melting zone. It does not make any difference how much or how little blast a cupola has. If it is given an even volume of blast throughout the heat, the cupola will melt a stream of iron of an even size and temperature throughout the heat except toward the end of a long heat, when the stream may get smaller. If the melter cannot run this kind of a stream from his cupola with an even blast, then he is at fault, and neither the blast, nor too much fuel, is the cause of the uneven melting.

We have watched the charging of cupolas in a great many

stove and machinery foundries, and as a rule more fuel is consumed in making dull iron in a machine foundry than is consumed in making hot iron in a stove foundry. This is simply because hot even iron cannot be produced with bad working of a cupola and too great a quantity of fuel, and the stove founder must have his cupola properly worked or he cannot use the iron to pour the work.

TAPPING BARS.

Tapping bars are made of round iron of from ½ to 1 inch diameter, and are from 3 to 10 feet long. The hand bars are made with an oval ring at one end to serve as a handle for rotating and withdrawing the bar when tapping. The other end is drawn down to a long sharp point for cutting away the bod and making the tap hole. The bars for sledging are made straight with a long sharp point at one end. This bar is only used in case the tap hole becomes so tightly closed that it cannot be opened with the hand bar, and seldom more than one is provided for a cupola. From three to six hand bars are provided for each cupola, and when the ladles are all of the same size the tap bars are all made of the same size, except one or two small ones which are provided for clearing the hole of any slag or dirt that may be carried into it by the iron.

When the iron is melted for different sized work and large and small ladles are used, the bars are of different sizes, so that a large or small hole may be made to suit the tap to be made or ladle to be filled. The bars are all straight except when the tapping is done from the side of a long spout. They are then slightly curved near the point, so that the hole can be made in a line with the spout. The bars are dressed and pointed at the forge before each heat, and are given any shape of point the melter may fancy. A square point cuts away a bod very rapidly when rotated and leaves a nice, clean hole, but is very difficult to keep a point square, for they generally become round after a few taps have been made and they come in contact with the molten iron a few times. For this reason they are generally made round at the forge.

Some melters have a short steel bar, with a sharp flat point, which they use for cutting away the bod before tapping, but never use it for opening the hole. This they do to remove the greater part of the bod from the spout before tapping, and prevent it getting into the ladles. A hammer and an anvil, or an iron block, should be placed near the cupola for straightening the points and breaking cinder or dross from the bars, and a rack should be provided within easy reach of the tap hole, in which to place the tap bars on end until wanted for use. There is nothing more slovenly and dangerous about a foundry than to have the tap bars lying around the floor when a heat is being run, and it is just as bad to set them up against a post from which they are all the time falling down.

BOD STICKS.

Two kinds of bod stick are used for stopping in a cupola; The wood stick and the combination wood and iron stick. The wood sticks are octagonal or round, from $1\frac{1}{2}$ to 2 inches diameter and from 5 to 10 feet long. They are made of both hard and soft wood and about an equal number of each wood, as some prefer one and some the other. When stopping in, the stick is held against the bod in the tap hole until it sets in the hole, and the stick generally takes fire from the heat of the spout. On this account they soon become small near the ends and have to be sawed off; for this reason they are always made longer than necessary and sometimes larger in diameter.

The combination stick is of the same diameter as the wood stick, and from 4 to 10 feet long. An iron ring is placed on one end of the stick, and a rod of round iron of from $\frac{1}{2}$ to $\frac{5}{8}$ inch diameter and 1 to 3 feet long is placed in the end of the stick. On the end of the rod is placed a round button of from $1\frac{1}{2}$ to 2 inches diameter, for carrying the bod. The object of the rod is to prevent the stick being burned by the heat of the spout every time the cupola is stopped in, and the length of the rod is made to correspond to the length of the spout. The ojection to the combination stick is that the button does not carry the bod as well as the wood stick, and the button and rod must be wet every time the stick is used to keep it cool, or the heat will dry out the bod and it will fall off. This repeated wetting rusts the buttom, and if the edges come in contact with the molten iron it makes the iron sparkle and fly; and for this reason most founders prefer the wood sticks, even at the extra expense of keeping them up.

An iron rod and button without the wood stick is also used in some foundries for stopping in, but they were not used by our grandfathers and are not popular with melters. Three or four bod sticks are provided for each cupola. They are placed on end in a rack alongside of the tap bars, within easy reach of the tap hole, and a bod is kept on each stick all the time the cupola is in blast.

BOD MATERIAL.

The bod is a plug used for closing the tap hole when it is desired to stop the flow of iron from a cupola, and the material of which the bod is composed has a great deal to do with the nice working of the tap hole. When the bod is composed of fire clay, or largely of fire clay, it does not give up the water of combination rapidly, and if a tap is quickly made after stopping in, the iron sputters and flies as it comes out of the hole. If the bod is permitted to become perfectly dry it bakes so hard that it cannot be cut away with the hand bar, and the heavy bar and sledge have to be used to make a hole of the proper size. If a friable sand is used it crumbles easily before the bar and a nice clean hole can be made; but it does not hold well, and if the cupola is stopped in for any length of time the bod may be forced out by the pressure of metal.

Some of the loams make an excellent bod that holds well and is easily cut away with the point of the bar, and leaves a clean hole. Some of the molding sands also make good bods in their native state, and there are several materials that are peculiar to certain localities that make good bods. When a suitable material cannot be found it must be made by mixing two or more materials. A good bod is made by mixing blue or yellow clay and molding sand. When these clays cannot be procured, a good bod can be made by mixing just enough fire clay with the molding sand to give it a little greater adhesive property, but not enough to make it bake hard. When a large body of iron is collected in the cupola before a tap is made, the bod material must be strong, and bake in the hole sufficiently hard to resist the pressure of the iron, and an entirely different material must be used for this kind of tapping than is used when the cupola is only stopped in for a few minutes at a time.

Small cupolas from which only a small hand ladle is drawn before it is stopped in also require a different bod, for the hole has hardly time to clear itself before it is stopped in again, and if the bod burns hard or sticks in the hole, the hole is so hard to open that the small amount of iron is chilled by the bar and slow tapping before it can be run out. This kind of cupola requires a bod that will crumble and fall out as soon as touched, or burn out as soon as the hole is opened. A nice bod is made for this kind of work by mixing clay, molding sand and sawdust and making it fully half sawdust. Blacking or sea coal is also mixed with bod material to make it more porous when burned and crumble more readily when tapping.

Horse manure was at one time considered to be one of the essentials of a good bod, but it has been replaced by blacking or sawdust, and is seldom used. A good bod should have strength to resist the pressure of molten iron in the cupola and at the same time break away freely before the iron and leave a clean hole. Such a material can be made suitable for any cupola, no matter how it is tapped, and a bod material should never be used that requires the sledging tap bar to open the tap hole.

TAPPING AND STOPPING IN.

When the blast is put on the tap hole is always open, and is left open until the iron melts and flows freely and hot from the hole. This is generally in from 5 to 20 minutes after the blast is on. While the melter is waiting for the iron, he arranges his tap bars, bod stuff and bod sticks, and places a bod on each stick to be ready for instant use. The bod material is worked a little wetter than molding sand to make it adhere to the end of the bod stick or button, but care must be taken not to have it too wet or it will make the iron fly when stopping in, and, furthermore, the bod does not hold well when too wet. The bod is made by taking a small handful of the bod stuff and pressing it firmly on the end of the stick with the hand. The size and shape the bod is made depend upon how the iron is tapped and the size of tap hole. When the hole is small and only stopped in for a few minutes at a time a small bod stick is used, and the bod made very small and shallow and only pressed into the hole a short distance, so that it can be quickly broken away when tapping. When the hole is large or has to be stopped in until a large body of iron collects in the cupola, the bod is made large, long and pointed, so that it may be pressed well back into the hole and stay in place until removed with the tap bar.

The first iron melted flows from the hole in a small stream, and generally chills in the hole or spout and has to be removed with the tap bar; but it soon comes hot enough to clear the hole, which is then closed, unless the first iron is used to warm the ladles and the hole is kept open through the heat. If the work is light, a small tap is made in a few minutes to remove any iron that has been chilled and dulled by the dampness in the sand bottom. But when the work is heavy this tap is not made and the molders go on with their regular pouring from the start. The tap is made by placing the point of the bar against the bod and giving it a half forward and back rotation and at the same time pressing it into the bod, or by carrying the handle end of the bar around in a small circle and at the same time pressing it in. As soon as the bod is cut through, the bar is run into the hole once or twice and worked around a little to remove any of the bod that may be sticking round the sides of the hole. The bar must always be held in such a position as will make the hole in a line with the spout, or the stream will not flow smoothly and may shoot over the sides of the spout and burn the men catching in.

When about to stop in, the bod is placed directly over the stream close to the tap hole and the other end of the stick is elevated at a sharp angle from the spout. The hole is closed by a quick downward and forward movement of the stick that forces the bod into the hole and checks the stream at once. The stick is then held against the bod for a few seconds until the force of the stream is stopped and the heat has set the bod fast in the hole. The part of the bod that does not enter the hole is then removed with the stick to keep the spout clean, and the stick is dipped in water to cool it, and another bod applied to be ready for the next time. There is a great knack in stopping in, that some melters never acquire. They hold the bod too far from the hole, and attempt to push it up, under or through the stream; they get nervous and are not sure of their aim and strike the stream too soon, or the side of the hole, and the iron sputters and flies in all directions.

The bod sticks are frequently made so long and slender or so heavy that it is impossible to accurately place the bod, and it is sometimes difficult to get the cupola stopped in with these long sticks. It is also difficult to stop in when the cupola is placed very high, for the melter cannot get up to place the bod stick at a proper angle for stopping in, and has to run the bod up through the stream, in place of cutting off the stream with the bod. An arm or bracket is sometimes placed over the spout near the tap hole, when long bars and sticks are used, upon which to rest the tap bars and bod sticks when tapping and stopping in. But a better plan is to construct a movable platform that can be placed alongside the spout for the melter to stand on. He can then use short bars and sticks, and has much better control of the tap hole than with the long bars and sticks.

At the tap hole is seen the skill of the melter in the results obtained from his labor. If the bed has been burned too much the first charge comes down fast and slack or dull. If the bed is too high the first iron is a long time in coming down. If too much fuel is used, the iron melts slowly and is dull toward the last if the heat is long. If the charges of iron are too heavy, the iron comes dull at the end of each charge. If the charges of fuel are too large, there is slow melting at the end of each charge. If the iron flows from the tap hole with great force and is difficult to control, the sand bottom has too much pitch. If slag flows freely from the tap hole with the iron, the hole is too large or the bottom has too much pitch. If the spout melts, crumbles or chips off, the material is poor or has not been properly mixed. If the tap hole cuts out, the material is poor or the tap hole has not been properly made. If the tap hole gums up and cannot be kept open, the front material is poor and is melted by the heat, and it may also be melted by the heat when it is good if rammed soft and ragged inside. If the tap hole cannot be opened without a sledge and bar, the bod bakes too hard and the material should be changed. If the bod does not hold, the material is not good or the bod is not put in right. If the cupola does not melt evenly throughout the heat and the same every heat, it is the fault of the melter and not of the cupola.

DUMPING.

As soon as the molders are through pouring their work, if there is no iron to be melted for other purposes, preparations are at once made for dumping the refuse from the cupola. The blast is first shut off by stopping the blower and the tuyere doors are at once opened to prevent the escape of gas from the cupola into the blast pipe, where it might do much harm. The melted iron in the cupola is then drawn off by the cupola men and poured in the pig bed. If there is a lot of small iron in the cupola that has not been melted time is given it to melt to save picking it out of the dump, but if there is a lot of pig or other heavy iron unmelted it is let fall with the dump, and the bottom is dropped as soon as the melted iron is drawn off.

The small props supporting the bottom are first removed and laid away. The main prop is then removed by striking it with a long bar at the top or pulling with a hooked bar at the bottom, and the instant it falls the doors drop. When the doors of a large cupola drop, the sand bottom and a greater part of the refuse of melting falls with them and a sheet of flame and dust instantly shoots out ten feet or more from the bottom of the cupola in all directions. But the flame disappears in an instant and the dust settles, revealing the white hot dump in a heap under the cupola. The cupola men then throw a few buckets of water on it to chill the surface and deaden the heat, and the melter puts a long bar into the tuyeres and tries to dislodge any refuse that may be hanging to the lining while it is hot. Small cupolas do not dump so freely, and the sand bottom has frequently to be started with a bar after the door drops. A long bar must be used for this purpose and the melter must be on his guard, for the dump may fall as free as from a large cupola the instant it is started and a sheet of flame shoots out.

Small cupolas frequently bridge over above the tuyeres, and only the sand bottom and refuse below the bridge is dumped when the door falls. The aim of the melter is then to break away the bridge or get a hole through it, so that the cupola will cool off in time to be made ready for the next heat. He puts a bar in at the tuyeres and breaks away small pieces at a time, and if there is not a large body of refuse in the cupola, a few short pieces of pig are thrown in from the charging door, so that they will strike in the center and break through the bridge. This bridging and hanging up of the refuse in a cupola when only run for a few hours is entirely due to mismanagement, for any cupola, no matter how small it may be, can be run for six or eight hours without bridging and be dumped clean if properly worked.

When the dump falls from a cupola it is a semi-fluid mass of iron, slag, cinder, dirt and fuel. This mass falls in a heap under the cupola, and if scattered or broken up when very hot it is more readily wet down and more easily removed when cold. In some foundries a heavy iron hook or frame is placed under the cupola before dumping and is withdrawn with a chain and windlass after the dump has fallen upon it and been partially cooled with water to harden it so that the hook will not slip through it without breaking it up and scattering it. In other foundries it is scattered with a long rake or hook worked by hand. In pipe foundries two or three short lengths of condemned pipe are placed under the cupola before dumping, and the dump is broken up by running a bar into the pipe and lifting it up after the dump has been slightly cooled. But in a great many foundries where the dump is small or where there is plenty of room to remove it when cold it is let lie as it falls and is wet down by the cupola men, or a few buckets of water are thrown on by the cupola men to deaden it, and it is left for the watchman to wet down during the night. Care must be taken not to put on too much water, or the floor under the cupola will be made so wet that there will be danger of the dump exploding when it falls upon it the next heat.

REMOVING THE DUMP.

A number of plans have been devised for removing the dump from under the cupola. Iron cars or trucks have been constructed to run under the cupola and receive the dump as it falls, but they cannot be used unless there is sufficient room for the doors to swing clear of the car, and few cupolas are so constructed. The dump must be removed from the car when hot to avoid heating and injuring the car, and considerable room is required for handling the car after it is taken from under the cupola. For these reasons cars are seldom used. Iron crates have been made to set under the cupola and receive the dump and be swung out with the crane, but they get fast under the cupola and are soon broken, and it is almost as much work to handle the dump from the crate as it is from under the cupola. A number of other plans have been tried, but the dump must be picked over by hand, and it is as cheap to pick it over at the cupola and remove it in wheelbarrows as by any way that has yet been devised. The dump is broken up with sledge and bar when cold and picked over. The large pieces of iron are picked out and thrown in a pile for remelting. The coke is thrown in a pile to be taken to the scaffold or core oven furnace. Anthracite coal that has passed through a cupola and been subjected to a high heat will not burn alone in a stove or core oven furnace, and it is very doubtful if it produces any heat when mixed with other coal and again put in the cupola, and only the large pieces are picked out, if any.

The cinder, slag and other refuse are shoveled into a wheelbarrow and taken to the rattle-barrels or dump. If the sand bottom is to be used over again it is riddled out in a pile and wetted. If not, it is removed with the cinder and slag. As soon as the bulk of the dump is removed the melter goes into the cupola and breaks down the ring of cinder over the tuyeres and chips off any that may be adhering to other parts of the lining. The dump is then all removed and the floor around the cupola is cleaned up preparatory to daubing up. Nothing is done with the dump after it is taken from the cupola but to recover the iron from it. This is done in two ways, by picking it over or milling it. The iron is often of the same color as the dump, and so mixed with it that it is almost impossible to recover it all by picking unless a great deal of time and pains be taken; and it is cheaper to throw out only the pieces of pig and shovel all the remainder into the tumbling barrels, where it is separated in a short time and all the iron recovered that is worth recovering.

CHIPPING OUT.

Before going into the cupola to chip it out the melter slushes one or two buckets of water around the lining from the charging door to lay the dust. He then goes in from the bottom if he can get in, but if the cupola is so badly bridged that he cannot get up into it he takes a long bar and endeavors to break down the bridge from the charging door, or goes down into it from the charging door, and with a heavy bar or sledge breaks it down. As soon as he gets a hole through large enough to work in, he goes down through it and with a sledge or heavy pick breaks down the shelf of slag and cinder that always projects from the lining over the tuyeres. He then takes a sharp pick and trims off all projecting lumps of cinder and slag and gives the lining the proper shape for daubing up. It is not necessary or advisable to chip off all the cinder and adherent matter down to the brick, for the cinder stands the heat equally as well as new daubing, and in some cases better. But all soft honeycombed cinder should be chipped off, and all projections of hard cinder that are likely to interfere with the melting or tend to cause bridging should be removed.

Some melters have a theory that to prevent iron running into the tuyeres they must have a projection or hump on the lining over the tuyeres, and they let the cinder build out from 3 to 6 inches thick and 6 to 12 inches deep at the base. These humps tend rather to throw iron into the tuyere than to keep it out, for the fuel becomes dead under the hump and the iron in its descent strikes the hump and follows it around into the tuyere. They also form a nucleus for bridging. The refuse of melting as it settles lodges upon these humps and is chilled by the blast. A small cupola with these humps over the tuyeres will not work free for more than an hour, while the same cupola with the humps removed and the lining straight would work free for two hours and dump clean. They also interfere with the melting and in large cupolas cause bridging. All humps that form on the lining above the melting point from bad charging or other causes should be removed, for they hang up the stock and retard melting.

The cupola picks generally used are entirely too light for the work to be done with them, and the handles are not firm enough. When the melter strikes a blow he cannot give the pick force enough to cut away the point desired and the handle gives in the eye, so that the pick glances off and cannot be held to the work. Repeated blows with a light pick turn the edge and render the pick worthless and the melter has to do two or three times the work really necessary in chipping out the cupola, and then he does not get it right. What the melter wants is a heavy pick with a firm handle. Then he can hold the pick where he strikes and prevent it glancing off. He can strike a blow that will cut away the cinder at one stroke and not jar and injure the lining nearly so much as he would by repeated blows with a light dull pick. The melter should be provided with three picks made of the best steel, weighing 4, 6 and 8 pounds each. They should be furnished with iron handles solidly riveted in, or should be made with large eyes for strong wood handles. The picks should be dressed, tempered and ground as often as they get the least bit dull.

DAUBING.

After the cupola is chipped out the lining is repaired with a soft plastic adhesive material known as daubing, with which all the holes that have been burned in the lining are filled up and thin places covered, and the lining given the best possible shape for melting and dumping. There are a number of substances used for this purpose, some of which are very refractory, and others possess scarcely any refractory properties whatever and are not at all suitable for the purpose. Molding sand is frequently used for a daubing. It is easily and quickly wet up and mixed, is very plastic and readily put on, but possesses none of the properties whatever requisite to a good daubing. It crumbles and falls off as soon as dry in exposed places, and a lining cannot be shaped with it. Furthermore, when put on in places from which it is not dislodged in throwing in the stock, it melts and runs down and retards the melting by making a thick slag that is readily chilled over the tuyeres by the cold blast.

Some of the yellow and blue clays are very adhesive and refractory, and make good daubing alone or when mixed with a refractory sand. Ground soapstone and some of the soapstone clays from coal mines make excellent daubing. But probably the best and most extensively used is that composed of fire clay and one of the silica sands known under various names in different sections of the country, and which we shall designate sharp sand. Fire clay is very plastic and adhesive when wet, but shrinks and cracks when dried rapidly. Sharp sand alone possesses no plastic or adhesive properties whatever and expands when heated. When these two substances, in exactly the right proportions, are thoroughly mixed, they make a daubing that is very plastic and adhesive, does not crack in drying, neither expands nor shrinks to any extent when heated, and resists the action of heat as well as fire-brick in a cupola. When not evenly mixed the fire clay cracks and the sand expands and falls out of the clay when heated, making an uneven and uncertain daubing.

Fire clay absorbs water very slowly, and it requires from 12 to 24 hours' soaking before it becomes sufficiently soft to be thoroughly and evenly mixed with the sand. A large soaking tub should be provided near the cupola and it should be filled with clay every day after the cupola is made up, and the clay covered with water and left to soak until the next heat. The clay and sand cannot be evenly mixed in a round tub with a shovel, therefore a long box and a good strong hoe should be provided for the purpose. The amount of sand a clay requires to make a good daubing varies from one-fourth to three-fourths, according to the qualities of the clay and sand, but generally one-half of each gives good results. The sand is added to the clay dry, or nearly dry, and the daubing is made as thick and stiff as it can be applied to the lining and be made to stick. The more it is worked in mixing the better, and if let lie in the mixing box for a day or two after mixing it makes a better daubing than if applied as soon as prepared.

Nothing is gained by using a poor, cheap daubing, for it

does not protect the brick lining, but falls off or melts into a thick tough slag which runs down and chills over the tuyeres and retards the melting by bunging up the cupola, and more fuel and time are required to run off the heat. The daubing is taken from the mixing box on a shovel when wanted for daubing and placed on a board under the cupola if the box is near at hand. When it is some distance from the cupola the daubing is placed in buckets or small boxes made for the purpose and conveyed to the cupola. The parts of the lining to be repaired are first brushed over with a wet brush to remove the dust and wet the lining so that the daubing will stick better. The daubing is then thrown on to the lining with the hands in small handfuls; it can be made to penetrate the cracks and holes better in this way than in any other, and stick better than when plastered on with a trowel. After the required amount has been thrown on in this manner, it is smoothed over with a trowel or wet brush and made as smooth as possible.

SHAPING THE LINING.

Daubing is applied to a lining for two purposes—viz., to protect the lining and to shape the cupola, the latter being by far the more important of the two. A great many melters never pay any attention to it, their only aim being to keep up the lining, and they pride themselves on making a lining last for one, two or three years. Nothing is gained by doing this if the melting is retarded by doing so and enough fuel consumed and time and power wasted every month to pay for a new lining. Besides, a lining will last just as long when kept in good shape for melting as when kept in a poor one, and the aim of the melter should be to put the lining in the best possible shape for melting and make it last as long as he can.

New linings are made straight from the bottom plate to the charging door when the cupola is not boshed. When it is boshed the cupola is made of a smaller diameter at and below the tuyeres, and the lining is sloped back to a larger diameter from about 6 inches above the tuyeres, with a long slope of 18 or 20 inches. In the straight cupola, slag and cinder adhere in every heat to the lining just over the tuyeres, and if not chipped off close to the brick after each heat, gradually build out and in time a hard ledge forms that is difficult to remove. It furthermore reduces the melting capacity of the cupola by increasing the tendency to bridge. Above this point at the melting zone the lining burns away very rapidly and in every heat a hollow or belly is burned in it at this point that requires repairing. Above the melting zone the lining burns away very slowly and evenly and seldom requires any repairing until it becomes so thin that it has to be replaced with a new one.

The cinder and slag that adhere to the lining just over the tuyeres must be chipped off close to the brick every heat, and the lining made straight from the bottom plate to 6 inches above the top of the tuyeres. No projection or hump of more than ½ or ¾ inch should ever be permitted to form or be made over the tuyeres to prevent iron running into them, and it should be placed right at the edge of the tuyere when it is thought necessary to make it. The upper edge of the tuyere lining should be made to project out a little further than the lower edge, and the brick lining should be cut away a little under each tuyere so that molten iron falling from the top of the tuyere will fall clear of the bottom side of the tuyere and not run into it.

It is not necessary or advisable to fill in the lining at the melting zone and make it perfectly straight, as it is when it is new, for a cupola melts better when bellied out at the melting zone. It must, however, be filled in to a sufficient extent for each heat to keep up the lining and prevent it being burned away to the casing. No sudden offsets or projections should be permitted to form or remain at the upper edge of the melting zone, for the stock lodges in settling upon projections and does not expand or spread out to fill a sudden offset, and so the heat passes up between the stock and lining and cuts away the lining very rapidly. No sudden offset or hollow should be permitted to form at the lower edge of the melting zone over the tuyeres, for the stock will lodge on it in settling and cause bridging of the cupola. The lining should be given a long taper from 6 inches above the tuyeres to the middle of the melting zone, and a reverse taper from there to the top of the melting zone. The belly in the lining should be made of an oval shape, so that the stock will expand and fill it as it settles from the top, and not lodge at the bottom as it sinks down in melting. As the lining burns away above the melting zone the straight cupola assumes the shape of the boshed cupola, and only the lower taper is given to the melting zone.

Daubing should never be put on a lining more than 1 inch thick, except to fill up small holes, and even then small pieces of fire brick should be pressed into it to reduce the quantity of daubing and make it firmer. All clays dry slowly and give up their water of combination only when heated to a high temperature. When daubing is put on very thick it is only skin dried by the heat of the bed before the blast goes on. The intense heat created by the blast glazes the outside of the daubing before it is dried through to the lining, and as there is no way for the moisture to escape, it is forced back to the lining, where it is converted into steam and in escaping shatters the daubing or tears it loose from the lining at the top.

In the accompanying illustration, Fig. 19, is shown a sectional view of a cupola that we saw at Richmond, Ind., in 1875. This cupola was a small one, about 35 inches diameter at the tuyeres, and the average heat was about 4 tons. The melter was a hard working German, who knew nothing about melting whatever, and his only aim was to keep up the lining in the cupola. With this object in view he would fill in the hollow formed in the lining every heat at the melting zone with a daubing of common yellow clay and make the lining straight from the tuyeres up. The daubing required to do this was from 2 to 4 inches thick all around the cupola, and was put on very wet. The heat dried and glazed this daubing on the outside before it was dried through. There being no way for the water to escape, it was converted into steam, and in escaping from behind the daubing tore it loose from the lining at the top. The fuel and stock in settling got down behind the daubing and pressed it out into the cupola from the lining until it formed a complete bridge, with only a small opening in the center through which the blast passed up into the stock. Before the heat was half

FIG. 19.

SECTION THROUGH BRIDGED CUPOLA.

over all the iron melted was running out at the tuyeres, and the bottom had to be dropped. When the cupola had cooled off, the daubing and stock were found in the shape shown in the illustration, and when the bridge was broken down it was found to be composed entirely of daubing that had broken loose from the lining in a sheet and doubled over.

This melter always had slow melting and difficulty in dumping. Some nights after dumping he would work at the tuyeres with a bar until eight o'clock before he got a hole through, so the cupola would cool off by morning. The lining was not protected by the thick daubing, but was cut out more by the repeated bridging than if it had been properly coated with a thin daubing. We daubed this cupola properly and ran off two heats in it and melted the iron in less than half the time usually taken, and had no difficulty in dumping clean.

The lining of the boshed cupola does not burn out at the melting zone in the same shape nor to so great an extent as in the straight-lined cupola, and in shaping the lining it is made almost straight from the top of the slope to the bosh up to the charging door. The taper from the bosh to the lining should start at 6 inches above the top of the tuyeres, and should not be less than 18 or 20 inches long, and must be made smooth with a regular taper so that the stock will not lodge on it in settling. Should the cupola be a small one with a thick lining and only slightly boshed and burn out at the melting zone similar to the straight cupola, it must be made up in the same way as the straight cupola. The great trouble with boshed cupolas is that the melter does not give a proper slope to the taper from the bosh, but permits a hollow to form in the lining over the tuyeres, in which the stock lodges in settling and causes bridging out over the tuyeres.

These directions for shaping the lining only apply to the common straight and boshed cupolas. Many of the patent and odd-shaped cupolas require special directions for shaping and keeping up the lining as it burns out, and every manufacturer of such cupolas should furnish a framed blue print or other drawing, to be hung up near the cupola, showing the shape of the lining when new and the shape it should be put in as it burns away and becomes thin. Full printed directions should be given for chipping out and shaping the lining. All the improvements in cupolas are based on the arrangement of the tuyeres and shape of the lining, and when the lining gets out of shape the working of the tuyeres is disarranged, and the cupola is neither an improved one nor an old style, and is generally worse than either. More of the improved cupolas have been condemned and thrown out for want of drawings showing the shape of the lining and directions for keeping it up, than for any other cause.

RELINING AND REPAIRING.

When a cupola is newly lined the lining is generally made of the same thickness from the bottom to the top except when the cupola is boshed. The casing is then either contracted to form the bosh or it is formed by putting in two or more courses of brick at this point. The lining varies in thickness from $4\frac{1}{2}$ to 12 inches, according to the size of the cupola, the heavier linings always being put in large cupolas. The greatest wear on the lining is at the melting zone, where it burns away very rapidly. From this point up it burns away more gradually and evenly, but the greatest wear is toward the bottom, where the heat is the greatest, and so a cupola gradually assumes a funnel shape with the largest end down and terminating at the melting zone, and the lining is always thinnest at about this point when it has been in use for some time.

At and below the tuyeres the destruction of the lining by heat is very slight, and the principal wear is from chipping and jarring in making up the cupola. At the charging door the principal wear is from the stock striking the lining in charging. In the stack the lining becomes coated with sulphides and oxides and is but little affected by the heat. A stack lined with good material properly put in generally lasts the lifetime of the cupola. The length of time a cupola lining will last depends upon the amount of iron melted and the way in which it is taken care of, and varies from six months to three or four years when the cupola is in constant use.

A lining burns away very rapidly at the melting zone, and if not repaired every heat would burn out to the casing in a few heats. Above the melting zone it burns away more slowly and evenly, and gets thinnest just above the melting point. From this point it gradually grows thicker up toward the charging door, where the wear is comparatively slight. The thickness of lining required to protect the casing where the heat is most intense depends upon the quality of the fire-brick and how the lining is put in. A lining of good circular brick made to fit the casing, and laid up with a good, well-mixed grout, remains perfectly solid in the cupola as long as it lasts, and may be burned down to $1\frac{1}{2}$ inches in thickness, and even less, for several feet above the melting point. When the brick do not fit the casing and large cracks or holes have to be filled in with grout, and daubing or the lining is poorly laid up, it becomes shaky as it burns out and in danger of falling out, and it cannot be burned down so thin as when solid.

It is therefore cheaper in the long run to get brick to fit the casing and have the lining well put in. It will then only be necessary to reline when the lining gets very thin almost up to the charging door. The lining at the melting zone, where it burns away the fastest, is often taken out for 2 or 3 feet above the tuyeres and replaced with a new one when it is not necessary to reline all the way up. In repairing a lining in this way the same sized brick are generally used as were used in lining. The lining has been burned or worn away above and below the point repaired, and the new lining reduces the diameter of the cupola to the smallest at the very part where it should be the largest. The result is that the new lining is cut away faster than any other part, and after a few heats it is as bad as it was before the new section was put in.

A better way of repairing a lining at the melting zone is to put in a false lining over the old lining. This is done by putting on a layer of rather thin plastic daubing over the old lining and pressing a split fire-brick into the daubing with the flat side against the lining. The brick are pressed into the daubing close together almost as soon as it is put on, and all the joints are filled up and the surface made smooth. A lining may be put in a cupola in this manner all the way around and to any height desired, or only thin places may be repaired, which is done without forming humps in the lining that interfere with the melting.

A split brick is an ordinary fire-brick, only 1 inch thick in place of 2 inches, and is now made by all the leading fire-brick manufacturers. We believe we were the first to repair a lining in this way, some 20 years ago. The split brick could not then be procured from fire-brick manufacturers, and they were made by splitting the regular sized brick with a sharp chisel after carefully nicking them all around. When the regular split brick cannot be procured they may be made in this way. Most of the new brick split very readily and true, but bats from old lining generally spall off and are difficult to split. A lining of split brick can be put in almost as rapidly as the cupola can be shaped with daubing alone. The diameter of the cupola is not reduced to the same extent as with a section of new lining put in in the regular way, and the best melting shape for the cupola is maintained with only a reduction in the diameter of from 3 to 4 inches. This lining, when put in with a good daubing well mixed, lasts as long as an equal thickness of lining put in in the regular way; and it can be put in at a great deal less expense for labor and material. It is, however, worthless if put in with a poor, non-adhesive and unrefractory daubing.

CHAPTER V.

EXPERIMENTS IN MELTING.

In visiting different foundries years ago, when the management of cupolas was not so well understood as at the present time, we found that there were many and different opinions held by foundrymen as to the point in a cupola at which the melting of iron actually took place. Some foundrymen claimed that melting was done from the tuyeres to the charging door, others that iron was only melted in front of the tuyeres by the blast and flame, on a similar principle to melting in an air furnace, and still others claimed that iron was only melted at a short distance above the tuyeres. These various opinions led to different ways of charging or loading cupolas. In some foundries one or two hundred-weight of iron were put in on the bed, then one or two shovels of fuel, then more iron and fuel in the same proportion, until the cupola was filled or loaded to the charging door. This way of charging mixed the fuel and iron together, and cupolas were charged in this way to melt from the tuyeres to the charging door. In other foundries from five to twenty-hundred weight of iron were placed on the bed and a layer or charge of fuel placed upon it to separate it from a second charge of iron of a similar weight; which was again covered with a second charge of fuel to separate it from the third charge of iron, and the cupola in this way filled. This charging was done upon the theory that a cupola only melted at a short distance above the tuyeres. Foundrymen who were of the opinion that the iron was melted by the flame and blast charged their cupolas in a similar way, but made the charges of iron light and those of fuel heavier, using an extravagant amount of fuel for each heat.

THE CUPOLA FURNACE.

To learn definitely at what point iron was really melted in a cupola, and also to ascertain something in reference to a number of other points in melting, as to which we had found there was a wide difference of opinion among foundrymen, we constructed a small cupola with a light sheet-iron casing and a thin lining, through which tuyere and other holes could be easily cut and closed when not required. This cupola we connected with a Sturtivant Fan placed at a short distance from the cupola. The fan was entirely too large for the size of the cupola, but it was arranged to regulate the volume of blast supplied by increasing or decreasing the number of revolutions of the fan. On the blast pipe, near the cupola, we placed a very accurate steel spring air-gauge to ascertain the exact pressure of blast in each experimental heat. The cupola was eighteen inches diameter inside the lining, and we first put in two round tuyeres of four inches diameter and placed them on opposite sides of the cupola, twenty-four inches above the bottom.

The first experiments made in melting in this cupola were for the purpose of learning at what point in a cupola iron melted, and at what point it melted first. To ascertain these facts we procured a number of small bars of No. 1 soft pig iron and placed ten of them across each other in the cupola, six inches apart from center to center, and fastened the ends of each pig in the lining so that they could not settle with the fuel as it burned away. At the ends of each pig we removed the brick lining and filled in the space between the ends of the pig and casing with fire-clay, and through this clay and the casing made a small hole through which the heat and blast would escape as soon as the iron melted and fell out of the lining. The first bar of iron was placed three inches above the bottom, and the others at intervals of six inches. When they had all been put in place the bottom door was put up, a sand bottom put in and the fire started in the usual way. As soon as the fire was burned up the cupola was filled with coke to the charging door, which was six feet from the bottom, and the blast put on. The fan was run very slowly during the heat and

EXPERIMENTS IN MELTING.

the air-gauge showed less than one ounce pressure of blast in the pipe at any time during the heat, and the greater part of the time showed no pressure at all. We attributed the light pressure of blast to the fact that no iron was placed in the cupola but the ten bars of pig iron, and the blast escaped freely through the fuel. The pressure of blast would probably have been greater if the fuel had been heavily weighted down with charges of iron closely packed in the cupola. The tap hole was made small and not closed during the heat, and the iron permitted to run out as fast as melted and a note made of the time at which it melted. Iron first appeared at the tap hole in three minutes after the blast was put on, and continued to flow freely until one pig was melted, as was shown by the weight of the iron when cold. The pig melted was the one placed six inches above the top of the tuyeres, as indicated by the escape of the blast from the holes placed in the casing at each end of the pig. After this pig had melted there was a cessation in the flow of iron from the tap hole for about three minutes, when it began to flow again and flowed freely until another pig was melted. The pig melted this time was the one placed twelve inches above the tuyeres, as indicated by the small holes at the ends of the pig. There was then a dribbling of iron from the tap hole for a short time, when it ceased altogether; but the blast was kept on until the appearance of the flame at the charging door indicated that the fuel was all burned up, and the bottom was then dropped.

When the cupola cooled off it was found that none of the four bars placed below the tuyeres had been melted or bent, and they showed no indications of having been subjected to an intense heat. The fifth bar, however, showed such indications and was partly melted, but was still in place. This bar was placed across the cupola almost on a level with the tuyeres, and at a point where the blast met in the center of the cupola from the two tuyeres. The iron that dribbled from the tap hole, as mentioned above, was melted from this bar. The sixth and seventh bars had melted as indicated by the escape of blast from the small holes in the casing at the ends of each bar and were entirely gone. The eighth bar was badly bent and showed evidence of having been subjected to an intense heat, but was not melted at all. The ninth and tenth bars were in place and showed less signs of having been highly heated than the eighth bar. The iron from the two pigs melted was a shade harder than when in the pig, and the iron from the pig partly melted was two or three shades harder, showing that iron melted very slowly or burned off was hardened in the process, and we afterward found this to be correct in the regular way of charging a cupola. This heat showed that with a light blast the cupola melted only from about the top of the tuyeres to twelve or fourteen inches above the tuyeres.

For the next heat the two bars melted out were replaced by new ones, and the bent one was also removed and replaced by a straight one. The cupola was made up and fired in the same manner as in the former heat, and filled with fuel to the charging door. The same sized tuyeres were used and the speed of the fan increased so as to give a four-ounce pressure of blast in the blast pipe, as indicated by the air-gauge. In this heat, as in the former one, the iron placed below the tuyeres did not melt, and the bars placed above the tuyeres at different heights melted at different times. The sixth bar placed six inches above the top of the tuyeres was the first to melt. Then in a few minutes later the seventh bar melted, and still a few minutes later the eighth bar, placed eighteen inches above the tuyeres. These three bars melted rapidly after they began, and were melted within a few minutes of each other. The iron from the first bar melted was a little dull, but the iron from the other bars was very hot. There was no dribbling of iron from the tap hole after the pigs were melted, as in the former heat, and one pig placed higher in the cupola, was melted in this heat. There was no fuel placed in the cupola after the blast was put on and when the fuel required to fill it to the charging door, or about twelve inches above the top of the last pig, was all burned out, the bottom was dropped, and the cupola permitted to cool off. When we went in to examine it, it was found that the fifth bar, placed opposite the tuyeres, which had to some extent melted in the former heat, showed no change and had not been subjected to so high a temperature in this heat. The sixth, seventh and eighth bars had been melted entirely out, as indicated by the escape of the blast through the small holes in the casing at the ends of each bar. The ninth bar was in place, slightly bent, and showed indications of having been subjected to a higher temperature than during the former heat. The tenth bar at that point showed no change from increase of heat. From this heat we learned that directly in front of the tuyeres and just above them, the heat was decreased by a stronger or greater volume of blast, and the melting temperature was raised to a higher level in the cupola; for the heat had been decreased at the fifth bar to an extent that prevented it from melting at all, and increased at the eighth bar to so great an extent that it was readily melted.

For the next heat we arranged the bars and cupola in exactly the same way, and increased the speed of the fan to give an eight ounce-pressure, as shown by the air-gauge. The melting in this heat was practically the same as in the last one just described. We had anticipated that the melting temperature would be raised to a higher level in the cupola by the increase of blast, and were very much disappointed when it was found that the results were the same as with a four-ounce pressure of blast. After thinking the matter over for several days, it occurred to us to put on the blast without charging the cupola and test the air-gauge with different speeds of the fan. In doing this it was found that with the fan running at the same speed that showed eight ounces pressure on the gauge when the cupola was in blast, the gauge showed six ounces pressure when the cupola was not in blast. We at once concluded that the tuyeres were too small to permit so great a volume of blast to pass through them, and the pressure of blast shown by the gauge was due to the smallness of the tuyeres, and not to the resistance offered to the blast by the stock in the cupola.

Since making this discovery, we have seen a great many cupolas when in blast show a high pressure of blast on the air-gauge when the pressure was almost wholly due to the size of the tuyeres and very little blast was going into the cupola.

After making the discovery that the tuyeres were too small to admit to the cupola the volume of blast produced by the fan, we placed two tuyeres in the cupola, four by six inches, laid flat. The tuyeres were made of this shape to increase the tuyere-area and at the same time neither raise the top of the tuyere nor lower the bottom, so that comparison of results in melting could be made with the former heat without rearranging the bars placed in the cupola.

For the next heat we replaced the bars, melted out and made up, and charged the cupola as before and ran the fan at the same speed that had shown eight ounces pressure on the gauge with the small tuyeres. The result was that the gauge only showed a pressure of four ounces of blast. The sixth bar placed six inches above the tuyeres, which had been the first to melt in former heats, was not melted at all in this heat, and the seventh, eighth and ninth bars were melted in the rotation named at about the same time apart as in former heats. The tenth bar was not melted, and none of the bars below the tuyeres were melted. The iron from the ninth bar, which was placed twenty-four inches above the tuyeres and was the last to melt, was accompanied by a good deal of slag as it flowed from the tap hole, and the iron when cold was white hard, although it was No. 1 soft pig iron when placed in the cupola. The slag and hardness of the iron we attributed to the strong or large volume of blast used in this heat, as there had been no hardening of the last pig melted in former heats with a lighter blast. But this pig had remained in the cupola unmelted during the three former heats and been subjected to the heat of the cupola, and it was afterwards found that the hardness and slag were due to the roasting and burning of the iron in these heats, and not to the strong blast as at first supposed. In this heat the melting temperature was raised to a higher level in the cupola, but only three bars were melted as before at a lower level.

EXPERIMENTS IN MELTING.

For the next heat we placed two more tuyeres in the cupola at the same level and of the same size as those used in the last heat, and arranged the cupola as before with a view to melting the tenth or top bar. The spread of the fan was the same as in the last heat, in which the gauge showed four ounces pressure of blast with two tuyeres. In this heat with double the tuyere-area, the gauge indicated a pressure of about one ounce, showing that the tuyere was still too small in the last heat to permit the blast to escape freely from the blast-pipe into the cupola. We were standing near the spout during this heat with our watch and note-book in hand, waiting to time the first appearance of iron at the tap hole and thinking it was a long time in coming down, when our assistant reported there was no flame or heat at the charging door, and the fire must have gone out. We at once examined the charging door and found that nothing but cold air was coming up. We then stopped the fan and removed the tuyere pipes, and found there was no fire in the cupola at the tuyeres. The front was then removed and plenty of fire was found in the bottom of the cupola, which immediately brightened up. The fire had been well-burned up as we supposed, above the tuyeres, before the blast was put on, and it had not been on more than fifteen minutes. We were not satisfied with the results in this heat, and as the fire showed signs of burning up when the front was out and the tuyeres were opened, it was determined to let it burn up and try it again with the strong blast. After the fire had burned up until there was a good fire at the tuyeres and we were quite sure the fuel was on fire to eighteen or twenty inches above the tuyeres, we put on the same volume of blast as before and watched the results at the charging door. At first the blast came up through the fuel quite hot, but the temperature gradually decreased until it became cold, and it was evident that the large volume of blast had put out the fire, and this was found to be the case when the tuyere pipes were removed.

When the bottom was dropped there was fire in the bottom of the cupola, and the coke around the tuyeres showed that it had been heated, but the coke in the upper part of the cupola showed no signs of having been to any extent heated.

The fuel used in this heat was hard Connelsville coke in large pieces. Large cavities were formed under the bars of iron supported by the lining in charging. The coke was not weighted down with iron in the cupola, and the blast escaped freely through the crevices between the large pieces.

We afterward made a heat in this cupola with the same tuyeres and blast, and charged the cupola in the regular way. The iron melted in this heat was pig and small scrap, that packed close in the charges and did not permit the blast freely to escape through the fuel. The gauge in this heat showed a blast pressure of three ounces and the fire was not blown out, but the cupola did not melt so well as with a less volume of blast, and the iron was harder.

These heats showed that iron is not melted in a cupola by the blast and flame of the fuel; for if it were, the bars directly in front and over the tuyeres, where the blast was the strongest, would have been melted first and been the only ones melted. But the one in front of the tuyeres was not melted by a mild blast, and the one just over the tuyeres was not melted by a strong blast.

The failure of the sixth and tenth bars to melt in the same heat, showed that iron is not melted in a cupola all the way from the tuyeres to the charging door, as it was years ago supposed to do by most foundrymen, when the fuel and iron were mixed in the cupola in place of being put in in separate charges, as is now commonly done.

The raising of the melting temperature to a higher level in the cupola by increasing the blast, showed that there is a certain limited melting space or zone in a cupola in which iron melts, and that this melting zone may be raised or lowered by an increase or decrease of the volume of blast. However the depth of the melting zone is not increased by a strong blast, but the zone is placed higher in the cupola. It was also shown that iron cannot be melted in a cupola outside of this zone, either

EXPERIMENTS IN MELTING.

above or below it, for the bars placed above and below it were not melted with either a light or a strong blast. The putting out of the fire in the cupola by a very large volume of blast and the subsequent poor melting done with a large volume of blast when the cupola was charged in the regular way, showed that too much blast may be given to a cupola and the iron thereby injured.

FUEL UNDER THE TUYERES.

In the first two heats it was noticed that considerable coke fell from the cupola when the bottom was dropped, although the indications of the charging door were that all the fuel in the cupola had been burned up. We determined to learn where this coke came from, and in the third heat we kept the blast on until the cupola was well cooled off, and we then turned a stream of water into it from a hose until the fire was out and the cupola cold. The ashes and cinder were then removed from the tuyeres, and it was found that there was no fuel above them except a few small pieces that had been buried in the ashes and cinder. The bottom was let down gradually, and the cupola found to be filled with coke and very little ashes from the bottom to the tuyeres.

The coke when examined showed that it had been heated through, and was soft and spongy like gas-house coke, and totally unfit for melting purposes. When put into the cupola it was hard Connelsville coke. We thought that all the ash found in the coke was made by the burning up of the bed before the blast was put on, and that the coke was not consumed at all after the blast was put on; but we had no means of accurately determining this point. We afterward put a number of peep holes in the cupola at different points below the tuyeres to observe the action of the fuel at this point. The holes were arranged with double slides, the inner one with mica and the outer one with glass. The mica was not affected by the heat, and could be withdrawn for a few minutes and the action of the fuel observed through the glass without the escape of the blast. Through these openings it was observed that the fuel was always at a white heat just before the blast was put on, but after the blast had been on for a short time it became a dull red and remained so throughout the heat. Molten iron could be seen falling through the fuel in drops and small streams. But the fuel was never seen to undergo any change or to settle down as it would do if it were burning away. From these observations it was concluded that the fuel placed under the tuyeres was not consumed during the time the blast was on, and that the only fuel burned in this part of the cupola was that consumed in lighting up before the front was closed.

LOW TUYERES.

After the failure to melt with the four large tuyeres, we placed two tuyeres, four by six inches, in the cupola, on opposite sides, three inches from the bottom or one inch above the sand bottom. The bars were placed in the cupola as before and the cupola filled with coke to the charging door, and a four-ounce pressure of blast put on, the same as in the heat with these two tuyeres when placed at a higher level, namely, twenty-four inches above the bottom. In this heat three bars were melted, but the quantity of slag that flowed from the tap hole with the iron was so great that we did not know where it came from and we were so afraid of the tuyeres being filled with slag or iron, that we failed to note the time the iron melted or the points we were looking for; but something else was learned.

We at first thought the slag came from tuyeres being placed so near the sand bottom, and when the coke with which the cupola had been filled was burned out and the heat over, we took out the front and raked out the fuel and ash in place of dropping the bottom, to see how badly the sand bottom had been cut up by the blast. It was found that it had not been cut at all and was as perfect as when put in and nicely glazed. The lining had not been burned out to any greater extent than in former heats when there was no slag, and we were at a loss to imagine where the slag came from. But when the iron that had been melted in this heat was examined, it was found where the slag came from. All the pigs melted were placed in the cupola at the beginning of the experiments and had remained there unmelted under the tuyeres during a number of heats, and the iron had been burned by the fire in the bottom of the cupola when lighting up and during the heats. When placed in the cupola this iron was No. 1 soft pig, but when melted it was as hard and brittle as glass, and fully two-thirds of it had been burned up and when melted converted into slag.

The results of this heat were so unsatisfactory that we replaced the bars melted out and repeated the experiment. The results in this heat were practically the same as in the heat with the tuyeres placed twenty-four inches above the bottom. Three bars placed six, twelve and eighteen inches above the tuyeres were melted in the same rotation and in about the same time. There was no trouble with slag, and the cupola melted equally as well as when the tuyeres were placed twenty-four inches above the bottom.

MELTING ZONE.

These heats established the fact that there exists a melting zone in a cupola when in blast, and that iron cannot be melted in a cupola outside of this zone. The location of a melting zone in a cupola is determined by the tuyeres and the distance or height of the zone above the tuyeres by the volume of blast, and the depth of the zone by the volume of blast and charging of the cupola. In these heats the melting zone was lowered in the cupola twenty-one inches by lowering the tuyeres to that extent without making any change in the character of the melting, and the zone could have been raised the same distance without making any differance in the melting. The zone was raised from one level to another above the tuyeres by increasing the volume of blast. In the first heat, with a light blast, a bar of iron placed on a level with the tuyeres was partly melted, and one placed eighteen inches above the tuyeres was highly heated and almost ready to melt. Bars placed above and below these two bars were very little affected by the heat, and bars between them were melted, showing that these two bars were on the edges of the melting zone, and the zone had a depth of about eighteen inches. In the next heat, with a larger volume of blast, the bar placed on a level with the tuyeres was not melted at all, showing that it was outside of the melting zone and the zone had been raised by the stronger blast. In the next heat, with a still larger volume of blast, a bar placed six inches above the tuyeres was not melted, showing that the zone had again been raised by the volume of blast. In each of these heats a bar placed higher in the cupola was melted, showing that the depth of the zone remained about eighteen inches and the entire zone was raised to a higher level in the cupola. We attributed the raising of the zone by increasing the volume of blast to the fact that the blast was cold when it entered the cupola, and it was necessary for the air to pass through a certain amount of heated fuel and become heated to a certain degree before its oxygen entered freely into combination with the carbon of the fuel to produce an intense heat; and the greater the volume of cold air, the greater the amount of heated fuel it must pass through before it became heated. With a hot blast this would not have been necessary, and the zone would probably have remained stationary and the depth of the zone been increased. In heats that were afterward made in this cupola with fuel and iron charged in the regular way, we found that the location and depth of the zone were somewhat changed by the weighting down of the fuel with heavy charges of iron. These tests were made by carefully measuring the fuel in the cupola from the charging door after the fire was burned up and the fuel settled, and we took care to have the fuel burned as nearly alike in each heat as possible, and to have the fire show through the top of the bed before iron was charged.

In a former heat, with only bars in the cupola, a bar was melted placed twenty-four inches above the tuyeres. We placed a bed of that height in the cupola and put a charge of three hundred weight of iron on it, and turned on the same blast with which we melted the bar at that height. The blast was on for half an hour before any iron melted, and the melting was very slow until about half the charge was melted, when it began to melt faster. This indicated that the iron was placed above the melting zone and supported there by the fuel, and the fuel had to be burned away before the iron was permitted to come within the zone by the settling of the stock.

In the next heat we placed the top of the bed two inches lower, and in each subsequent heat two inches lower, until it was lowered to ten inches above the tuyeres, and made the charges of iron the same, or three hundred weight.

With a twenty-two inch bed, iron came down in twenty minutes and was hot, but melted slowly throughout the heat.

With a twenty-inch bed, iron came down in ten minutes, melted *hot* and faster than in previous heats.

With an eighteen-inch bed, iron came down in five minutes, and melted fast and hot throughout the heat.

With a sixteen-inch bed, iron came down in four minutes, melted hot and fast at first, but toward the latter end of the charge the iron was a little dull, and as each charge melted the first part of it was hot and the latter part dull.

With a fourteen-inch bed, iron came down in four minutes. Melted fast, but was too dull for light work.

With a twelve-inch bed, the iron was very dull, and with a ten-inch bed it was so dull that it could not be used for general foundry work. With a light blast and low melting zone, the iron in these two heats would probably have been hot.

In these experiments we obtained the best general results with a bed of eighteen to twenty inches, and we adopted this bed for further experiments.

Our next experiments were to learn the depth of the melting zone in practical melting, and the amount of iron that should be placed in each charge to melt iron of an even temperature throughout a heat. In these experiments we made the charges of fuel placed between the charges of iron at a ratio of one pound of fuel to ten pounds of iron.

THE CUPOLA FURNACE.

For the first heat we put in a bed of eighteen inches, on this bed four cwt. of iron on this iron forty pounds of coke, on the coke four cwt. of iron, and so on until the heat was all charged. The blast was the same as before, four ounces pressure with two large tuyeres. In this heat the iron melted hot and fast, and of an even temperature throughout the heat.

For the next heat we made the charges of iron five cwt., and charges of coke fifty pounds. The results in melting were practically the same in this heat as in the former one.

For the next heat we made the charges of iron six cwt., and coke sixty pounds. In this heat there was a slight change in the temperature of iron as the last of each charge melted.

For the next heat the charges were, of iron seven cwt., and coke seventy pounds. The iron in this heat was a little dull when the last of each charge melted, and hot when the first of the next charge melted, making the iron of a very uneven temperature throughout the heat. How often have we seen cupolas melt in this way. In fact it is a common thing in the majority of machine and jobbing foundries for a cupola to melt iron of an uneven temperature, and moulders may be seen almost every heat standing round the cupola watching their chance to catch a ladle of hot iron to pour a light pulley or other piece of light work. The uneven melting is never attributed to improper charging, but to the mysterious working of the cupola.

For the next heat the charges were, of iron eight hundred weight, and coke eighty pounds. In this heat the iron was hot until the last of the first charge, when it became dull. The first of the second charge was hot, but it soon became dull, and before the charge was all melted it was very dull. At the beginning of the third charge the iron livened up a little, but soon became too dull to pour the work and had to be put into the pig bed. In this heat we used exactly the same percentage of fuel (one to ten) between the charges as in the former heats, which should have raised the top of the bed to its former height after melting a charge of iron; but it did not do so, as shown by the melting, and the iron became duller as the melting of

EXPERIMENTS IN MELTING.

the heat progressed. Had another charge of iron been put in, it probably would not have melted at all. The failure of the cupola to melt well in the latter part of the heat was not due to the heat being too large for the cupola, for we afterwards melted heats double the size of this one in the same cupola, and had hot iron to the end of the heat. The top of the bed was reduced to a lower level in this heat in melting the heavier charges of iron, and the fuel in the bed must have burned away more rapidly when the bed was low, or the charges of fuel would have restored it to its former height, as with the light charges of iron. We tried to determine this point more accurately by placing a vertical slot in the cupola at the melting zone in order to observe the settling of the charges, but the heat was so intense at this point that the heat could not be confined within the cupola, and the slot had to be closed up.

In these experiments the most even melting was done with four and five hundred-weight charges. With these charges the fuel kept the top of the bed at a proper height in the melting zone, while with heavier charges it became lower after the melting of each charge, until it became too low to make hot iron, and if the charges had been continued, too low to melt at all. We afterward tried a number of heats with a twenty-inch bed and six hundred-weight charges, and did good melting. With a twenty-four inch bed and six hundred-weight charge the melting was even, but slow.

By the experiments in this cupola it was found that it was necessary to pass the blast through a certain amount of heated fuel before a melting zone was formed in a cupola, and that the amount of heated fuel required for the blast to pass through depended upon the volume of the latter. This heated fuel must be above the tuyeres, for the blast passes upward from the tuyeres, and the melting zone is located at a point dependent upon the amount of heated fuel the blast must pass through before it becomes heated and forms the zone. The blast does not pass downward from the tuyeres except when it may be permitted to escape from the tap or slag hole, and fuel placed below the tuyeres takes no part in the melting of iron in a cupola. When the tuyeres are placed high, the fuel grows deader as the heat progresses and becomes a dull cherry red. We believe the fire would go out in this part of a cupola in a long heat were it not for the molten iron dropping through the fuel, and the occasional escape of blast from the tap and slag holes.

Iron melted high in a cupola is made dull by passing through a large amount of fuel below the tuyeres. With the tuyeres in this cupola placed three feet above the bottom and iron properly charged to make hot iron, it was found impossible to get hot iron at the tap hole for light work. This was undoubtedly due to the iron being chilled in its descent through the fuel under the tuyeres, for the same charging and blast produced hot iron with low tuyeres. The amount of fuel under the tuyeres makes no difference in the location of the zone, and it is the same distance above the tuyeres with high tuyeres as with low ones, when the blast is the same. No iron is melted outside of the zone, and fuel placed above the zone takes no part in melting until it descends into the zone. If too large a quantity of fuel is placed in a bed, the iron charged upon the bed is placed above the zone and cannot be melted until fuel in the zone is burned away and the iron settles into the zone, and iron is a long time in melting after the blast is put on. If too great a quantity of fuel is placed in the charges, the top of the bed is raised above the zone after the melting of each charge of iron, and fuel must again be burned away before the iron can settle into the zone to be melted, and there is a stoppage in melting at the end of each charge of iron. If the charge of iron is made too heavy, the bed is lowered to so great an extent in melting the charge that the top of the bed is not raised to the top of the zone by the charge of fuel; and as each succeeding charge is melted, the bed sinks lower until it gets near the bottom of the zone and iron melts dull, or sinks below the bottom of the zone, and melting ceases. Scarcely any two cupolas have the same tuyere area or receive the same volume of blast, and for this reason scarcely any two cupolas can be charged exactly alike. To do good melting in a cupola it is necessary for the melter to vary the amount of fuel in the bed until he finds the top of the melting zone, and to vary the charges of fuel until he finds the amount of fuel that will raise the bed to the top of the zone after melting a charge of iron. He must vary the weight of the charges of iron until he finds the amount of iron that can be melted in a charge without reducing the bed too low to be properly restored by a charge of fuel.

After twenty years' active experience in melting in different cupolas, the above are the only practical instructions we can give for charging and managing a cupola; and no table of charges for cupolas of different sizes, with different tuyere-area and volume of blast, would be of any practical value to a melter. Fuel placed in a cupola above the zone to replenish the bed is heated by the escaping heat from the zone, and prepared for combustion in the latter, and iron placed above the zone to be melted is heated and prepared for melting in it, and the more fuel and iron brought into a cupola at one time the greater the amount of heat utilized. And the charging door in a cupola should be placed at a sufficient height to admit of a large amount of stock, or the entire heat, being put into the cupola before the blast is put on.

The melting zone is developed above the tuyeres by permitting the blast or carbonic oxide to escape upward after passing through the zone, and it may be developed below the tuyeres by permitting it to escape downward. A cupola has been constructed with the tuyeres placed near the top, and provision made for the escape of the blast through flues arranged near the bottom of the cupola. It was hoped by this plan that all the heat produced by the fuel would be utilized in melting, and the entire heat placed in a cupola melted very quickly and economically. But these hopes were not realized, for the depth of the melting zone was not increased by being below the tuyeres, but remained the same as above the tuyeres. Iron could not be melted outside of the zone, and the cupola was a failure.

MELTING WITH COAL.

All the experiments just described were made with Connelsville coke, but we also made a number of similar ones in this same cupola with anthracite coal. In these experiments it was found that the melting zone was not so high above the tuyeres with the same volume of blast as with coke, nor was the depth of the zone so great, but the coal did not burn away so rapidly in the zone as coke and heavier charges of iron could be melted. In these experiments the best results were obtained with a bed of about fourteen inches above the tuyeres and charges of coal of one to eight, and charges of iron from one-half to two-thirds heavier than with coke. An opinion prevails among foundrymen that the tuyeres in a cupola must be especially adapted for coke, or coke cannot be used. In these experiments we used the same tuyeres as with coke, placed them at the same heights, and found no difficulty in melting with them; and iron may be melted in almost any cupola with either coal or coke, if charged to suit the fuel and tuyeres.

SOFTENING HARD IRON.

In experimenting with iron in a crucible, we found that the hardest iron could be softened by melting it, or subjecting it to a prolonged heat in a closed crucible with charcoal. We thought the same results might be obtained in a cupola by passing molten iron through charcoal in its descent from the melting zone to the bottom of the cupola. It had been found that fuel below the tuyeres was not consumed during a heat, and we decided to try permitting the iron after melting to drop through a bed of charcoal under the tuyeres. The tuyeres were placed twenty-four inches above the bottom and the cupola was filled with charcoal to the tuyeres, and above the tuyeres coke to do the melting was placed. We were afraid the charcoal would all be burned up before the coke above the tuyeres was ready for charging, and to prevent this we put in a wood fire to dry the bottom and warm the cupola. When this was burned out we filled the cupola with charcoal to the tuyeres, put in shavings and wood, and lit the fire at the tuyeres above the charcoal. The charcoal was only burned a little on top when the coke was ready for charging, and not on fire at all in the bottom of the cupola. When the cupola was ready for charging we put in one charge of five cwt. of hard pig and scrap, and put on the blast. The iron melted hot, but in its descent through the charcoal to the bottom of the cupola was cooled to such an extent that it would scarcely run from the tap hole, and the heat was a failure. This was not the only failure in our experimental melting, and we are afraid if we attempted to write up all our experimental heats more failures than successes would be recorded. Experiments in a cupola are not always a success, no matter how much care may be taken in making them. Experimenters generally report only their successful experiments, but if they would report their failures also, they would give much valuable information and save other experimenters much time and expense in going over the same ground.

For the next heat we placed shavings over the bottom, filled the cupola with charcoal to the tuyeres, and put shavings and wood on top of the charcoal for lighting the coke. There was a great deal of trouble in getting the two fires to burn at the same time, and the results were not at all satisfactory.

For the next heat we filled the cupola with charcoal to a short distance above the tuyeres to allow for burning away, and settling, and lit the fire from the front in the ordinary way, and as soon as it was burned up to the tuyeres put in the front to shut off the draught at the bottom. This worked very well, and we found we had a good bed of hot charcoal up to the tuyeres when the cupola was ready for charging. On the bed of coke was placed a charge of five cwt. of pig and scrap, all white hard, and the blast put on. The charcoal bed did not appear to burn away at all during the heat, and the iron melted well and came down hot. When tapped almost as fast as melted, the iron was very little softened by the charcoal. But when allowed to remain in the cupola for some time after melting, it was softened to the extent of becoming a mottled iron when run into pigs or heavy work. But when held in the cupola for a sufficient length of time to soften it to this extent, the iron became very dull and not fit to run light work. This experiment was repeated a number of times with different grades of hard iron, but we never found any marked change in the iron when tapped almost as fast as melted and hot. When held in the cupola a sufficient length of time to soften it to a limited extent, it was too dull to run light work, for the flowing properties of the iron were not to any extent increased by the charcoal. As there is no difficulty in making mixtures of iron soft enough for heavy work into which dull iron can be poured, we could see no advantage in using charcoal in this way.

TIME FOR CHARGING.

There is a wide difference of opinion among foundrymen as to the proper time for charging iron on the bed and putting on the blast after charging. Some claim that if iron is charged several hours before the blast is put on, fuel in the bed is burned up and the heat is wasted, and others claim that heat is wasted by putting on the blast as soon as iron is charged. In some foundries the cupola is filled with fuel and iron to the charging door before lighting the fire. In others, iron is charged after the fire is burned up and permitted to remain in the cupola two or three hours before the blast is put on, and in some foundries the blast is put on as soon as charging of iron begins.

We made a number of experiments in the heats just described to ascertain the proper time for charging and putting on the blast after charging. Iron charged before the fire was lit was very uncertain as to the time at which it melted after the blast was put on. In some heats it melted in five minutes and in others in thirty minutes.

Iron charged before the fire was burned through the bed was a long time in melting after the blast was put on, and the time of melting was very uncertain; in some heats it melted in ten minutes, and in others not for thirty minutes. Iron charged after the bed was burned through and the heavy smoke burned off, melted sooner after the blast was on and was more regular in time of melting, and generally melted in ten minutes when the bed was of a proper height.

Iron charged two or three hours before the blast was put on, melted in from three to five minutes after it was put on.

Iron charged and the blast put on as soon as charging began, melted in from fifteen to twenty minutes.

In these heats it was found that time and power to run the blower were saved by charging the iron two or three hours before putting on the blast, for iron melted in from three to five minutes after the blast was on, and melted equally as fast during the heat as when the blast was put on soon after the iron was charged. We do not think that any fuel was wasted by this manner of charging, for we shut off the draught from the bottom of the cupola by putting in the front and closing all the tuyeres but one as soon as the bed was ready for charging. The bed burned very little after the front was put in, and the heat that arose from it was utilized in heating the first charge of iron preparatory to melting, or iron would not have melted in less time than when the blast was put on as soon as the iron was charged. There is great risk in charging iron before the fire is lit or has burned up, for the fire may go out or not burn up evenly, and we prefer to have the bed burned through before charging the iron.

DEVICES FOR RAISING THE BOTTOM DOORS.

A number of devices have been used for raising the bottom doors of cupolas into place, and thus avoiding the trouble and labor of raising them by hand. One of the oldest of these devices is a long bar, one end of which is bolted to the under side of the door, on the other end is cast a weight or ball almost sufficient to balance the door upon its hinges when raised. When the door is down the bar stands up alongside of the cupola, and when it is desired to raise the door the bar and weight are swung downward. As the weight descends the door is balanced upon its hinges and swings up into place, where it is supported by a prop or other support. This device, when properly arranged and in good order, raises the door very easily and quickly into place, but it is continually getting out of order. The sudden dropping of the door in dumping and the consequent sudden upward jerk given to the heavy weight on the end of the bar, frequently breaks the bar near the end attached to the door or breaks the bolts by which the bar is attached to the door, and the door is sometimes broken by the bar. For these reasons this device is very little used.

Another device, and probably the best one for raising heavy doors, is to cast large lugs with a large hole in them, on the bottom and the door, and put in an inch and a half shaft of a sufficient length to have one end extend out a few inches beyond the edge of the bottom plate. The door is keyed fast upon the shaft, and the shaft turns in the lugs upon the bottom when the door is raised or dropped. An arm or crank is placed upon the end of the shaft, pointing in the same direction from the shaft as the door. When the door is down the arm hangs down alongside of the iron post or column supporting the cupola and is out of the way in removing the dump, and when the door is up the arm is up alongside of the bottom plate, out of the way of putting in the bottom props. The door is raised by a pair of endless chain pulley blocks attached to the under side of the scaffold floor at the top and the end of the arm at the bottom, and it is only necessary to draw up the arm with the chain to raise the door into place. This is one of the best devices we have seen for raising heavy doors.

Another one, equally good for small doors and less expensive, is to make the end of the shaft square and raise the door by hand with a bar or wrench five or six feet long, placed upon the end of the shaft. The bar is placed upon the shaft in an upright position, and by drawing down the end of the bar the door is swung up into place by the rotation of the shaft on to which it is keyed. When the door is in place the bar is removed from the end of the shaft, and is not at all in the way of handling the iron or managing the cupola.

CHAPTER VI.

FLUXING OF IRON IN CUPOLAS.

FLUX is the term applied to a substance which imparts igneous fluidity to metals when in a molten state, and has the power to separate metals contained in metallic ores from the non-metallic substances with which they are found in combination; also to separate from metals when in a fluid state any impurities they may contain. Fluxes are also used for the purpose of making a fluid slag in furnaces to absorb the non-metallic residue from metals or ores and ash of the fuel, and removing them from the furnace to prevent clogging and to keep the furnace in good working order for a greater length of time. The materials used as fluxes for the various metals are numerous and varied in nature and composition, but we shall only consider those employed in the production of iron and the melting of iron for foundry work.

The substances employed for this purpose are numerous, but they consist chiefly of the carbonate of lime in its various forms, the principal one of which is limestone.

In the production of pig iron from iron ore in the blast furnace, limestone is used for the two-fold purpose of separating the iron from the ore, and for liquefying and absorbing the non-metallic residuum of the ore and ash of the fuel, and carrying them out of the furnace. For this purpose large quantities of limestone are put into the furnace with the fuel and ore. The stone melts and produces a fluid slag, which absorbs the non-metallic residuum of the ore and ash of the fuel in its descent to the bottom of the furnace. Thence it is drawn out at the slag hole, and carries with it all those non-metallic substances which tend to clog and choke up the furnace. By this process of fluxing the furnace is kept in good smelting order for months, and even years. Were it not for the free use of limestone, the furnace would clog up in a few days.

The blast furnace is a cupola furnace, and is constructed upon the same general principle as the foundry cupola. Foundrymen long ago conceived the idea of using limestone as a cupola flux. In many foundries it is the practice to use a few shovelfuls or a few riddlefuls of finely broken limestone in the cupola on the last charge of iron, or distributed through the heat, a few handfuls to each charge of iron. The object in using limestone in this way is not to produce a slag to be drawn from the cupola, but to make a clean dump and a brittle slag or cinder in the cupola, that can be easily broken down and chipped from the lining when making up the cupola for a heat.

Limestone used in this way does not produce a sufficient quantity of slag to absorb the dirt from the iron and ash of the fuel and keep the cupola open and working free, but rather tends to cause bridging and reduce the melting capacity of the cupola.

The making of a brittle cinder in a cupola by the use of limestone depends to a great extent upon the quality of the stone. Some limestones have a great affinity for iron and combine with it freely when in a molten state, while others have but little affinity for iron and do not enter into combination with it at all. In the cinder piles about blast furnaces we find cinder almost as heavy and hard to break as iron, resisting the action of the atmosphere for years; while at others we find a brittle cinder that crumbles to pieces after a short exposure to the atmosphere, or even slacks down like quicklime when wet with water. In a cupola we may have a hard or brittle cinder produced by limestone. The results obtained from the use of limestone in small quantities in a cupola are so uncertain that we do not think they justify the foundryman in using it.

LIMESTONE IN LARGE QUANTITIES.

The tendency of slag or cinder in a cupola is to chill and

FLUXING OF IRON IN CUPOLAS. 137

adhere to the lining just over the tuyeres and around the cupola at this point, and prevent the proper working of the furnace. So great is this tendency to bridge that a small cupola will not melt properly for more than two hours, and a large one for more than three hours. To overcome this tendency to clog and bridge, foundrymen in many cases have adopted the blast-furnace plan of using a large per cent. of limestone as a flux in their cupolas, and tapping slag.

When a large per cent. of limestone is charged with the iron in a cupola, it melts when it settles to the melting point and forms a fluid slag. This slag settles through the stock to the bottom, and in its descent melts and absorbs the ash of the fuel and dirt or sand from the iron and carries them to the bottom of the cupola, where the slag and dirt it contains may be drawn off and the cupola kept in good melting order and in blast for days at a time. The amount of limestone required per ton of iron to produce a fluid slag depends upon the quality of the stone and the condition of the iron to be melted. It is the custom in some foundries, where the sprews and gates amount to from thirty to forty per cent. of the heat, to melt them without milling to remove the sand, and to use enough limestone in the cupola to produce a sufficient quantity of slag to absorb and carry out of the cupola the sand adhering to them. In this case a larger per cent. of limestone is required than would be necessary if the sprews and gates were milled and only clean iron melted. Poor fuel also requires a greater amount of slag to absorb the ash than good fuel, and a lean limestone must be used in larger quantities than a stone rich in lime. The quantity required to produce a fluid slag, therefore, varies with the quality of the limestone and the conditions under which it is used, and amounts to from 25 to 100 pounds per ton of iron melted.

The weight of the slag drawn from a cupola when the sprews and gates are not milled, and the cupola is kept in blast for a number of hours, is about one-third greater than the weight of the limestone used. When the sprews and gates are milled, the weight of the slag is about equal to the weight of the limestone. When the cupola is only run for a short time and slag only drawn during the latter part of the heat, the weight of the slag is less than the weight of the limestone.

The slag drawn from a cupola has been found, by chemical analysis, to contain from 4 to 7 per cent. of combined iron and numerous small particles of shot iron mechanically locked up in the slag. These cannot be recovered except at a greater cost than the value of the metal. In a number of tests made in the same cupola, we found the loss of iron to be from 3 to 4 per cent. greater when the cupola was slagged.

EFFECT OF FLUX UPON IRON.

Many of the limestones and other mineral substances employed as cupola fluxes contain more or less finely divided oxides, silicates, etc., in combination with earthy materials, The flux is often reduced in a cupola and its component parts separated, and in minute quantities they alloy with the iron and injure its quality. The conjoined effect upon iron of these diffused oxides, silicates, etc., liberated in a cupola from their native elements in fluxes, is to prevent the metal running clean in the mould or making sharp, sound castings, and the tensile and tranverse strengths are frequently impaired by them. When the oxides, silicates, etc., are not separated in the cupola from their native elements, they do not impair the quality of the metal, nor do they improve it. The tendency of the cupola furnace is to clog and bridge over the tuyeres, and concentrate the blast upon the iron through a small opening in the center and injure its quality. If by the free use of limestone we prevent bridging and keep the furnace working open and free, we avoid injuring the iron in melting by the concentration of a strong blast upon it. The effect, therefore, of limestone in a cupola is not to improve the quality of iron, but to prevent its deterioration in melting.

THE ACTION OF FLUXES ON LINING.

Limestone and other minerals employed as fluxes frequently contain impurities which enter into combination with the lining material of a furnace and render it fusible. This was illustrated at the foundry of John D. Johnson & Co., Hainesport, N. J., in 1893. The cupola front had been put in with new moulding sand for a long time, and no flux used in the cupola. The sand made an excellent front that resisted the action of the heat and molten iron upon it. As the heats enlarged, it became necessary to use flux and tap slag to run off the heat. Oyster shells were used and produced a slag that flowed freely and had no effect upon the sand in the front. When the supply of shells became exhausted, a limestone was used in place of them. Trouble then began with the front. It was melted by the flux into a thick, tough slag that settled down and closed up the tap hole, and iron could only be drawn by cutting away a large portion of the front to enlarge the tap hole. Mr. Johnson called at our office to learn what could be done to keep the tap hole open. We advised that the front material be changed and a mixture of fire-clay and sharp sand be used in place of moulding sand. This was done, and there was no further trouble in keeping the tap hole open and in good order to run off the heat. This serves to illustrate the effect of fluxes upon lining material. With no flux and with oyster shells the moulding sand resisted the heat and pressure of molten iron and slag upon the front; but with limestone it melted into a thick, tough slag. This was due to some property in the limestone entering into combination with the sand and making it fusible. Had the cupola been lined with this moulding sand, the entire lining would have been cut out in one heat, while it would have stood many heats with shells or no flux at all.

From the various qualities of cupola brick and lining material now in the market, a lining may be selected that will resist the action of almost any flux or slag, and foundrymen may select a flux to suit the lining or a lining to suit the flux, which ever they find to be the most profitable in their locality.

HOW TO SLAG A CUPOLA.

Foundrymen sometimes experience trouble in slagging their cupolas. This is largely due to a lack of knowledge in charging the limestone and drawing the slag, for any cupola can be slagged if properly worked. To draw slag from a cupola, a sufficient quantity of limestone or other slag-producing material must be charged in the cupola with the iron to make a fluid slag. The exact amount required can only be learned by experimenting with the fluxing material used, but it is generally from fifty to sixty pounds of good limestone per ton of iron, when the remelt is not milled. The limestone is generally charged on top of the iron and put in with each charge after the melter begins using it. No limestone is used with the iron on the bed or first few charges of iron. In small cupolas limestone is generally charged with the second or third charge of iron. In large cupolas, when the charges of iron are light, six or eight charges, or generally about one-sixth of the heat, are charged without limestone. This is the way limestone is used when the cupola is run in the ordinary way for a few hours. When the cupola is run for some special work, the limestone is charged in a number of different ways.

The slag is drawn from the cupola through an opening known as the slag-hole. This opening is made through the casing and lining under the lower level of the tuyeres and at a point in the cupola where it will be out of the way in removing iron from the spout and convenient for removing the slag. The height the slag hole is placed above the sand bottom depends upon how the iron is drawn from the cupola. When it is desired to hold iron in a cupola until a sufficient quantity is melted to fill a large ladle, the slag hole is placed high, and when the iron is drawn as fast as melted the slag hole is placed low. When the slag hole is placed high, slag can only be drawn as the cupola fills up with iron and raises it to the slag hole. When the iron is withdrawn from the cupola, the slag falls and the slag hole is closed with a bod to prevent the escape of blast. When the iron is drawn from the cupola as fast as melted, the slag hole is placed low and when opened it is permitted to remain open through the remainder of the heat. This is the best way of drawing slag from a cupola, for the flow is regulated by the amount of slag in the cupola, and if the hole is not made too large, there is no escape of blast.

The slag in the bottom of a cupola takes up impurities from the fuel and iron, and if permitted to remain in the cupola for too long a time, it may become so thick and mucky it will not flow from the slag hole. Or it may be filled with impurities, become over-heated, boil up and fill the tuyeres with slag; and when boiling, it will not flow from the cupola through a small slag hole. The time for drawing the slag from a cupola is therefore a matter of great importance. The slag hole is generally opened in from half an hour to an hour after the cupola begins to melt, and when placed low is permitted to remain open throughout the remainder of the heat. When placed so high that slag can only be drawn when the cupola fills up with molten iron, it should be opened as soon as the slag begins to rise and closed as soon as it falls below the opening.

DOES IT PAY TO SLAG A CUPOLA?

Nothing is gained by slagging a cupola when the sprews and gates are milled and the heat can be melted successfully in the cupola without slagging; but a great saving in labor and wear and tear of machinery can be effected in many foundries by melting the sprews and gates with the sand on, and slagging to carry the sand out and keep the cupola working free. A cupola can not be made to melt iron faster by slagging, but it can be kept in blast and in good melting condition for a greater length of time and a much larger amount of iron melted by slagging. Foundrymen who find their cupolas temporarily too small to melt the quantity of iron required for their work, can overcome the difficulty by slagging the cupola and keeping it in blast for a greater length of time.

In endeavoring to make an estimate of the cost of slagging a cupola, we found that the cost of limestone in different localities varied from 50 cents to $3 per ton. The amount used varied from 25 to 100 pounds per ton of iron melted. The amount of slag drawn varied from 25 to 100 pounds per ton of iron. The iron combined with the slag varied from 4 to 7 per cent. With these wide differences in the cost and quantity of limestone used, and the difference in the quantity of slag drawn and per cent. of iron it contained, we found it impossible to make an estimate that would be of any practical value to foundrymen. Such an estimate must be made at each foundry to be of any practical value.

SHELLS.

Oyster, clam and other shells are largely composed of lime, and are frequently used as a flux in place of limestone in localities where they can be procured at a less cost than limestone. The shells are charged in the same way as limestone and in about the same proportion to the iron. They may be used in place of limestone either in large or small quantities, and have about the same effect upon the iron and cupola as limestone. When used in large quantities, they produce a fluid slag that keeps the cupola working free and flows freely from the slag hole, carrying with it the refuse of melting that clogs the cupola. When the heat first strikes shells in a cupola, they produce a crackling noise and flakes of shell may be seen to pass up the stack, and the foundry roof, when flat, is often covered with flakes of shell after a heat, when shells are used in large quantities. The crackling is due to the destruction of the hard inner surface of the shell; the flakes thrown from the cupola are entirely of this surface, and the loss of shell is not as great as it would appear to be at first sight. The remainder of the shell melts and forms a fluid slag that absorbs the refuse of melting, becomes thick and helps to clog up a cupola when the shells are used in small quantities, or assists in keeping it open when used in large quantities.

MARBLE SPALLS.

Marble is another of the carbonates of lime, and the spalls or chippings from marble quarries or works are quite extensively used in some localities as a cupola flux. Their action in a cupola and their effect upon iron is very similar to that of limestone, and they are used in the same way and in about the same proportions. There are a number of other substances, such as fluor-spar, feld-spar, quartz-rock and a number of chemical compounds that are used as cupola fluxes.

In 1873, when engaged in the manufacture of malleable iron, we began experimenting with mineral and chemical materials with the view of making a cheap malleable iron, and changing the nature of iron in a cupola furnace so that it might be annealed at a less cost, and produce stronger iron. In this we succeeded to some extent, and then drifted off into improving the quality of iron in a cupola for grey iron castings; this we have followed for nearly twenty years. During this time we have melted iron in foundries all over the greater part of the United States and Canada, and have constructed and worked a number of experimental cupolas of our own, to learn the effect of different mineral and chemical substances upon iron and cupola linings. In these investigations we have used all the mineral and chemical fluxes known to metallurgical science, and observed their effect upon the various grades of iron employed for foundry work.

In these experiments it was found that iron can be improved or injured when melted in a cupola furnace, and is often ruined as a foundry iron by improper melting and fluxing. The point at which iron is melted in a cupola has a great deal to do with its quality. Iron melted too high in a cupola is burned and hardened; melted too low, it runs dirty in a mould; melted with too strong a blast, it is hardened. Iron melted dull does not make a sound casting. Iron melted with poor coal or coke is injured by the impurities in the fuel. Iron melted with oyster shells, limestone and other mineral fluxes may take up oxides, sulphides, phosphides, silicates and other impurities contained in the flux and be ruined by them for foundry work.

The per cent. of iron lost in melting is increased by improper melting and fluxing, and may be double or treble what it should be. We have made a great many experiments to ascertain the effect of silicon on iron, and have found that silicon enters freely into combination with cast iron and has a softening effect upon it. Iron as hard as tempered steel may be made as soft as lead by combining it with silicon. But silicon is an impurity having a deleterious effect upon iron. An excess of it destroys cohesive force and crystallization, and reduces transverse and tensile strength. So great is the destruction of cohesive force in cast iron by silicon that the strongest iron may be reduced to a powder when combined with an excess of silicon. Silicon in any proportion is a detriment to cast iron, as an iron. The nature and form of crystallization of a pure cast iron is changed by sudden cooling in a mould, and a soft iron in the pig may become a hard iron in a casting. This chilling property in cast iron is destroyed by silicon, and an iron high in it is not hardened when run into a sand mould or upon an iron chill. The destruction of the chilling tendency in cast iron is very desirable in the manufacture of light castings, and for this reason silicon irons are largely used in foundries making this class of work.

The per cent. of silicon an iron may contain and yet retain sufficient cohesive force for the work, depends upon the amount of other impurities in the iron and the work the iron is employed to make. For heavy work, requiring great strength, it should contain none at all. For light machinery it may contain from one-half to one per cent.; and for stove plate, light bench work, etc., it may contain from two to three per cent. This amount is sufficient to reduce the chilling tendency of the iron, without impairing its strength to any great extent in this class of work. But a larger amount destroys the strength of the iron and also injures its flowing property in a mould.

At the present time there is a large amount of high silicon cheap Southern iron being used in stove foundries for the purpose of making a cheap mixture and a soft casting. At one of these foundries we recently visited, the foreman informed me that they were using a mixture that cost $14 per ton, and said their breakage in the tumbling barrels and mounting shop was very large, and he never made a shipment to their warehouse in New York, a distance of 25 miles, but a lot of stoves were broken in transit and sent back to be remounted and repaired.

At another stove foundry in Troy, N. Y., they informed us they were using a mixture of Pennsylvania irons that cost them $20 per ton. They had scarcely any breakage at their works, and shipped their lightest stoves and plate to their warehouse in Chicago without boxing or crating, and never had any breakage in transit or in handling. They had found by experience that a mixture of Pennsylvania irons at a cost of $20 per ton was cheaper in the long run than a mixture of cheap Southern irons at $14 per ton.

In a number of other foundries we visited, they all complained of heavy breakage when using high silicon irons as softeners. Another matter to be considered in using these high silicon irons for stove plate, is, how long will a stove last, made of such weak iron, and can a reputation for good work be maintained by foundries using them? A stove made of this kind of iron will certainly not last as long as one made of good iron.

Carbon has the same effect upon cast iron as silicon, in softening and reducing the chilling tendency. The hardest of cast iron can be made the softest by the addition of carbon, without destroying its cohesive force and rendering it brittle or rotten, and carbon can be added to iron in a cupola as readily as silicon. Before the high silicon Southern irons were put upon the Northern market, highly carbonized irons were used as softeners for stove plate and other light work, and a far better grade of castings were made then than now are made from the silicon irons.

It is difficult to remove silicon from iron when melted in a cupola, but free carbon is readily removed by the oxidizing flame in a cupola produced by a strong and large volume of blast; and a soft iron may be hardened in melting to such an extent as to make it unfit for the work. This can be prevented to some extent by using a mild blast and melting the iron low in the cupola, and it can also be prevented by the use of chemicals in the cupola to produce a carbonizing flame.

We have spent a great deal of time and money in experimenting on the production of such a flame in a cupola as would not only prevent the deterioration of iron in melting, but would improve its quality, and at the present time are engaged in the manufacture of a chemical compound for this purpose.

FLUOR SPAR.

Fluor spar is extensively used as a cupola flux, in sections of the country where it is found native and can be procured at a moderate cost, and it has also been used to a considerable extent in other sections of the country, but the expense of transporting this heavy material has greatly retarded its use as a flux at any great distance from the mines. Fluor spar when used in sufficient quantities in a cupola, produces a very fluid slag that absorbs and liquefies the non-metallic residue of melting with which it comes in contact; keeps the cupola open and working freely, and causes it to dump clean. But it also fluxes the cupola lining, causing it to burn out in a very short time, and for this reason it can only be used in large quantities with certain grades of lining material that are only affected to a very limited extent by it. This quality of lining material can generally be procured in the vicinity of the mine, but it cannot always be had at a moderate cost in other parts of the country, and for this reason it is frequently used with limestone to increase the fluxing properties of the limestone and reduce the injurious effect of the spar upon the cupola lining. When used in this way, fluor spar greatly increases the efficiency of a poor limestone, and often enables a founder to use a cheap limestone that could not be employed alone as a flux, while the limestone reduces the injurious effect of the spar upon the lining, and the two combined make an excellent flux for tapping slag in long heats.

We have used fluor spar in a number of cupolas and with a

great many different brands of iron. We never found it to harden or soften any of these irons to a noticeable extent, but it improved the melting very materially in a number of cases where the cupola was run beyond its melting capacity, melted slow in the latter part of the heat, and could not be dumped without a great deal of labor.

CLEANING IRON BY BOILING.

Before the use of fluxes in cupolas was so well understood as at the present time, it was a common practice in many foundries to cleanse iron of impurities in a ladle by agitating or boiling the molten metal. This caused a large amount of dross to collect on the surface, from which it was skimmed off and the iron was considered to be purer after the boiling. A favorite way of agitating iron in a ladle was to place a raw potato or apple on the end of a tap bar and hold it in the molten metal, near the bottom of the ladle, for a short time. The potato or apple contained a sufficient amount of moisture to agitate or boil the metal gently without exploding it, and the metal was said to be greatly benefited by this gentle boiling; but practice has demonstrated that nothing is gained by boiling iron in a ladle, and the practice has long since been discontinued in this country.

A ball of damp clay placed upon the end of a tap bar was also used for boiling iron in a ladle, but this was not considered as good or as safe as an apple or potato, for if the clay chanced to be too damp, it caused the iron to boil violently and sometimes to explode.

Another favorite way of cleansing and mixing irons years ago was to pole the molten iron. This was done with a pole two or three inches in diameter, of green hickory or other hard wood. The pole was thrust into the molten metal in a ladle or reverberatory furnace, and the metal stirred with it. The effect of the green wood thrust into the metal was to cause it to boil around the pole, and as the pole was moved through the metal all parts of the metal were agitated, and a better mixture of the different grades of iron melted was effected and a more homogeneous casting produced. The poling of iron was a common practice in many foundries twenty-five years ago, but we have not seen iron poled in a ladle for many years, and we believe the practice has been entirely discontinued with cupola-melted iron; but poling is still practiced in many foundries in the mixing of iron in reverberatory furnaces for rolls and other castings requiring a very strong homogeneous iron.

CHAPTER VII.

DIFFERENT STYLES OF CUPOLAS.

OLD STYLE CUPOLAS.

BEFORE describing the construction of the cupolas now in use, a short account of the old-fashioned cupolas may be of interest to many founders who have not had an opportunity of seeing them or observing their defects, all of which defects should be avoided in modern ones.

In Fig 20 is seen the old style cupola in general use throughout the country many years ago, many of which are still in use in some of the old-time small foundries. A square cast-iron bottom plate, with opening in the center and drop door, is placed upon a brick foundation at a sufficient height above the floor for the removal of the dump. An iron column is placed upon each corner of the plate, and upon these columns is placed another cast-iron plate, having an opening in the center for the top of the cupola. Upon this plate a brick stack is constructed to carry off the flame and unconsumed gases from the cupola. The stack plate was sometimes placed upon brick columns or brick walls, built on each side of the cupola, through which openings were made for manipulating the tuyere elbows. The stack was built square and of a much larger size than the inside diameter of the cupola. It was not subjected to a very high heat, and was built of common red brick. These large stacks were not built very high and threw out very few sparks at the top, which was due to their size. The cupola was placed between the bottom and stack plate, and the casing was formed of cast-iron staves, which were held together by wrought-iron bands, drawn tight by draw-bolts placed through the flanged

150   THE CUPOLA FURNACE.

ends of the bands. When the casing was made tapering, the bands were placed in position when hot and shrunk on. The cupolas were only from six to eight feet high, and those of

FIG. 20.

OLD STYLE CUPOLA.

small diameter were generally made larger at the bottom than at the top, to facilitate dropping, and that a large quantity of molten iron might be held in the cupola for a heavy casting. The charging door was placed in the stack just above the stack plate. From two to four tuyeres were put upon each side of the cupola, one above the other, and from eight to ten inches apart. The tuyeres were supplied from a blast pipe on each side, to which was attached a flexible leather hose and tin or copper elbow for conducting the blast into the tuyeres. A small hole was made at the bend of the elbow for looking into the tuyere, and closed with a wooden plug. The tuyeres were frequently poked with an iron bar through these openings.

When light work was to be cast, the upper tuyeres were closed with clay or loam, and the blast sent through the lower tuyere. When it was desired to accumulate a large amount of molten iron in the cupola for a heavy piece of work, the lower tuyeres were used until the molten iron rose to the lower edge. The tuyere elbows were then withdrawn and shifted to the next tuyere above, and the lower tuyere closed with clay or loam rammed in solid. The shifting of the tuyere elbows was continued in this way until the necessary amount of molten iron for the work to be cast was accumulated in the cupola. When a heavy piece of work was to be cast, a sufficient quantity of fuel was placed in the cupola to bring the top of the bed some distance above the top of the highest tuyere to be used; on the bed two cwt. of iron was charged, and a shoveful of coke and a cwt. of iron charged throughout the heat. The charging was raised a little in different sized cupolas, but the fuel and iron were always mixed in charging. The large body of molten metal frequently pressed out the front and sometimes the plugging of the tower tuyeres. After the iron was tapped, the stock in the cupola dropped so low that no further melting could be done with the blast in the upper tuyeres, and frequently the lower tuyeres were so clogged that they could not be opened, and the bottom had to be dropped.

In practice it was found that in a cupola constructed large at the bottom and small at the top for the purpose of retaining a large amount of molten iron, the stock did not spread to fill the cupola as it settled, and a great deal of heat escaped through the space made between the lining and stock by the settling of the stock. It was also found that the shifting of tuyeres required such a high bed that the cupola melted slowly, and a greater per cent. of fuel was consumed in large than in small heats. .

THE RESERVOIR CUPOLA.

To overcome the objections to the tapering cupola and shifting of the tuyeres, and still be able to hold a large amount of molten iron in a cupola, the reservoir cupola, Fig. 21, was designed.

The casing of this cupola was made of wrought iron, and the bottom section, to a height of from twelve to twenty-four inches, was constructed of one-third greater diameter than the upper section or cupola proper. This arrangement admitted of a large body of molten iron being held in the cupola without shifting the tuyeres. The metal was spread over a larger surface, which reduced the pressure on the breast, and did not leave the stock in so bad a condition for melting after a large tap was made as in the taper cupola, and melting could be continued after a large body of iron was tapped. The reservoir cupola did faster and more economical melting in large heats than the tapered cupola, but in small heats the amount of fuel required for the bed was too large for economical melting.

At the present time cupolas are made of the same diameter from the bottom to six or eight inches above the tuyeres. The tuyeres are placed at a height to suit the general run of work to be done, and when a heavy piece is to be cast, the iron is held in ladles and covered with charcoal or small coke to exclude the air. The molten iron can in this way be kept in almost as good condition for pouring as in the cupola, and the cupola is kept in better condition and melts faster and longer.

DIFFERENT STYLES OF CUPOLAS.

FIG. 21.

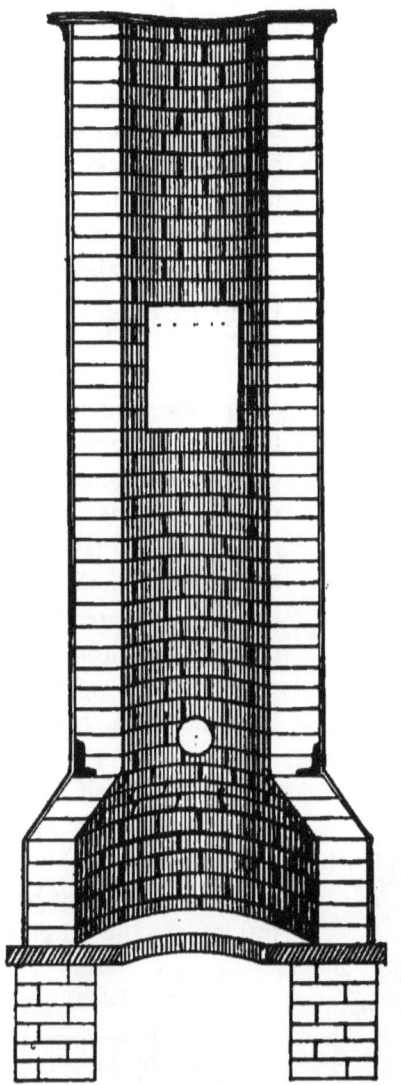

RESERVOIR CUPOLA.

THE CUPOLA FURNACE.

STATIONARY BOTTOM CUPOLA.

In Fig. 22 is shown the old style English cupola. This cupola. is constructed upon a solid foundation of stone or brick work and has a stationary bottom of brick, upon which is made a sand bottom. The refuse, consisting of ash, cinder and slag, remaining in the cupola after the iron is melted, is drawn out at the front in place of dropping it under the cupola, as is now generally done with the drop-bottom cupola. These cupolas are generally of small diameter. The opening in front for raking out is about two feet square, and when the cupola is in blast, is covered with an apron of wrought iron. When the cupola has been made up for a heat, shavings, firewood and a small amount of coke are placed in it and ignited with the front open; when the coke is well alight, a wall is built up with pieces of coke even with the inside of the cupola lining.

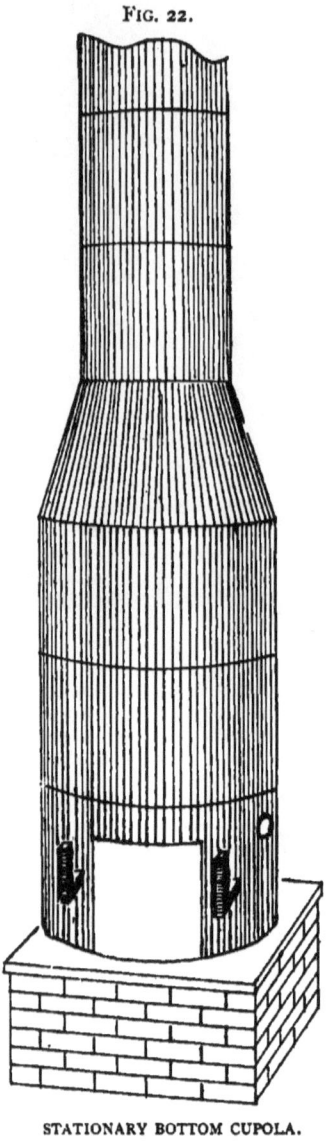

FIG. 22.

STATIONARY BOTTOM CUPOLA.

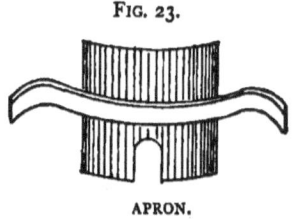

FIG. 23.

APRON.

The bed of coke is then put in, a round stick is placed in the spout to form the tap hole, and the front is then filled in with new molding sand or loam even with the casing, and rammed solid. The apron, Fig. 23, is then placed in position over the loam and wedged tight against it, to prevent it being forced out by the pressure of molten iron in the cupola. After the breast-plate is placed in position, the tap hole and spout are made up in the ordinary way. Some melters prefer to place the apron in position before lighting the fire, and put the breast in from the inside when making up the sand bottom. It is then rammed solid against the apron and made up to the full thickness of the brick lining of the cupola. When the heat has been melted the breast-plate is removed and the loam front dug out. After the loam front has been broken away, a sheet-iron fender is placed in front of the cupola to protect the workmen from the heat, and the raking out process begins. This is done by two men with a long two-pronged rake. If the refuse hangs in the cupola, it is broken down from the charging door with a long bar or by throwing in pieces of pig iron. These cupolas were extensively used in England, but never to any extent in this country. We saw one in Baltimore a few years ago, and believe this is the only one in use in this country; but they are still in general use in England.

EXPANDING CUPOLA.

Fig. 24 is a sectional elevation of the expanding cupola, which is said to have melted very rapidly and with very little fuel. This peculiar form was designed to admit of the charging of a large quantity of iron before putting on the blast, for the purpose of utilizing all the heat produced by the combustion of the fuel. These cupolas were built of common brick, banded with wrought-iron bands and lined with firebrick. The diameter at the charging door was sixty inches and at the tuyeres thirty inches, or one-half the diameter at the charging door. Below the tuyeres the lining expanded to forty or even fifty inches, to give room for molten metal. The bottom was

156   THE CUPOLA FURNACE.

FIG. 24.

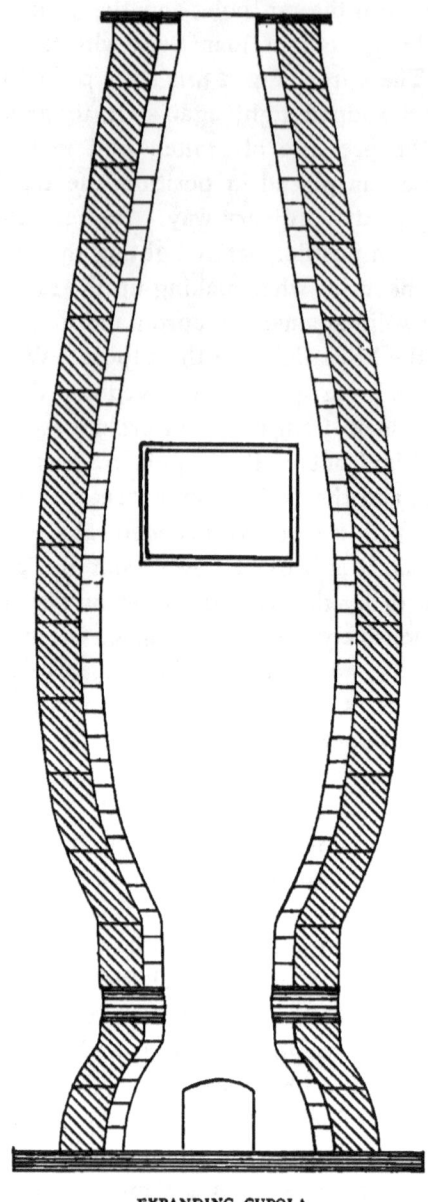

EXPANDING CUPOLA.

stationary, and the refuse after melting was drawn at the front. The cupola expanded from a level a little above the tuyeres to the bottom of the charging door, thence to the top of the stack it gradually contracted.

The greatly increased diameter at the charging door certainly admitted of a large quantity of iron being placed in the cupola at one time, and the utilization of a very large per cent. of the heat in melting. The even taper of the lining insured the even settling of the stock, so that good melting should have been done in this cupola; but the best results obtained appear to have been about six and a half pounds of iron to the pound of coke.

This old form might be used to advantage in the construction of very large cupolas; but in the ordinary sized cupola, practically the same results are obtained by boshing or contracting the lining at the tuyeres, and making it straight from the top of the boshes to the charging door.

IRELAND'S CUPOLA.

Ireland's cupola, for which the inventor took out a number of patents in England about 1856, and which was largely used there about that time, was constructed of a variety of shapes and sizes, but probably the best design is that shown in sectional view Fig. 25. It is built with a bosh and contraction of the diameter at the tuyeres, and has a cavity of enlarged diameter below them to give increased capacity for retaining molten metal in the cupola.

The cupola, of which a section is shown, was twenty-five feet high from bottom plate to top of stack, twelve feet from bottom plate to sill of charging door. The shell was parallel and fifty inches diameter to the charging door, thence it gradually tapered to two feet three inches at the top. There were two rows of tuyeres eighteen inches apart, eight in the upper row two inches diameter, and four in the lower row six inches diameter. The cupola was constructed with stationary bottom and draw front.

158   THE CUPOLA FURNACE.

FIG. 25.

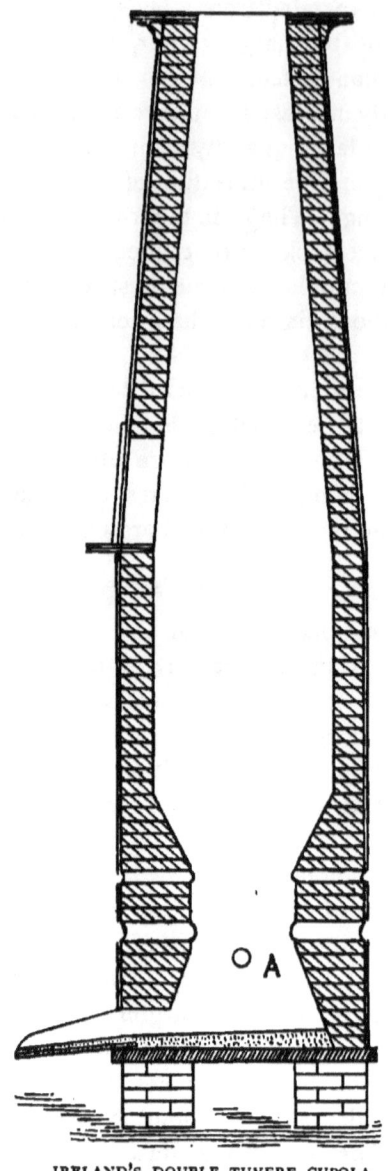

IRELAND'S DOUBLE TUYERE CUPOLA.

DIFFERENT STYLES OF CUPOLAS.   159

It was at first proposed to use a hot blast in the top row of tuyeres, but it was found to be difficult and expensive to heat the blast, and that nothing was gained by using the upper row with a cold blast, and they were closed and the cupola constructed with only the lower row of tuyeres. The interior shape was slightly modified to give more space for retaining molten metal, while, at the same time, retaining the boshes and increasing the diameter of the bottom of the cupola, as seen in the Fig. 25. Two of these cupolas were used by the Bolton Steel and Iron Company in England, in melting the iron for a large anvil block weighing two hundred and five tons, for which two hundred and twenty tons of metal, including eight tons Bessemer steel, were used.

The cupolas were each seven feet outside diameter, three feet nine inches diameter below the boshes in the crucible, and five feet diameter above and below the crucible. The blast was supplied from an external air-chamber, extending round the casing and delivered into the cupolas through two rows of tuyeres placed eighteen inches apart, sixteen in the upper row of three inches diameter, and four in the lower row of eight inches diameter. The metal was melted in ten hours and forty-five minutes from the time of putting on the blast until the mold was filled, and only one hundred and twenty-five pounds of coke consumed per ton of metal. Slag was tapped from the slag hole A below the tuyeres throughout the heat.

IRELAND'S CENTER BLAST CUPOLA.

In Fig. 26 is seen a sectional elevation of Ireland's cupola with bottom tuyere. The height from bottom plate to top of stack is twenty-seven feet, from bottom plate to sill of charging door twelve feet. The casing is parallel from the bottom plate to charging door, and thence it gradually tapers to the top; diameter of casing up to charging door four feet six inches, tapering to two feet six inches at the top of stack. The inside diameter at bottom of crucible, on the cupola hearth L is two feet six inches, contracting to two feet three inches at spring of

160   THE CUPOLA FURNACE.

Fig. 26.

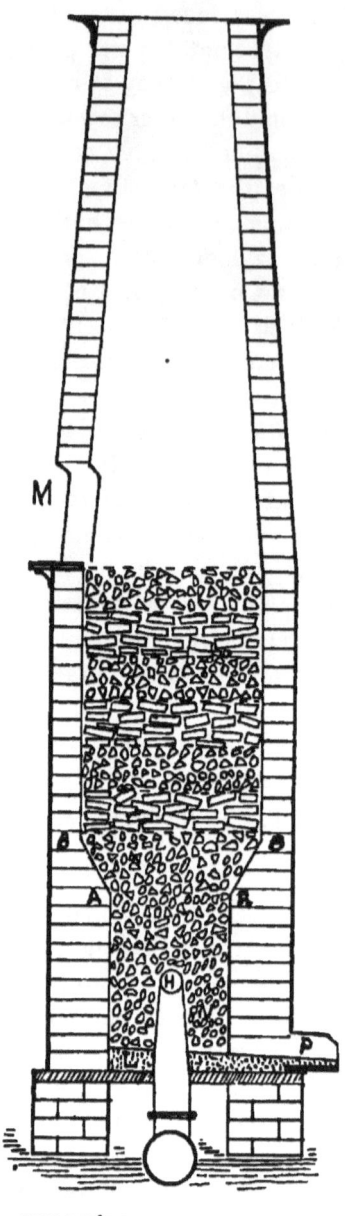

IRELAND'S CENTER BLAST CUPOLA.

the bosh AA, and three feet nine inches diameter from top of bosh to charging door, whence it tapers to one foot nine inches at top of stack. Height of crucible four feet five inches, length of boshes from AA to BB, eighteen inches; height from top of bosh to charging door, six feet seven inches. The blast is supplied from one tuyere placed in the center of the bottom of crucible.

The tuyere hole through the iron bottom is nine inches diameter, into which is passed a seven and a half-inch water tuyere, the mouth of which, H, is two feet above the sand bottom L. A slag hole N, five inches diameter, is placed just below the level of the mouth of the tuyere. P is the tap-hole and spout.

This cupola melted three tons of iron per hour with two and a-half cwt. of coke per ton, but it does not appear to have given satisfaction, for it never came into general use in England or this country, and Mr. Ireland changed his plans and constructs his cupolas with side tuyeres.

VOISIN'S CUPOLA.

In illustration Fig. 27 is seen a sectional elevation of Voisin's cupola, in which very good melting has been done. The shell is constructed of boiler plate with an external air chamber of the same material, extending all the way round the body of the cupola. This air chamber is supplied from two pipes, one on each side of the cupola. Two sets of tuyeres lead from the air belt into the cupola. The lower set are oblong, four in number, placed at equal distances apart and at right angles to the air belt. The upper set are round, of less capacity than the lower set, are placed horizontally through the lining and diagonally to the lower set, so that they are between them at a higher level.

Mr. Voisin claims through this arrangement of the tuyeres, that the escaping gases are burnt in the cupola, creating a second zone of fusion with those gases alone, and the second set of tuyeres obviates to some extent the evil effect of the formation of carbonic oxide in the cupola.

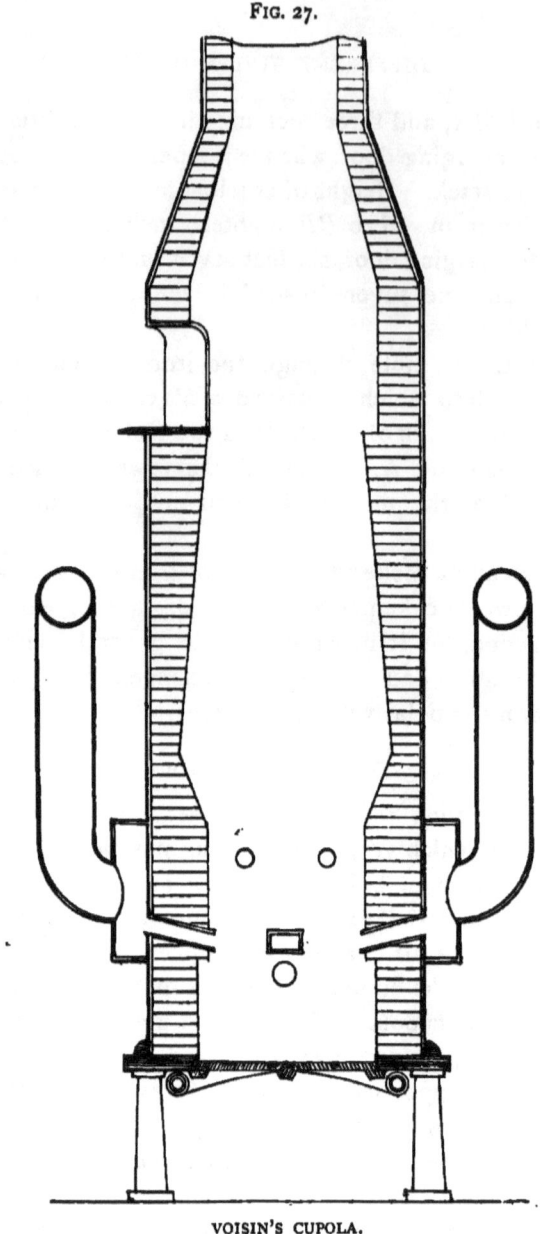

Fig. 27.

VOISIN'S CUPOLA.

This cupola is constructed in slightly varying shapes inside the lining, but the following dimensions give a general outline of it: Vertical dimensions from bottom to offset below tuyeres, one foot ten inches; offset below tuyeres to lower end of bosh, two feet four inches; length of bosh, one foot two inches; top of bosh to charging door, six feet ten inches; bottom of charging door to bottom of stack, two feet seven inches; taper to stack, three feet ten inches. Horizontal dimensions: Below tuyeres, two feet; at tuyeres, one foot eight inches; at top of bosh, two feet four inches; at bottom of charging door, one foot ten inches; at charging door, two feet seven inches.

The casing is made straight from the bottom plate to taper to the stack, and to get the above dimensions it has to be lined with brick made specially for this cupola.

Mr. Voisin has invented a number of different cupolas, but this one is said in melting to give the best results.

WOODWARD'S STEAM-JET CUPOLA.

In Fig. 28 is seen a sectional view, showing the construction of the Woodward steam jet cupola, in use to some extent in England. This cupola is worked by means of an induced current or strong draught caused by a steam-jet blown up the cupola stack, which is very much contracted just above the charging door. There are several different modes of applying the steam-jet, but the general principle will be at once understood from the figure (28). The cupola is constructed upon the general plan of the English cupola, with a stationary bottom and draw front. Two rows of tuyeres or air-inlets, as they are termed, are placed radially at two different levels. In the lower row there are four openings, varying in size from five to eight inches in diameter, according to the size of the cupola. In the upper row there are eight, varying in diameter from three to five inches. Each of the air-inlets is provided with a cover outside, which can be closed when it is' desired to shut off the draught. The upper row of air-inlets is placed from ten to

164   THE CUPOLA FURNACE.

Fig 28.

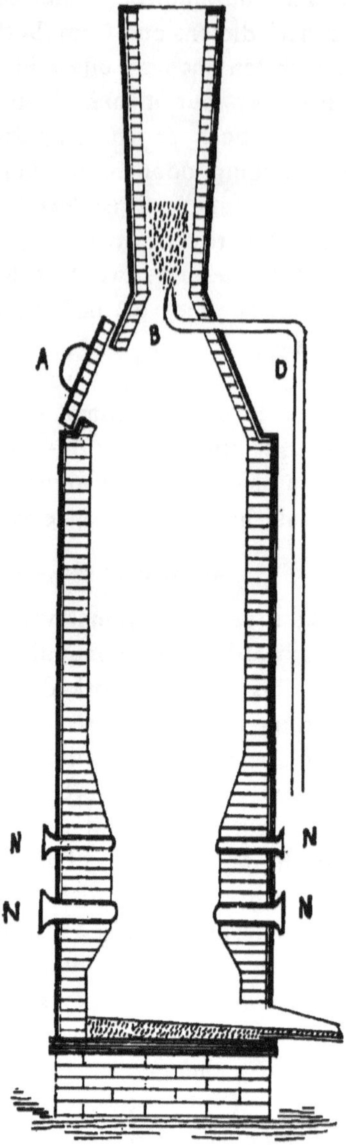

WOODWARD'S STEAM-JET CUPOLA.

fifteen inches above the lower row. The lining is contracted at the air-inlets to throw the air to the center of the stock, and enlarged below the air-inlets to admit of the retention of a large amount of molten iron in the cupola.

The charges of fuel and iron are put in at the charging door A in alternate layers in the ordinary way, and the door tightly closed and luted to prevent the admission of any air. The steam is then turned on through the nozzle B connected with the boiler by steam-pipe D, and the air-inlets N opened for the admission of air. When the cupola is working, the draught has to be regulated by the melter and care taken to close any air-inlets near which iron is seen to accumulate in a semi-fluid state. The temperature at the spot where the iron chills will soon rise to a degree that will cause the iron to run freely, when the air-inlet may be again opened. All the iron to be melted is put in and the door closed before the steam is turned on. The charging may be continued throughout the heat, but the opening of the door has the same effect on the stock as shutting off the blast in the ordinary cupola, and the melting stops. The repeated opening of the door soon gets the cupola into bad working order and it bungs up in a short time.

When it is desired to use the cupola for continuous melting or for a larger amount of iron than can be put in at one time, it is constructed with a side flue and feeding hopper, as shown in Fig. 29. The general construction and air-inlets are the same as those shown in Fig. 28. The stack is removed and the feeding hopper A with a sliding door B at the bottom, to be worked by the lever D, is placed on top of the cupola. The flue H near the top of the cupola connects it with the stack M, and the draught is induced by a steam-jet from the nozzle N attached to the steam-pipe P. When filling the cupola, the bottom of the hopper is left open and the charges put in in the ordinary way until the cupola is filled. The bottom door of the hopper is then closed, and when the cupola is melting the charges of fuel and iron are put into the hopper and dropped into the cupola as the stock settles, and the door is at once closed to exclude the air at the top of the cupola.

166   THE CUPOLA FURNACE.

FIG. 29.

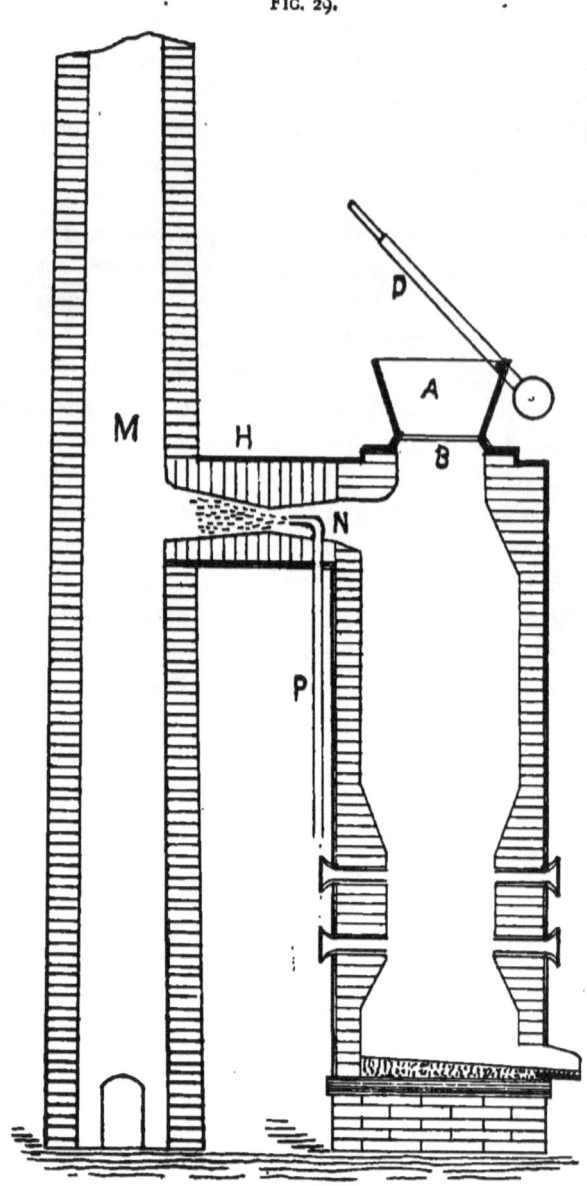

WOODWARD'S STEAM-JET CUPOLA.

It is asserted by those interested in this cupola that it effects a great saving in fuel over the ordinary blast cupola. The consumption of coke in melting a ton of iron is placed at one hundred and fifty pounds, a very low rate of fuel; but the same results are also claimed to have been obtained in blast cupolas of good design when properly worked.

The steam required to create the draught is only equal in quantity to what would be required by an engine for driving a fan or blower of sufficient power to work an ordinary cupola of the same size. Considerable saving is effected in the first cost of engine and fan or blower, besides the saving in wear and tear of machinery.

The objection to this style of cupola is the slow melting, for it cannot be forced beyond a certain point, and when a large amount of iron is to be melted the cupola must be kept working all day. This does not meet the views of the foundrymen of this country, who desire to melt their heats in from one to two hours from the time the blast is put on until the bottom is dropped, and with that object in view construct their cupolas.

TANK OR RESERVOIR CUPOLA.

In Fig. 30 is seen a sectional elevation of a reservoir cupola. This cupola was designed for the purpose of making soft iron for light castings. It only differs in construction from the ordinary type in the reservoir or tank placed in front, which may be attached to any cupola.

The cupola is set high and the tank A is placed in front of it, with the cupola spout leading into it near the top. The molten iron is run from the cupola into the tank as fast as melted, and drawn from the tank-spout into the ladles as it may be required for pouring. The tank is made of boiler plate and lined with fire-clay or other refractory material, and is covered with an iron lid, lined likewise with same material. The spout and breast are made up the same as for an ordinary cupola. Before putting on the blast, the tank is filled with charcoal and closed with the cover; and as the iron melts, it is run into the

FIG. 30.

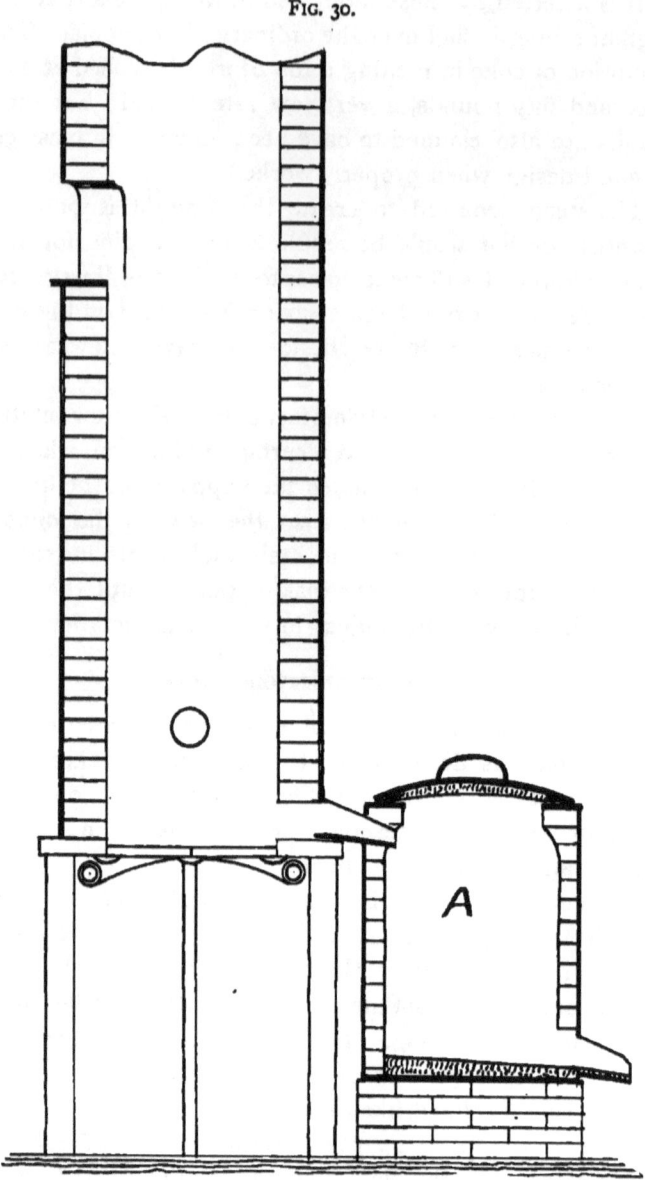

TANK OR RESERVOIR CUPOLA.

DIFFERENT STYLES OF CUPOLAS. 169

tank, where it is allowed to remain a sufficient length of time to be carbonized and softened by the charcoal.

These cupolas have been constructed in a number of different ways; the tank has been made of sufficient size to hold the entire heat of molten iron before pouring, so that the iron might be of an even grade throughout the heat and softened to a greater extent; and they have been riveted to the cupola casing and the lining continued from the cupola to the tank. In this latter case, the top is bolted or clamped to the tank and a tight joint made to prevent the escape of the blast, which has the same pressure in the tank as in the cupola.

The tank cupola produces a softer iron than the ordinary cupola, but there is considerable additional expense attached to it in keeping up the tank and supplying it with charcoal. Another objection is the change made in the shrinkage of the iron; that taken from the tank shrinks less than the same grade of iron when taken from the cupola, and when some parts of a machine or stove are made from the tank and other parts from the cupola, allowance must be made in the patterns for the difference in shrinkage.

It is claimed by some founders that soft iron can be produced by putting a quantity of charcoal on the sand bottom, and placing the shavings and wood for lighting the bed on top of the charcoal. In lighting up, the charcoal is not burned, but remains in the cupola during the heat and may be found in the dump. This is the case if the tuyeres are high and the front is closed before lighting up, but if the tuyeres are low or the front and tap-hole are not closed, the charcoal will be burned out in lighting up the bed, the same as the wood.

Tanks are, in England, used in connection with cupolas to some extent at the present time for mixing irons or to enable the founder to run a large casting or heat from a small cupola. The iron for an entire heat, requiring several hours to melt in a small cupola, is melted and run into the tank and drawn from the tank into the ladles at casting time. This makes a well-mixed and even grade of iron in all the castings and saves considerable time in casting, as the moulders are not obliged to wait for iron to melt, as is often the case.

MACKENZIE CUPOLA.

In Fig. 31 is shown a sectional elevation of the Mackenzie Cupola, designed by Mr. Mackenzie, a practical foundryman, and manufactured by Isbel-Porter Co., Newark, N. J. When this cupola was designed the only one in use was the common straight one with a limited number of very small tuyeres and low charging doors, and it melted very slowly. It was the custom in foundries at that time, to put on the blast at one or two o'clock and blow all the afternoon in melting a heat. Moulders generally stopped moulding when the blast went on and a great deal of time was lost in waiting for iron. To save this time and get a few hours' more work from each moulder on casting days, Mr. Mackenzie conceived the idea of constructing a cupola that would melt a heat in two hours from the time the blast was put on until the bottom was dropped. He had discovered that the tuyeres in common use were too small to admit blast freely and evenly, and cupolas did not melt so well in the center as near the lining and tuyeres. To overcome this fault in the old cupola, and admit the blast to the stock evenly and freely, a belt tuyere was put in extending around the cupola, and to place the blast nearer to the center of the cupola at the tuyeres, the lining was contracted or boshed at this point. To avoid reducing the capacity for holding molten iron below the tuyeres, the lining just above the tuyeres was supported by an apron riveted to the cupola casing and the bosh made to overhang the bottom, leaving the cupola below the tuyeres of the same diameter as before boshing.

This cupola, when first introduced, was known as the two-hour cupola and wrought a great revolution in melting and in foundry practice. Heats that had required half a day to melt were melted in two hours, the quantity of fuel consumed in melting was reduced, the number of moulds put up by each moulder increased, and the cost of producing castings greatly reduced

DIFFERENT STYLES OF CUPOLAS.

FIG. 31.

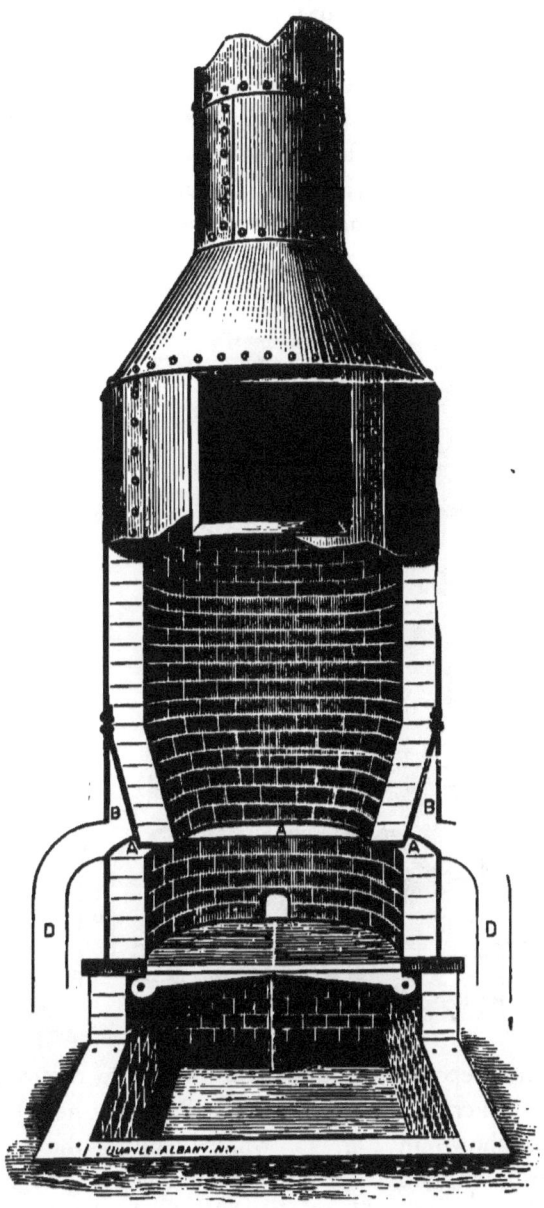

MACKENZIE CUPOLA.

Many of these cupolas are still in constant operation, and for short heats of one or two hours, are probably the most economical melting ones now in use. In long heats the tendency of the cupola to bridge at the bosh is so great, that it melts slowly toward the end of a heat and is frequently difficult to dump, especially if the cupola is a small one.

We have had much experience in melting in these cupolas, and have found that slag and cinder adhere to the lining over

FIG. 32.

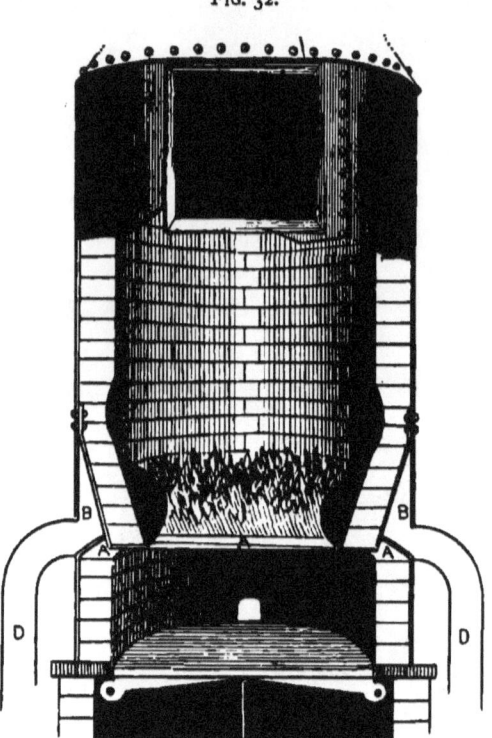

the tuyeres and become very hard and difficult to remove, and if care be not taken to remove them after every heat it soon builds out, as shown in Fig. 32, which reduces the melting capacity very much, and increases the tendency of the cupola to bridge and hang up. The lining should be kept as near the shape shown in Fig. 31 as possible, and all building out over the tuyeres and bellying out in the melting zone, as far as possible, prevented.

In the illustration (Fig. 31) is shown the cupola pit, commonly placed under cupolas when they are set very low for hand-ladle work. The outlet to the pit may be placed at the front, back or side of the cupola as found most convenient for removing the dump.

THE HERBERTZ CUPOLA.[*]

The cupolas generally used either for melting iron or for any other purpose, are cupolas blown through one or several rows of tuyeres inserted at some distance above the hearth. The pressure of the blast varies in most cases from $\frac{1}{4}$ pound to 1 pound, and the blast is obtained by blowing engines or blowers driven through belting and shafting by special steam engines. Such a plant, requiring as it does many mechanical appliances, consequently subject to continual care and repairs, is expensive. The Herbertz cupola, instead of being blown by blast forced from below through the melting material, is provided at its upper end with a steam-jet pipe, which in action creates a vacuum of from 3 inches to 4 inches of water in the upper region of the cupola, while the air is allowed to enter freely at the lower part through an annular opening between the movable hearth and the upper shaft.

The movable hearth, as shown in Fig. 33, is mounted on four screws, which by their common action lower or raise it at will, and thereby allow of a complete and easy regulation of the quantity of blast introduced through the annular opening. The screws work either in the standards of the cupola, as seen in Fig. 33, or are carried on a special car together with the hearth, so that this latter can be removed at any moment from underneath the shaft. The steam-jet is applied in the center line of the smoke pipe, which connects the cupola either directly with a special stack or is built like the down-comer of a blast furnace,

[*] By J. B. Nau, New York.

174 THE CUPOLA FURNACE.

FIG. 33.

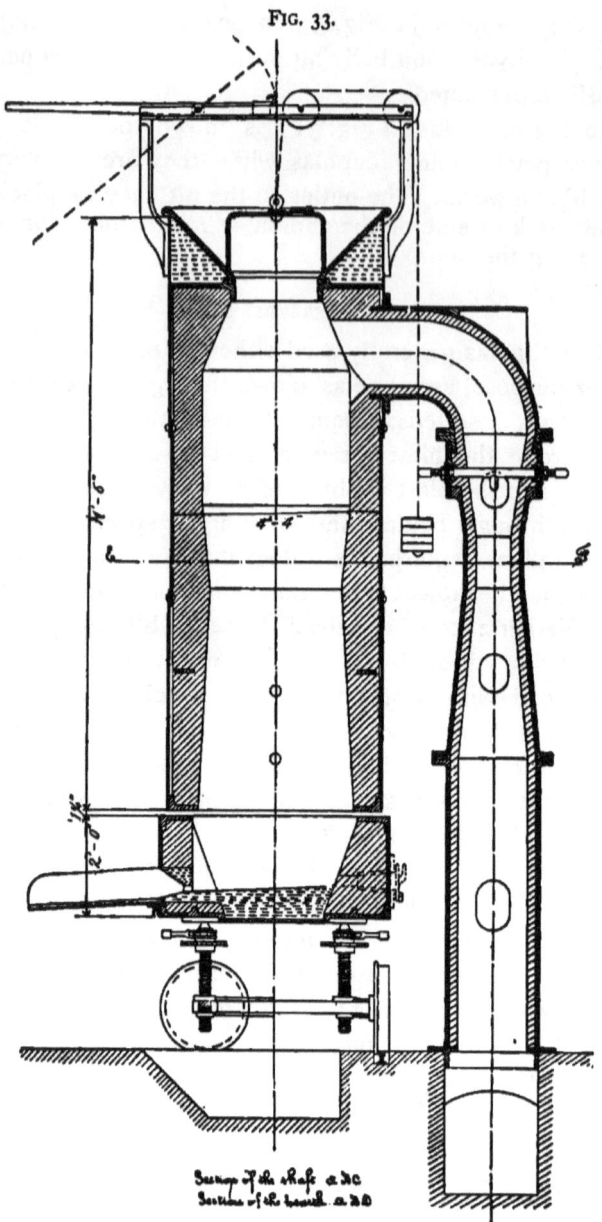

SECTION OF HERBERTZ'S IRON-MELTING CUPOLA.

and connects the cupola with a horizontal underground flue leading to any chimney. The top of the cupola is provided with a hopper hermetically closed while the melting is proceeding, and only open at regular intervals and for a very short time, when the charge is being introduced.

The bottom of the hearth is provided with a door turning on hinges and kept tight by a lock. This door, once lowered after the melting is done, turns around the hinges and the contents of the cupola are dropped into an ash pit, where, after having been cooled with water, the unburnt coke can be collected and saved for the lighting of the cupola in the next melting.

Three tuyeres are placed all around at the level of the bottom of the hearth. These tuyeres, as we shall see later on, are plugged up with sand during the melting, but are used before the melting in the kindling of the fire and to give access to the air necessary for combustion.

The shaft is provided at two different levels with bull's-eyes, through which the fire can be watched.

The application of a steam-jet to create draft in the cupola has many advantages. The only mechanical appliance required is a small boiler supplying the necessary steam for the ejector. No blowing engine or steam engine with blower, shafting, pulleys, belts, no blast pipe connecting the blower with the cupola, is necessary. The only repairs are those on the boiler and steam pipe, very light indeed, without mentioning the fact that oil for lubricating will be entirely dispensed with. But besides these already important advantages, some other features are met with. Most of the blowers, running at a speed of from 1000 to 1200 revolutions or more per minute, produce sometimes a noise, which often can be heard at a great distance. The Herbertz cupola runs without any appreciable noise, and can be established in any populous center without the slightest inconvenience to the neighborhood. Its top being closed, no sparks or flame are thrown out. The repairs to the movable hearth are very easy and can be done outside.

In the United States it is as yet little known. For some

time, however, tests have been made with it at a car-wheel foundry in Elizabethport, N. J.

This cupola is very well adapted for the melting of pig iron. The very reduced consumption of coke, claimed by the inventor to be as low as 4 to 5 per cent. of the weight of iron (or in other words, 1 pound of coke would be enough to melt 20 pounds of iron), leads to the conclusion that the combustion of coke must be complete, or that the coke must be burnt completely to carbonic acid, and thus generate the greatest possi-

FIG. 34.

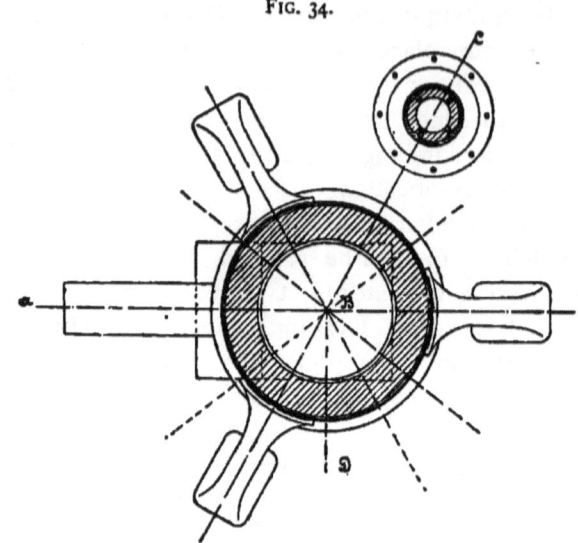

HORIZONTAL SECTION.

ble amount of heat. In order to prove this, test-heats have been made in Europe, and the analyses of the escaping gases showed that in most cases the whole amount of carbon was burnt to carbonic acid, while in a few other cases a very small proportion of carbon burnt to carbonic oxide. In one of these test-heats the mixture in the cupola was 1050 kg.* of Luxemburg foundry iron No. 3 and 450 kg. of foundry scrap, a total of 1500 kg., or 1.5 tons.

* 1 kilogramme = 2.2 lbs.

The melting coke was air dry and contained but 3 per cent. of water, 6.8 per cent. of ash and 1.037 per cent. of sulphur; 190 kg. of filling coke was put in the cupola and on top of it 1000 kg. of pig iron. The total amount of coke used, including lighting coke, was 215 kg. to 1500 kg. pig iron. After the fusion was done, 67 kg. of coke were taken out and could be used again for the next day's charge, so that the real amount of coke used was only 215—67 = 148 kg., or 9.9 per cent., whereas the real amount of melting coke was only 5 per cent.

A careful weighing of the iron cast showed that 1460 kg., or 97.33 per cent. of the original iron charged, was obtained, constituting a loss of only 2.66 per cent. The temperature of the molten metal was high, and amounted in part to 1300°. The escaping gases had the following composition:

	Carbonic acid.	Carbonic oxide.	Oxygen.	Nitrogen.
Before the steam-jet was acting	7.1	0	7.1	85.8
Five minutes after steam-jet was acting	13.1	0	6.5	80.3
Twenty-five minutes after steam-jet was acting	9.25	0	7.0	83.75
At the end of the cast (after 35 minutes	13.3	0	6.3	80.4
Average	10.71	0	6.73	82.60

Another test heat with thoroughly wet gas coke was made. This coke contained nearly 20 per cent. of water and 7.5 per cent. of ash. About 12.7 per cent. of it was used (lighting and melting coke together). The loss in iron in this charge was only 3.45 per cent. The average composition of the gases was 11.5 of carbonic acid, 3.4 of carbonic oxide, 8.2 of oxygen, 76.9 of nitrogen. It will be seen that in this last heat carbon did not burn entirely to carbonic acid, which was probably due to the increased amount of coke that had been charged intentionally.

Nevertheless, the composition of these gases is still far more favorable than would be obtained with an ordinary blown cupola, where a certain number of analyses have shown that the escaping gases contain from 12.50 to 19.90 per cent. of carbonic acid and 4.80 to 11.73 per cent, of carbonic oxide. The analyses show, furthermore, that in the case of the Herbertz cupola, the fuel is thoroughly utilized and yields the maximum of heat.

To obtain such complete combustion it is necessary that the air should be in slight excess, and that this actually happens is shown by the presence in the gases of a certain amount of free oxygen. Several reasons have been advanced to explain this complete combustion of carbon to carbonic acid. The first is that the air enters the cupola all around the circumference in a thin sheet and gives rise to very uniform combustion. Another reason is the very reduced velocity with which the gases rise.

In the ordinary blown cupola these gases are pushed upward with great pressure and velocity, and the combustion under such conditions cannot be obtained entirely in the lower regions, but some of the air will reach the upper regions unburnt, where it causes the reduction of part of the carbonic acid.

The presence of free oxygen in the escaping gases of the Herbertz cupola, might lead to the supposition that it has a pernicious influence on the composition of the iron. Some of the elements in the pig iron, such as carbon, silicon and manganese, for instance, might be oxidized, and by their partial elimination deteriorate the quality of the iron. Not only is this not the case, but it seems that actual practice has shown that less carbon and silicon are eliminated from the iron in the Herbertz cupola, than in the ordinary blown cupola. This has been explained in the following manner: The combustion in this cupola takes place a little above the annular opening, and no flame is seen in the upper regions of the cupola, whereas in the ordinary cupola, combustion takes place through the entire length of the shaft and continues in a blue flame on top. In this case all the pig iron is more or less heated and pasty before it reaches the melting zone, and surrounded by an oxidizing atmosphere, the elimination of part of its elements is easy.

In the Herbertz cupola, where the combustion takes place almost entirely in the lower regions, and where the upper regions are less heated up, the pig iron better resists the influence of the ascending gases.

It must be stated at once that the above tests extended only over one single charge of 1.5 tons, lasting in the first heat 35 minutes. Had the work been continued for a certain length of time and had a greater number of charges been made, the consumption of fuel would have been considerably lowered, as for the following charge only melting coke would have been put in the cupola without any further addition of lighting coke. Then, if five consecutive charges had been made, we should have 190 kg. of lighting coke and 5×75, or 375 kg. of melting coke (at 5 per cent. of the weight of iron), or a total of 565 kg. And as 67 kg. of coke have been taken out of the cupola after the charge was over, this would leave 498, say 500 kg. of coke really burnt, which is equal to 6.6 per cent. In other words, 1 pound of coke would melt 15.15 pounds of iron.

In the cupola working for a few weeks at Elizabethport, the consumption of coke for the melting proper amounted during the tests to 6 per cent. The cupola is rated as a 2 ton, melting 2 tons an hour. The outside diameter of the hearth is 4 feet 7 inches, whereas the shaft has an outside diameter of only 4 feet 4 inches, and the total height from bottom plate of hearth to top of cupola is 13 feet 9 inches, when hearth and shaft are in contact with each other. The castings made during the tests were car wheels, and the mixture of iron put in the cupola was the same mixture of pig and scrap iron that had been used previously in the old ordinary cupola of the foundry, viz.: One-third No. 1 foundry iron and two-thirds car-wheel scrap. It was melted down with only 6 per cent. of coke, not counting the filling coke. Notwithstanding this very reduced amount of fuel, the iron began to melt rapidly. Ten to fifteen minutes after the steam was put on to create the draft, the first iron was tapped off. It was very good, and so hot that the men had to wait a few minutes before casting it into the molds. Though the iron mixture used in this test was the same as has always been made in the old cupola, it must be stated that the castings obtained were too soft for car wheels and presented very little chill on the tire. In order to obtain a better chill it was deemed advisable to use nothing but car-wheel scrap on the second day. The result showed a marked improvement on the first day's work; the chill was deeper and better. On the third day the mixture was one-fourth No. 3 foundry iron and three-fourths average foundry scrap. The use of this mixture constituted a large economy over what had been done in the ordinary cupola, and with it better results were obtained than with a mixture of one-third No. 1 foundry and two-thirds car-wheel scrap. The castings obtained were very tough and dense with $\frac{7}{18}$ inch chill. The metal, too hot to be cast immediately after tapping, was very pure throughout the cast.

The charge on this day was as follows;

Filling coke, 576 pounds.
Melting coke, 6×72 = 432 pounds.
Limestone, 6×15 = 90 pounds.
No. 3 foundry, 6×300 = 1800 ⎫
Foundry scrap, 6×900 = 5400 ⎭ 7200 pounds.

The hearth in the cupola used at Elizabethport is mounted on a small car. This hearth, prepared and dried outside, was filled with wood shavings, wood and coke on top to the upper level. It was then pushed under the shaft, and raised by means of the screws until it came in contact with its lower rim. Filling coke was then charged to the level of the highest bull's-eye, and fire started through the three tuyeres at the bottom of the hearth. After this the cupola was left to itself, working under natural draft through the three tuyeres at the bottom. When the filling coke was fully ignited the above-named charge was put on in alternate layers of iron, coke and limestone. When the filling was done, the hearth was lowered enough to form an annular opening of about 1¼ inches between the lower rim of the shaft and the top of the hearth. The tuyeres at the bottom were plugged with molding sand, and the cupola again allowed to work with natural draft through the annular opening, until the first iron was melting down. At this moment the steam-jet was put in action. The draft, which, when no steam was applied, had been equal to about $\frac{1}{16}$ inch water column, rose at once to between 3 and 4 inches of water. From that moment on, the melting was regular, hot and rapid. The top of the cupola was kept tight and only opened at regular intervals to introduce the raw materials. Lighting coke was only used at the start; all the subsequent charges were made with not more than 6 per cent. of coke, and continued regularly without any other addition. When the steam was applied, its pressure was 80 pounds. The entire charge of 7200 pounds was melted in 1 hour and 24 minutes, which corresponds to 5140 pounds of pig iron melted per hour, say $2\frac{3}{8}$ tons, instead of 2 tons. The vacuum created in the cupola remained between 3 and 4 inches of water as long as the level was kept constant. Toward the end of the charge, when this level became lower, the vacuum fell somewhat below 3 inches. No disagreeable noise was heard while the melting was going on, nor was any spark or flame seen at the top of the cupola.

As soon as the melting was done and the last iron run out from the cupola, the bottom door of the hearth was opened and the ignited mass fell down into the ash pit, where, once the hearth pulled out, the entire content was cooled with water, and the remaining coke gathered to be saved as lighting coke in the following melting. After careful weighing of the iron cast from the cupola, it was found that the entire loss amounted to only $3\frac{1}{2}$ per cent., a very low figure when compared with the ordinary loss of at least 6 per cent. in an ordinary foundry cupola. It is remarkable also that a lower grade of iron can be taken and still the same results as in the ordinary cupola be obtained. Thus the tests at Elizabethport have conclusively shown that as good results were obtained with a mixture of one-quarter No. 3 foundry iron and three-quarters foundry scraps when melted in the new cupola, as had been obtained with a mixture of one-third No. 1 foundry and two-thirds scrap iron when melted in the ordinary cupola. This may be explained by the reason that less carbon and silicon are eliminated from the iron when melted in the Herbertz cupola.

HERBERTZ CUPOLA USED FOR MELTING STEEL.

The first tests made in the Herbertz cupola to melt steel were attended with success. Rail ends, old files and other iron or steel crop ends were melted together with a small amount of foundry iron, and with only 8 to 10 per cent. of lighting coke. The molten metal was liquid enough to be cast easily. However, when small steel castings had to be made it was soon discovered that the metal lacked fluidity, and in order to obtain better results, it was deemed advisable to work with heated air. Figs. 35 and 36 illustrate the construction of a cupola especially adapted to this kind of work. The cupola in all its parts is entirely similar to the steam-jet cupola used for the melting of pig iron, with the exception, however, that a certain number of wrought-iron pipes are laid in the brick work. The air enters at the top in a circular space, and from there is sucked down at once through the iron pipes to the lower part, where it enters the cupola. By its passage through the pipes it is heated to a temperature ranging from 500° to 1100° F. and consequently the temperature in the melting zone will be sufficiently increased to obtain a thoroughly liquid steel. Bessemer steel, as well as wrought-iron crops, were melted in this way, each separately and with the greatest success.

For the casting of heavy steel castings especially, this cupola is better adapted than a crucible. On account of its direct contact with the fuel at a high temperature, the percentage of carbon in the metal seems to slightly increase. No steel-melting cupola is as yet working in the United States, but on the Continent of Europe their value seems to be more and more appreciated.

For melting other metals or alloys, such as lead, copper, brass, etc., the cupola has recently been introduced in some European works. Especially bronze has been melted in an ordinary foundry cupola with the best results. The tempera-

DIFFERENT STYLES OF CUPOLAS. 183

FIG. 35.

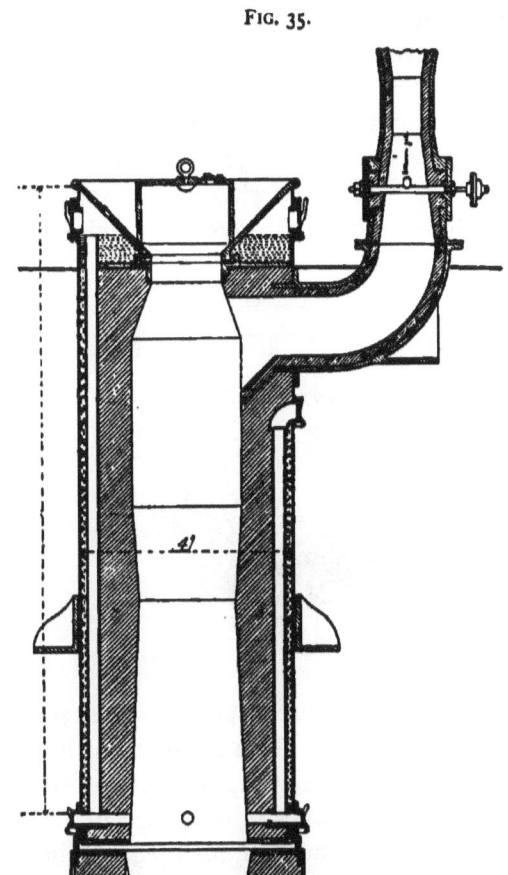

HERBERTZ STEEL MELTING CUPOLA.

ture required in this case being lower than in the case of pig iron, the cupola worked with natural draft and without any use of the steam-jet. The fuel economy obtained seems to be very large. The total consumption amounted to only 12 per cent., whereas in crucibles this consumption is sometimes as high as 40 per cent. The fusion, even without the application of the steam-jet, is claimed to be nearly five times quicker than in the

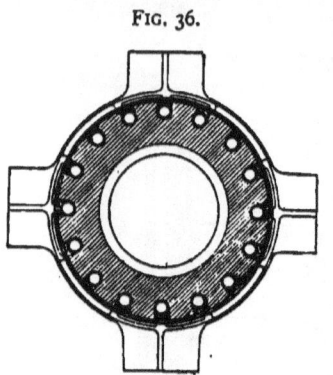

FIG. 36.

crucible furnace, and the bronze obtained is even purer than the bronze remelted in crucibles. This is said to be due to the fact that during the fusion, the small amount of tin often found in bronze is burnt out and escapes in the shape of a white smoke. The loss of metal during the melting seems to be slightly increased on account of the elimination of the tin. The cupolas used in the remelting of these metals are the same as those used for the remelting of foundry iron.

PEVIE CUPOLA.

In Fig. 37 is seen the Pevie cupola, designed by Mr. Pevie, a practical foundryman of the State of Maine. The small cupolas, 18 to 24 inches, of this design are built square, with square corners in the lining, and larger ones are made oblong with square corners and 24 to 30 inches wide inside the lining, and any increase in the melting capacity of the cupola desired,

DIFFERENT STYLES OF CUPOLAS.

FIG. 37.

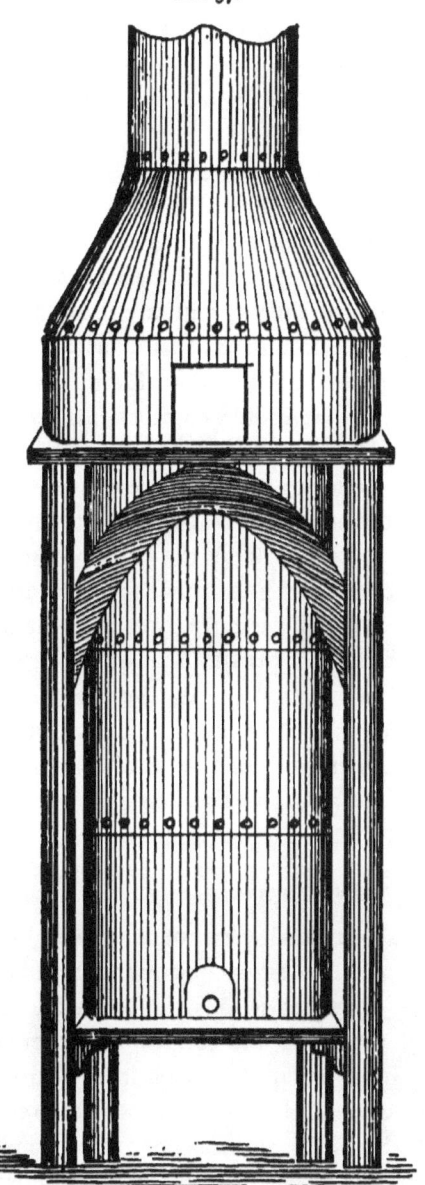

PEVIE CUPOLA.

is obtained by increasing the length of the cupola in place of increasing the diameter; as is done with the round cupolas.

Blast is supplied on two sides from an inner air chamber, through a vertical slot tuyere extending the full length of the sides of the cupola.

The object of Mr. Pevie in constructing a cupola upon this plan was to supply an equal amount of blast to all parts of the stock and to produce even melting. This theory was correct, for blast was certainly more evenly distributed to the stock than with the small round tuyere then commonly used, and we saw excellent melting done in cupolas of this construction in the foundry of Mr. Pevie, in a small town in Maine (the name of which is forgotten), which we visited some twenty years since. But in cupola construction an even distribution of blast is not the only matter of importance to be considered; for if it bridges and clogs up, the blast cannot do its work, no matter how evenly it may be distributed by tuyeres or by the construction of a cupola, and the peculiar construction of this cupola made the tendency to bridge very great. It was only by careful management that it could in long heats be prevented from bridging, when the lining was kept in its original shape, and for this reason it never came into general use. We know of only three of them at the present time in operation, one at Smithville, N. J., and two at Corry, Pa., and the shape of the linings in these cupolas has been greatly altered from their original form.

STEWART'S CUPOLA.

In Fig. 38 is seen a sectional view of a cupola in use at the Stewart Iron Works, Glasgow, Scotland. This cupola, which is one of large diameter, is boshed to throw the blast more to the center of the stock and reduce the amount of fuel required for a bed. Blast is supplied from a belt air-chamber extending around the cupola, through a row of tuyeres passing horizontally through the lining and a second row placed above and between the tuyers of the first row and pointing downwards, as shown in the illustration. The object of this second row of

DIFFERENT STYLES OF CUPOLAS.

FIG. 38.

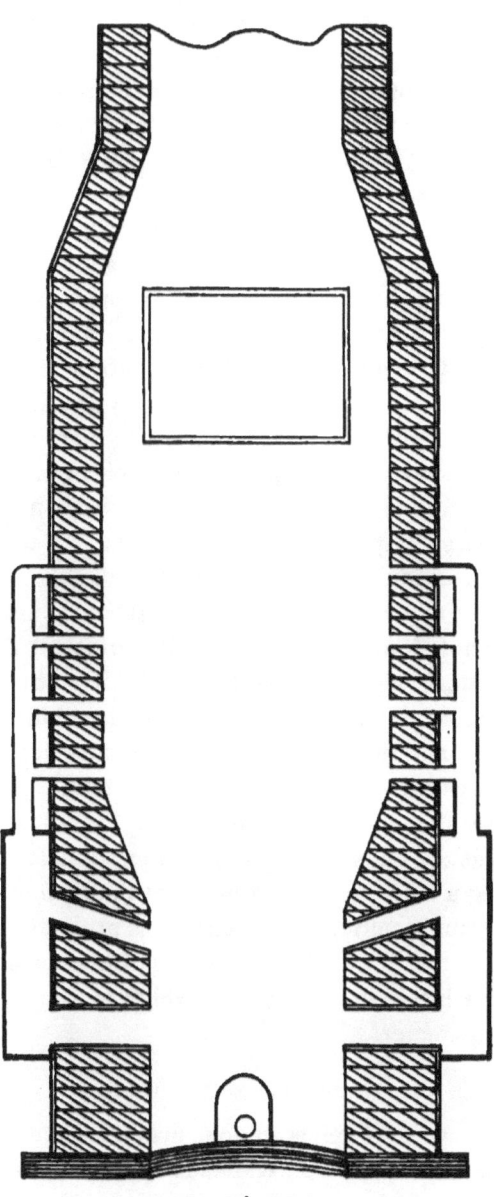

STEWART'S CUPOLA.

tuyeres is to increase the depth of the melting zone and increase the melting capacity of the cupola per hour. Attached to the top of the air-chamber at intervals of about two feet, is placed a vertical gas-pipe of two inches diameter, and from this pipe four branches of one-inch pipe lead into the cupola, about twelve inches apart. The object of these pipes is to supply a sufficient amount of oxygen to the cupola above the melting zone to consume the escaping unconsumed gas, namely carbonic oxide (CO), above the melting zone, and utilize it in heating and preparing the iron for melting before entering the zone. The cupola melts very rapidly, and is said to be the best melting one in Glasgow. But it is very doubtful if the one-inch gas-pipe tuyeres contribute anything toward the rapid melting, for it is absurd to suppose that one-inch openings placed twelve inches apart vertically and two or more feet apart around the cupola, would supply a sufficient amount of oxygen to fill a large cupola to such an extent as to ignite escaping carbonic oxide in the center of the cupola. While they might supply oxygen for combustion of carbonic oxide near the lining, we do not think they would admit a sufficient amount to be of any practical value in melting, even if they admitted a volume of blast equal to their capacity when placed in the lining. This they do not do, for they are frequently clogged by fuel or iron, filled with slag from melting of the lining, and as a lining burns away the ends of the pipes are heated and frequently collapse at the ends, and it is almost impossible to keep them open during a heat or to open many of them after a heat is melted. The rapid melting in this cupola is probably due to the arrangement of the first and second rows of tuyeres and the shape given to the inside of the cupola, which is excellent for cupolas of large diameter.

THE GREINER PATENT ECONOMICAL CUPOLA.

In Fig. 39 is shown the Greiner cupola, manufactured by The Greiner Economical Cupola Co., Kankakee, Ill., for which the following claims are made:

DIFFERENT STYLES OF CUPOLAS. 189

In placing the Greiner Patent Economical Cupola before the foundrymen and steel manufacturers in this country, we have the advantage of the splendid results already obtained with this cupola in Europe, where more than three hundred are in daily use.

The adoption of the Greiner system of melting iron there has

FIG. 39.

THE GREINER PATENT ECONOMICAL CUPOLA.

met with the most satisfactory results. In no case has the saving of fuel been less than twenty per cent., and in some instances it has reached forty and even fifty per cent.

The novelty of the invention consists in a judicious admission of blast into the upper zones of a cupola, whereby the combustible gases are consumed within the cupola and the heat utilized to preheat the descending charges, thereby effecting a saving in the fuel necessary to melt the iron when it reaches the melting zone. In order to fully explain the principle of its workings, we illustrate in Fig. 40 a cupola of the ordinary

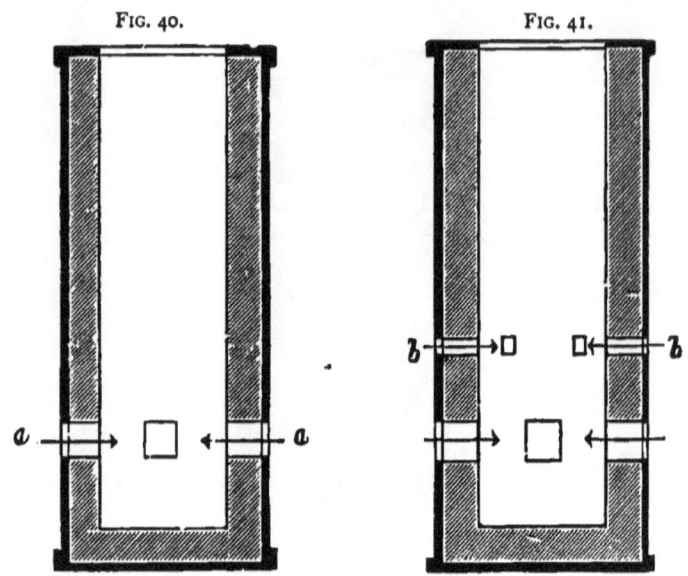

FIG. 40. FIG. 41.

SECTION OF ONE ROW TUYERE CUPOLA.   SECTION OF DOUBLE ROW TUYERE CUPOLA.

design, with a single row of tuyeres or air inlets, AA. The incoming air burns the coke in front of the tuyeres to carbonic acid gas, a combination indicating perfect combustion. As this gas ascends through the incandescent coke above, most of it is converted into carbonic oxide by the absorption of an equivalent of carbon. The result of the combustion is, therefore, a gas mostly composed of carbonic oxide (CO), indicating an imperfect utilization of the fuel, as one pound of carbon burned to carbonic acid (CO_2) will develop 14,500 heat units; whereas, the same amount of carbon burned to carbonic oxide (CO) will only develop 4480 heat units, or less than one-third of the heat given by perfect combustion.

DIFFERENT STYLES OF CUPOLAS.

To avoid this loss of heat, additional tuyeres have been placed at a short distance *bb* (Fig. 41) above the lower tuyeres to introduce air to consume the carbonic oxide (CO), but such arrangement does not have the desired effect, because the material at that place in the cupola has a very high temperature, consequently the entering air also ignites the coke, so that the action at the lower tuyeres is simply repeated, and carbonic oxide (CO) again formed at a short distance above *bb*.

This led Mr. Greiner to the following conclusions:

In every cupola there must be a point *cc*, (Fig. 42) above which the descending materials have not yet reached the tem-

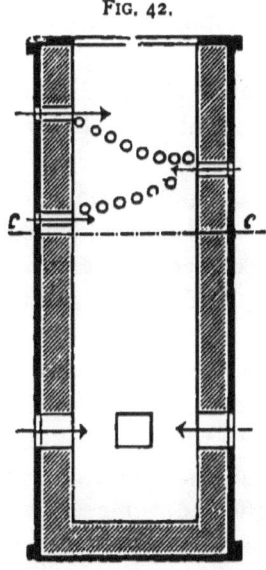

FIG. 42.

SECTION OF GREINER CUPOLA

perature necessary for the ignition of the solid fuel, while the ascending combustible gas is still warm enough to ignite when brought into contact with air. It is clear that air, if properly admitted above that point, will cause the combustion of the carbonic oxide (CO) without igniting the coke.

But if all the air necessary for the combustion of the carbonic oxide (CO) be admitted at one place or in one horizontal row of tuyeres, the heat developed will very soon raise the temperature so as to set fire to the coke, producing loss of carbon as before. Hence the upper blast must not be introduced on a horizontal plane, but through a number of small tuyeres, arranged (either in the form of a spiral or otherwise) so as to embrace the higher zones of the cupola, and must be regulated, both as to pressure and arrangement and dimensions of pipes, according to the capacity of each particular cupola.

The combustible gases are thus burned without heating the coke to incandescence, and the heat thus developed is utilized to preheat the iron and the coke, so that they reach the melting zone at a higher temperature and require less heat to effect the melting.

Another point in favor of the Greiner economical cupola, and which is very important in most foundries and steel works, is that the application of the Greiner system will increase the melting capacity of the cupola, owing to the more rapid melting in the fusion zone and to the additional room in the cupola that previously was occupied by the extra amount of coke not now required. Owing to the more rapid melting, a purer and better iron is obtained.

As will be understood, the number, size, position and arrangement of the upper tuyeres vary considerably, according to the capacity of the cupola to which the system is to be applied.

It can be readily adapted to existing cupolas, without material alteration being effected, while the only additional fittings necessary generally consist of a circular pipe connected by branches with the main blast box of the cupola, valves to regulate the blast, and connecting pipes for the small tuyeres.

COLLIAU PATENT HOT BLAST CUPOLA.

In Figs. 43 and 44 are seen external and sectional views of the Colliau patent hot blast cupola, designed by the late Victor Colliau, a civil engineer, who devoted a great deal of time to the study of cupola construction and management. The cu-

DIFFERENT STYLES OF CUPOLAS. 193

pola is at the present time extensively used, and has many good points in its construction. The following history and descrip-

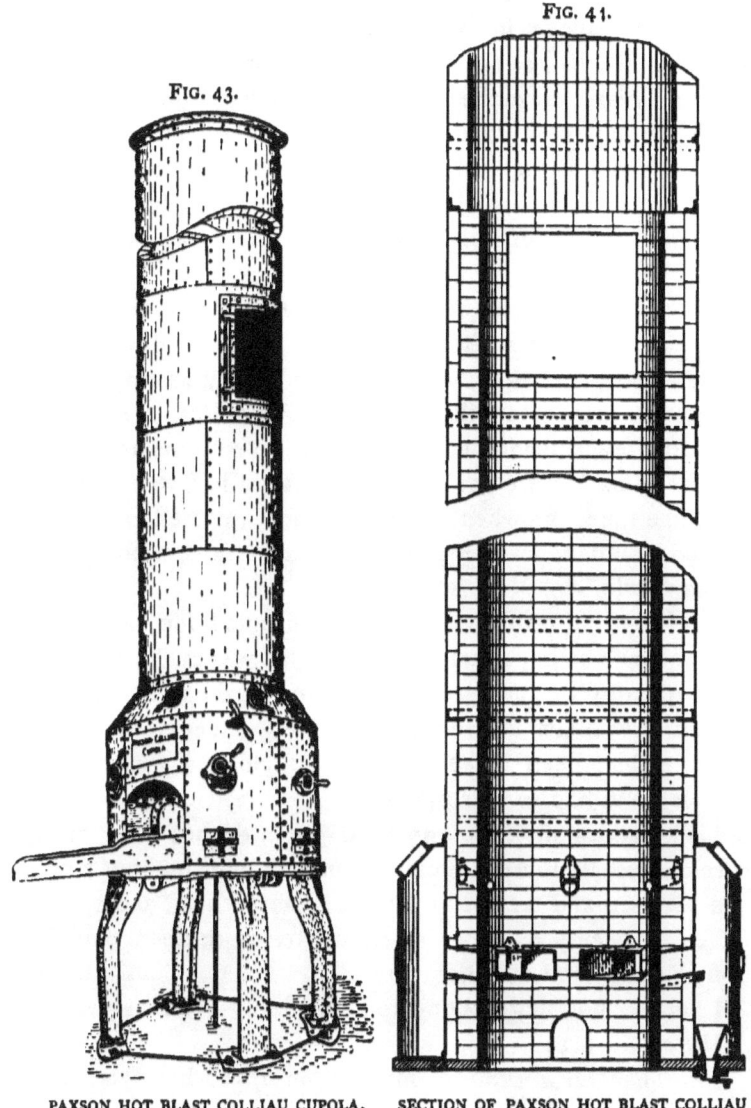

FIG. 43. FIG. 41.

PAXSON HOT BLAST COLLIAU CUPOLA.   SECTION OF PAXSON HOT BLAST COLLIAU CUPOLA.

tion of the cupola and results obtained in melting are furnished by his son, Victor Colliau, Detroit, Mich., who is engaged in the manufacture of it. It is also manufactured with some improvements, shown in Figs. 43 and 44, by J. W. Paxson & Co., Philadelphia, Pa., as the Paxson Hot Blast Calliau Cupola.

Some years since, the cupola for melting iron was very incomplete and ineffectual—the melting of twenty-five tons at one heat and a rate greater than three to four tons per hour was unknown, and a melting of three to four pounds of iron with one pound of coke was considered a very satisfactory result.

Large castings could not be made, and it was considered a large foundry that melted five to six tons per day, and later (only a few years ago), when large and heavy castings became necessary, such as anvils, steamboat bed-plates, cannon, etc., requiring ten, fifteen, and still later on, thirty tons at one time, several cupolas were used and were placed in a row, lighted at the same time, and when the iron in each was melted they were tapped simultaneously, the metal running in a common channel to the mould.

All these old-fashioned cupolas consumed too much fuel in consequence of imperfect combustion, as was evidenced by the large quantity of gas burning at the top of the chimney, which should have been utilized in the melting process; and after a few tons had been melted the cupola clogged with cold iron and slag and had to be stopped.

I have, with my new improved patented hot blast cupola, surmounted all these difficulties, and am now melting sixty to one hundred and ten tons a day in some of them, at a speed of fifteen to twenty tons per hour, and ten to thirteen pounds of iron to the pound of coke.

I am now building a cupola to melt twenty-five tons per hour.

My claims are:

1st. That the working of my new improved hot blast cupola has never been equaled.

2d. A saving of from 25 to 50 per cent. of fuel. I have replaced cupolas which were melting five pounds of iron with one pound of coke by one of my cupolas of the *same size*, and melted ten or twelve pounds of metal with one of coke.

3d. Great rapidity of fusion. With the same diameter inside of lining of the old model, I am melting in my cupola *double* the quantity of iron per hour.

4th. One very important feature is the saving of iron in the melting process. In common cupolas the loss is from 6 to 10 per cent., in my new improved cupola 5 per cent. is the maximum; the loss is as low as $3\frac{1}{2}$ per cent. in large meltings.

5th. The iron melted is improved compared with the old system, in which the slow process of melting exposed the iron for too long a time to the action of the blast, which, by its oxidizing influences, burned the carbon combined with the iron, and thus lowered its grade and value.

6th. Hot iron from the beginning to the end of the melting, and increasing the rapidity of the fusion as the operation advances—for instance, on a melting of forty-seven tons of iron in a cupola forty-eight inches in diameter inside the lining at the Detroit Car Wheel Works, Detroit, the first ten tons took one hour and fifteen minutes to melt; the second ten tons one hour and ten minutes; the third ten tons one hour; the fourth ten tons fifty minutes, thus showing a decrease of time as the operation advanced—that is to say, a better working of the cupola at the end of the operation than at the commencement. This is exactly the reverse of what generally occurs in other cupolas.

7th. By following my instructions, providing the quality of coke and iron is preserved, the same result will always be obtained; that is to say, that with the pressure of blast indicated in my instructions, it will take exactly the same time to melt a given quantity of iron with the same proportion of fuel to the iron melted.

8th. I claim that my cupola is built in the most substantial and workmanlike manner; that neither expense in material or labor is spared to make it stronger and more durable than any cupola hitherto constructed.

9th. There is a metallic shell surrounding the entire base, forming an annular air-chamber, which is provided with an inlet at the top, connected with the blower or fan, by means of which cold air is driven into the annular chamber. To compel a circulation of this air around the inside lining in order to take the caloric from it, thereby highly heating such air, and to prevent the passage of such air directly from the inlet at the top of the air-chamber to the outlets through the tuyeres into the furnace, I provide a diaphragm, the width of which equals the distance between the wall of the furnace and the outer shell. One end of this diaphragm is secured just above the inlet at the top of the air-chamber and extends spirally and downwardly, making at least one entire turn around the furnace and terminating at a point just above the tuyeres and immediately below the said inlet. This circulation, forced by the blower and compelled to take its course around the furnace, cools the latter, while the air becomes heated on reaching the tuyeres, through which it finds an outlet, thus performing the double function of cooling the furnace and supplying a hot blast.

THE WHITING CUPOLA.

In Fig. 45 is seen the Whiting patented cupola, designed by Mr. Whiting, a practical foundryman, and manufactured by the Whiting Foundry Equipment Co., Chicago, Ill., of which the following description is given by them:

The universal satisfaction given by the Whiting cupola is largely due to the patented arrangement and construction of the tuyere system, which is so designed as to distribute the blast most efficiently, carrying it to those portions of the cupola where it will do the most good, under a reduced pressure, and through an increased area.

There are two rows of tuyeres. The lower ones are arranged to form an annular air inlet, distributing the blast continuously around the entire circumference of the cupola.

This system of tuyeres is also arranged to be adjusted vertically. This provides for adjustment to the class of work,

DIFFERENT STYLES OF CUPOLAS.   197

kind of fuel, and changes in the inside diameter of the cupola. These tuyeres are flaring in shape and admit the blast through a small area which is expanded into a large horizontal opening on the inside of the cupola, thus permitting the air to reach the fuel through an area nearly double that through which it enters

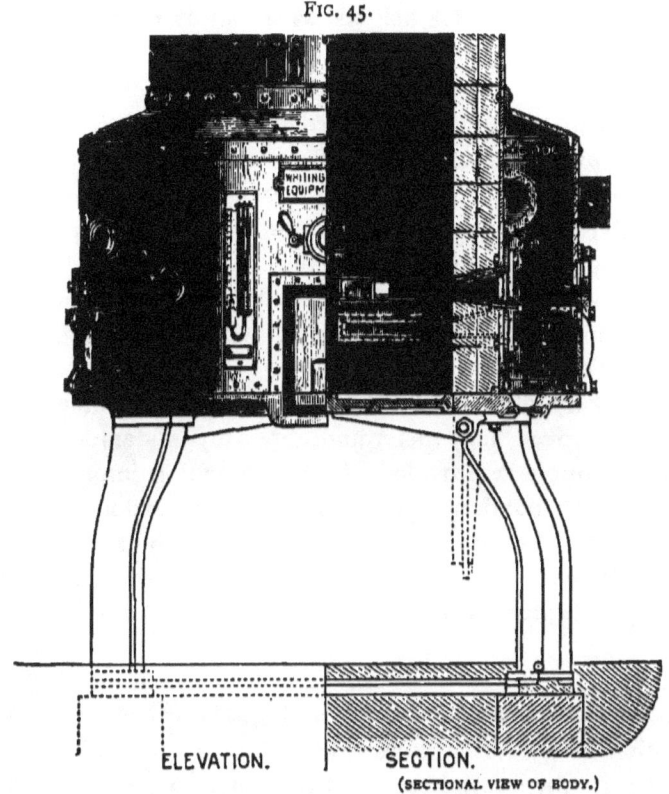

FIG. 45.

SECTION OF WHITING CUPOLA.

the tuyeres—admitting the same volume of blast, but softening its force.

There is an upper row of tuyeres of similar construction to supply sufficient air to utilize to the fullest extent the escaping carbon gas. These tuyeres are of great service in melting and in large heats—for small heats they may be closed by means of our improved tuyere dampers.

Fig. 45 represents the latest type of the Whiting patent cupola. A half vertical section is represented, showing the arrangement of the improved tuyeres and the method of adjusting them vertically. These tuyeres are arranged on slides and can be placed at various heights, as shown by dotted lines.

It sometimes happens that the operator finds the cupola too large for his needs. When this is the case, a thicker lining can be used and the tuyeres adjusted accordingly, and for small heats the proper ratio of coke to iron can be maintained; otherwise a large cupola running small heats will decrease this ratio materially, adding considerably to the cost of castings.

A change can be made from coke to coal fuel, and the bed made of suitable depth, by simply adjusting these tuyeres.

No other cupola has this device. It practically gives the operator two cupolas in one.

This figure also shows the safety alarm attachment, side plates, improved blast meter and upper tuyere dampers, etc.

Every cupola is provided with the foregoing improvements, together with foundation plate, bottom plate and doors, columns (three to five feet long), slag and tapping spouts and frames, peep holes with fittings, patent tuyeres and charging doors and frames. All fitted ready to erect.

JUMBO CUPOLA.

In the accompanying illustration, Fig. 46, is shown a sectional elevation of the large cupola known as Jumbo, in use in the foundry at Abendroth Bros., Port Chester, N. Y., to melt iron for stove plate, sinks, plumbers' fittings, soil pipe and other light castings, all requiring very hot fluid iron. The cupola, which was constructed for the purpose of melting all the iron required for their large foundry in one cupola, is of the following dimensions: diameter of shell at bottom to height of 24 inches, 7 feet 6 inches; diameter in body of cupola, 9 feet; taper from large to small diameter, 5 feet 6 inches long; diameter of stack, 6 feet;

taper from cupola to stack, 6 feet long; height from bottom plate to bottom of taper to stack, 20 feet; height to bottom of charging doors, 18 feet; two charging doors placed in cupola on opposite sides. Wind box inside the shell extending around the cupola, 5 feet 6 inches by 9 inches wide. Height of tuyeres, first row, 24 inches; second row, 36 inches; third row, 48 inches. Size of tuyeres, first row, 8×5 inches; second row, 6×4 inches; third row, 2×2 inches. Number of tuyeres in each row, 8; total number of tuyeres, 24. Slag hole, 17 inches above iron bottom, 11 inches above sand bottom. Two tap holes. Lining, 18 inches thick; over air belt, 9 inches. Diameter of cupola at bottom, inside the lining, 4 feet 6 inches. Diameter above taper, 6 feet. Cupola supplied with blast by No. 6 Baker blower.

It is charged as indicated in the table as follows:

Date, August 17, 1894.
No., Jumbo.
Cupola, No. 3.

	Pounds coal.	Pounds coke.	No. 1.	No. 2X.	No. 2X.	No. 2X.	Sprues and scrap.	Pounds at each charge and total.	No. 2 Foundry Pioneer.	Recorded by .... Remarks.
Charge........	2,100	900	2,500	2,500	1,500	1,500	4,000	12,000	....	
Added	125	600	2,000	2,000	1,000	1,000	3,000	9,000	....	
Added	125	600	2,000	2,000	1,000	1,000	3,000	9,000	....	
Added	125	600	2,000	2,000	1,000	1,000	3,000	9,000	....	
Added	125	600	2,000	2,000	1,000	1,000	3,000	9,000	....	
Added	125	600	2,000	2,000	1,000	1,000	3,000	9,000	....	
Added	125	600	2,000	2,000	1,000	1,000	3,000	9,000	....	
Added	125	600	2,000	2,000	1,000	1,000	3,000	9,000	....	
Added	200	700	2,000	2,000	1,000	1,000	3,000	9,000	1,000	
Added	300	800	2,000	2,000	1,000	1,000	3,000	9,000	1,000	
Added	300	800	2,000	......	2,000	1,000	3,000	9,000	1,000	
Added	300	800	2,000	......	2,000	1,000	3,000	9,000	....	
Added	100	500	2,000	2,000	2,000	1,000	3,000	9,000	....	1000 pounds iron over.
Added	....	150	800	800	400	400	1,100	3,500	....	
Pounds of each	4,175	9,350	29,300	21,300	18,900	14,900	44,100	132,500	4,000	

	Time.	Time.	Time.	Time.	Time.	Time.	Time.	Time.	Time.	
Blast put on	12.30									
Iron flowed......	12.38 M.									
Blast to iron	8	1.00	......	3.00	......	4.00	......	4.30	......	4.50
Dropped bottom	5.15	Oz. 14	Oz.	Oz. 16	Oz.	Oz. 17	Oz.	Oz. 14	Oz.	Oz. 12
Blast to drop of bottom.	H. M. 4.53									

Pressure of blast in ounces.

Three hundred and fifty pounds of limestone are placed on each charge of iron, except the last charge, and the slag hole opened after the blast has been on about three-quarters of an hour and permitted to remain open during the rest of the heat.

FIG. 46.

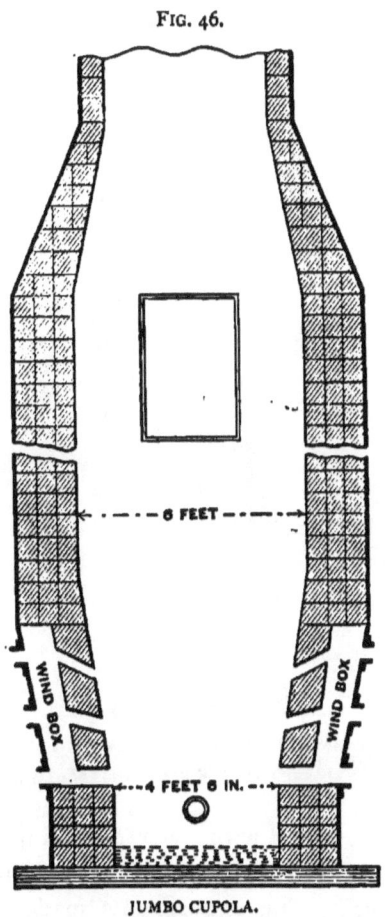

JUMBO CUPOLA.

The sprues, gates and foundry scrap are not milled before charging, and the large amount of limestone placed on each charge is required to liquefy the quantity of sand charged into the cupola on the scrap, and prevent clogging and bridging of the cupola. Sixty tons of iron have been melted in this cupola in four hours from the time the blast was put on until the bottom was dropped.

THE CRANDALL IMPROVED CUPOLA WITH JOHNSON PATENT CENTER BLAST TUYERE.

In Fig. 47 is shown the above-named cupola and tuyere manufactured by the Foundry Outfitting Co., Detroit, Mich., a description of which is furnished by them as follows:

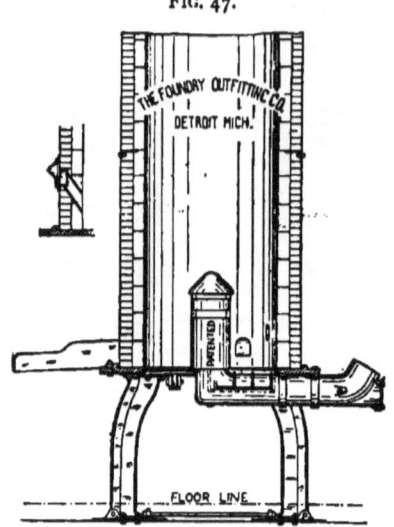

FIG. 47.

THE CRANDALL IMPROVED CUPOLA WITH JOHNSON PATENT CENTER BLAST TUYERE.

The cupola is designed with a view of getting a more efficient action of the blast than is possible to attain with the methods now in general use. The experiments made in this new departure have finally led to a very simple and durable construction, which we place before the foundrymen and request that they make a thorough investigation of it. It is a well-known fact that the matter of forcing blast to the center of a cupola and obtaining a complete combustion of fuel at that

DIFFERENT STYLES OF CUPOLAS. 203

point, has been to many a puzzle, and various means have been tried to accomplish this end. But it has been found in all cases, that a large portion of the blast when taken in at a high pressure through outside tuyeres, in striking the fuel is forced back against the brick lining, cutting it out very rapidly just above the tuyeres and then escaping up along the brick wall, doing no good, thereby requiring a greater volume of blast to melt the same amount of iron than is used when the blast is taken in at the center of the cupola. In the illustration (Fig. 47) is clearly shown the general arrangement.

The air, instead of being forced into the cupola-furnace from the outside, is applied from the inside by means of a center blast tuyere attached to the under side of the bottom plate. This tuyere terminates at about the same height as outside tuyeres, and a continuous annular opening is formed for the blast by putting on a loose section of pipe and spacing it apart by means of pins that can be varied in height, so as to get any desired opening. On top of this loose section a cap is set; also spaced apart from it by means of pins, so that a second opening is formed for the blast to enter, and by taking in more air at this point the carbonic oxide, which would otherwise go to waste, is changed into carbonic acid gas, forming the whole interior into a melting zone, insuring complete combustion. Both the loose pipe section and the cap can be removed to have the lining on them repaired. The horizontal part of the center blast pipe has an opening at the elbow which enables it to be cleaned out, in case any obstructions should fall through the tuyere opening above. The drop doors close over this tuyere and can be opened without in any way deranging it. No belt air-chamber is required, as the tuyere may be connected direct to the main blast pipe; but in cases where such air-chambers already exist, the center blast tuyere may be attached to them without in any way disarranging the blast pipe. We would draw special attention to the fact that but little expense need be incurred in making this change outside of the price charged for the center blast tuyere and piping.

Claims are made as follows:

1st. A saving in brick lining.
2d. A saving in fuel.
3d. More rapid melting with less volume of blast.
4th. A more uniform temperature of iron than can be attained by the outside tuyere.

BLAKENEY CUPOLA.

In Fig. 48 is seen a sectional view of the Blakeney cupola furnace, the following history and description of which are furnished by The M. Steel Co., Springfield, Ohio.

FIG. 48.

SECTIONAL VIEW OF BLAKENEY CUPOLA FURNACE.

By the Blakeney cupola furnace, the air is so distributed or projected into the furnace as to produce a uniform heat, giving the iron a uniform strength for all kinds of castings. The features peculiar to it are as follows:

The introduction of a combination of curved tuyeres or chutes placed upon the wall or lining of the cupola, and forming a part of the wall, a proper distance from the bottom, and nearly surrounding the inner and outer sides of the wall. The tuyeres are made of cast iron and in sections for convenience of handling. A blank space is left in the rear of the cupola two feet wide, through which the slag is blown, if required.

A chamber or base extending around the cupola and enclosing the space in which the air is conducted to the tuyeres. The bottom of this chamber, made irregular in form, hollows at suitable intervals to allow the metal to flow to the escape openings, in case it overflows through the tuyeres. The openings are closed with fusible plugs of lead or other material, to be melted out by the molten metal.

The blast is conducted to this cupola through one pipe, and striking the blank space sidewise in rear of chamber, passes all around through the curved tuyeres into the center of the furnace, the blast striking into the cupola every seven-eighths of an inch horizontal, and $3\frac{3}{4}$ inches perpendicular, or according to diameter of cupola.

As a producer of a uniform grade of iron for the purpose of casting car-wheels, it is just what is needed for the different grades of iron to prevent chill cracking.

This cupola, with its many superior advantages, has also rows of shelves bolted to the shell four feet apart up to the top of the charging door, so that it will not be necessary to tear out any of the lining except that which is burned out. These cupolas have run eighteen months with heavy heats without being relined.

These various cupolas are shown and described, not that we endorse all that is claimed for them, but to give our readers some idea of what has been done in design and coustruction of them, and what kinds may at the present time be obtained from cupola manufacturers. We have by no means exhausted the different varieties at hand, but have probably given sufficient examples to indicate the direction in which inventive genius has gone, and the objectionable points in construction which it has been their aim to overcome.

CHAPTER VIII.

ART IN MELTING.

THE melting of iron in a cupola is an art that is by many foundrymen and foundry foremen but little understood, and they never begin the melting of a heat without a dread that something will happen to prevent the iron being hot enough for the work, or that they may not be able to melt the entire heat. In many foundries it is almost an every-day occurrence to have something happen in or about the cupola to prevent good melting. The sand bottom cuts through, the front blows out, the tap hole cannot be opened without a heavy bar and sledge, slag flows from the tap hole with the iron and bungs up the spout and ladles, iron and slag get into the tuyeres, daubing falls off the lining and bungs up or bridges the cupola, stock lodges upon the lining in settling, and only part of the heat can be melted. Iron melts so fast in one part of the heat that it cannot be taken care of; in another part it melts so slowly that a ladle cannot be filled before the iron is too dull for the work; or, iron is not melted of an even temperature throughout a heat, and has to be watched in order to get hotiron to pour light work; the first iron is dull, or the last is dull, or the whole heat is dull. Some of these troubles to a greater or less extent occur almost daily, and it is a rare occurrence in a great many foundries that a perfectly satisfactory heat is melted. In foundries in which these difficulties occur, the foundryman or his foreman, or both, do not understand melting. The cupola is in charge of an old professional melter who always ran it in this way, or a foundry laborer or helper has been selected for a melter and given a few instructions by some one who has seen a cupola prepared for a heat, or perhaps has melted a few heats. He is instructed until he melts a heat successfully, and then he "knows it all" and is left to himself, and perhaps he knows as much as his instructor. If he is a practical man, he learns the cause of all the troubles in melting and in time becomes a fair melter; but at what an expense to his employer!

If he is not a practical man, he bungles along from day to day until he gets disgusted with his job and quits, or is discharged, and another man of the same kind is tried, with about the same result, for there is no one about the foundry who understands the art of managing a cupola to instruct him, and he must learn it himself or as a melter be a failure. The foundryman or foreman of a foundry in which this kind of melting is done, will tell you a cupola is a very hard thing to manage, and it cannot be made to melt evenly throughout a heat or the *same every heat*. If this were really the case, foundries making very light work, requiring hot fluid iron, would lose half their castings every heat or be compelled to pour large quantities of iron into the pig bed and wait for hot iron. But this is not the case in stove, bench and other foundries making very light castings. Heats of many tons are melted every day, and as many pounds of iron are melted in one minute as in another from the beginning to the end of a heat, and there is not a variation of fifty degrees in the temperature of the iron from the first to the last tap.

There is no chance work in nature, and there is no chance work in art when the scientific principles are understood and applied to practice, and there is no chance work in melting iron in a cupola when the cupola is scientifically managed, and there is no furnace used for melting iron more easily managed than the cupola furnace; but it is necessary to understand its construction and mode of operation to do good melting.

In the first place, the cupola must be properly constructed and of a size suitable for the amount of iron to be melted, and the time in which this melting is to be done. For fast melting, a cupola of large diameter is required, and for slow melting one of small diameter. There are those in use at the present time in which sixty tons of iron are melted in four hours, and those in which one ton of iron is melted in four hours and a half, and each of these cupolas melts iron as fast as it can be taken care of after it is melted. The large cupola would be useless in one foundry, and the small one in the other. So it follows that a cupola must be so constructed as to be suitable for the melting that it is desired to do.

To melt iron hot and of an even temperature, the tuyeres must be placed low, made of a size to admit the blast freely to the cupola and arranged to distribute the blast evenly to the fuel, and the latter must be of a proper volume for the size of cupola. To utilize the greatest possible amount of heat from the fuel, the charging door should be placed high and the cupola kept filled to the door until the heat is all in. When preparing a cupola for a heat, it must be properly chipped out and the lining given the best possible shape for melting, by the application of daubing. The daubing material must be of an adhesive and refractory nature, and not put on so thick that it will fall off when dried or heated. The bottom door must be put up and supported by a sufficient number of props to make it rest perfectly solid against the bottom plate. The bottom sand must be of a quality that will not burn or be cut up by the molten iron, and it must be of a temper that will neither wash nor cause the iron to boil. It must be carefully packed around the edges and rammed evenly, and no harder than the sand for a mould, and given a proper pitch to cause the iron to flow to the tap hole as fast as melted. A front and spout lining material must be selected or prepared that will not cut or melt. And the front must be put in solid with a proper sized tap hole, and the spout given the right shape and pitch. The cupola having been thus prepared, it is ready for melting. Shavings and wood are put in for lighting the melting fuel or bed, and a sufficient quantity of coal or coke is put in to fill the cupola to the top of the melting zone after it has settled. As soon as this fuel is well on fire and the heavy smoke is burned off so that the top of the bed can be seen, it is leveled up with a few shovelfuls of fuel, and charges of iron and fuel are put in until the cupola is filled to the door. The weight of the bed fuel, and charges of iron and fuel, must be learned for each cupola, for scarcely any two are charged exactly alike.

It will thus be seen that the melting of iron in a cupola is very simple. But all these things and many more must be learned and practiced to make it so, and they cannot be learned in one or in a dozen heats. Slag and cinder adhere to the lining at one point to-day and at another to-morrow, and the chipping out must be different. The lining is burned away more at one point to-day than it was yesterday. A new lining requires a different shaping than an old one, as a lining burns out and the diameter of the cupola increases. More fuel is required for a bed, and the weight of charges of fuel and iron must be increased. All brick are not suitable for a cupola lining, and a good brick may be laid up in such a way that a lining will not last half so long as it would do if properly put in. All daubing material is not suitable for repairing the lining of a cupola, and the best daubing is worthless when not properly applied. Bottom sand when used over and over again becomes worthless, and all sands are not suitable for a bottom. The front may be put in with material that melts, and the tap hole cannot be kept open and free of slag; or the front made of a shape that iron chills in the tap hole between taps. The spout lining material may not be suitable, and may melt and bung up the spout with slag, or the lining may be made of a shape that two or three ladles are required to catch the many streams that fall from it at the same time.

To learn to manage a cupola perfectly, a close study of all the materials used in melting and their application to melting are necessary, and months of careful observation are required to learn them, but by an intelligent man they can be learned. A moulder, when serving his time as an apprentice, is seldom given an opportunity to learn melting, and when he becomes foreman of a foundry knows nothing whatever about the management of a cupola and is completely at the mercy of the melter. The time has passed in many localities when the entire force employed in a foundry was subject to the whims of a melter and compelled to take a day off whenever he did not see fit to work, and a foreman who does not fully understand the management of cupolas is no longer considered a competent man to have charge of a foundry. It should be the aim of every moulder who aspires to be a foreman or foundryman to learn melting, and when he takes charge of a foundry he should at once learn all the peculiarities of the cupolas of that foundry, and be able to run off a heat as well as the melter, or instruct the melter how to do it. In conversing with foremen, we have frequently remarked to them that the foreman of a foundry should be the melter, and many of them have replied that they would give up the foremanship before they would do the melting. To be a melter does not imply that the melter should perform the labor requisite to melting, for a melter may direct the melting of a heat without ever touching the iron to be melted or any of the material required to melt it. By going inside for a few minutes and giving directions how it must be done, any intelligent man can be employed to do the work, and he can be instructed from the charging door how to pick out and daub a cupola or repair a lining. He can be shown how to put up the doors and support them in place; how to prepare daubing, front and spout material, select and temper bottom sand, and instructed from the charging door and front, how to put in a bottom front and spout lining; how to light up and burn the bed, and given a slate of charges and directions for putting them in the cupola. After he has been directed by a competent melter in this way for a few heats, it is only necessary for the melter or instructor to inspect his work from time to time, to see that it is properly done and prevent the lining getting out of shape or other things occurring, in which a new melter cannot be instructed in a few days; and his work should be inspected to prevent him getting into a rut, as melters so frequently do when left to themselves.

CHAPTER IX.

SCALES AND THEIR USE.

THERE is nothing more essential to the good melting and mixing of iron, than an accurate weighing scale upon the scaffold near the cupola charging door. The best for this purpose are the platform scales mounted on large wheels, with the platform about two feet above the floor or on a level with the charging door. For foundries that make a large quantity of gates, sprews and light scrap to be remelted, an iron box made of boiler plate and open at one end for shoveling out iron and fuel should be placed upon the scales. A one or two ton scale is sufficient for charging almost any cupola, for the iron and fuel are weighed in charges or drafts that seldom exceed this weight, and when large pieces are charged they are generally weighed on the scales in the yard. Scales placed in the floor on the scaffold upon which barrows of iron and fuel are weighed as they are brought on the scaffold, give the weight of the stock used, but they are of no value in dividing it into charges if the stock is not charged direct from the barrows into the cupola, which is seldom done.

The melting of iron in a cupola, when reduced to an art, consists in melting the greatest possible amount with a given amount of fuel; and this can only be done by first learning the amount of iron that can be melted with each pound of fuel, and placing that amount upon the fuel in the cupola at the proper place to be melted. If all the fuel required to melt ten or twenty tons of iron were first placed in a cupola and all the iron put upon it in one lot, it would be so high above the melting zone that none of it could be melted until fully one-half of the fuel had been burned away and the iron permitted to settle to the melting zone, in which case all the fuel consumed before melting began would be wasted, and the iron would not have a sufficient amount of fuel to melt it.

For this reason the fuel and iron are divided into charges and placed in a cupola in layers, each layer of fuel being only sufficient to melt the layer of iron placed upon it, when it descends into the melting zone. If the charge of fuel be too heavy, the excess must be consumed before the iron can be melted by it; and if the charge of iron be too heavy, all of it cannot be properly melted with the charge of fuel. It is, therefore, necessary that the layers of fuel and iron should be of exactly proper proportions to do economical melting.

There are many foundrymen who do not understand this theory of melting, but think fuel placed in a cupola melts iron no matter how it is put in, and trust to their melter to guess the weight of fuel consumed and iron melted in an entire heat. Others have the fuel and iron weighed in the yard or upon the scales placed in the floor of the scaffold, and permit the melter to guess the respective weights of fuel and iron in charging. In the first case an excess of fuel is always consumed, the melting is slow and the amount of iron charged is often more than required to pour off the work; or it is insufficient, and more iron has to be charged after the stock is low in the cupola, and the destruction of cupola lining is greater than if the iron had been charged at the proper time. In the second case the melting is irregular, and the temperature of the iron uneven, even if only a proper amount of iron and fuel to melt it is placed upon the scaffold, for the melter cannot in charging divide it evenly. No melter can guess the weight of a promiscuous lot of scrap, sprews, gates, etc., or accurately estimate the weight of pig iron by counting the pigs. The counting of shovels, riddles or baskets of fuel in charging is the greatest fallacy of all; for riddles and baskets always hold more the longer they are in use, and shovels hold less. The melter makes no allowance for the increase in size of riddles or baskets, but always puts in a few extra shovelfuls to make up for reduced size of the shovel, as it wears down. Even when these articles are new, a few pounds more or less may be put on, so that it is simply guess-work at best.

Placing upon a scaffold old worn-out scales that are unfit for use in other parts of the works and frequently only weigh correctly on one side or end, is a mistaken economy frequently practiced by foundrymen. The weighing of cupola stock upon such scales is only guess-work, and the saving in fuel and improvement in melting would soon pay the cost of accurate scales.

CHAPTER X.

THE CUPOLA ACCOUNTS.

IN all well regulated foundries a cupola account of melting is kept and an accurate record made of each heat, and preserved for future reference. In this way, the melting is reduced to a system and the foundryman knows what is being done in his cupola each day and is able to make an estimate of the cost of melting. These records are also of value in showing the amount of fuel required for a bed and in charges when the cupola is newly lined, and the amount they should be increased as the lining burns out and the cupola is enlarged. Mixtures of various brands and grades of iron are recorded, with the result of the mixtures upon the quality of castings, and a great deal of experimental work in melting and mixing of irons is saved and better results are thus obtained. The manner of keeping these accounts varies in different foundries. In some they are kept very simply, showing only the amount of fuel and iron in each charge and total fuel consumed, iron melted, and time required in melting. Others show kind and amount of fuel used, in bed and charges, and amount of each brand or quality of iron placed in charges, total amount melted, time of lighting up, time of charging, putting on blast, first iron melted, blast off, pressure of blast, etc.

Others are still more elaborate, and not only show all the details of the cupola management, but also a report presenting cost of various castings produced, good and bad, the cost of the bad ones being charged to the good ones made off the same pattern or for the same order, and the average found.

To give foundrymen who have never used such reports an idea of how they are made out, we here give a few blank reports from leading foundries. That of Abendroth Brothers, Port Chester, N. Y., and Byram & Co., is filled in to show the manner of placing the various items in the blank report.

THE CUPOLA ACCOUNTS.

ABENDROTH BROS., PORT CHESTER, N. Y.

Date, August 17, 1894. No., Jumbo. Cupola, No. 3.	Pounds coal.	Pounds coke.	No. 1.	No. 2X.	No. 2X.	No. 2X.	Sprues and scrap.	Pounds at each charge and total.	No. 2 Foundry Pioneer.	Remarks. Recorded by ......
Charge........	2,100	900	2,500	2,500	1,500	1,500	4,000	12,000		
Added	125	600	2,000	2,000	1,000	1,000	3,000	9,000		
Added	125	600	2,000	2,000	1,000	1,000	3,000	9,000		
Added	125	600	2,000	2,000	1,000	1,000	3,000	9,000		
Added	125	600	2,000	2,000	1,000	1,000	3,000	9,000		
Added	125	600	2,000	2,000	1,000	1,000	3,000	9,000	1,000	
Added	125	600	2,000	2,000	1,000	1,000	3,000	9,000	1,000	
Added	125	600	2,000	2,000	1,000	1,000	3,000	9,000	1,000	
Added	200	700	2,000		1,000	1,000	3,000	9,000	1,000	
Added	300	800	2,000	2,000	2,000	1,000	3,000	9,000		
Added	300	800	2,000	2,000	2,000	1,000	3,000	9,000		
Added	300	500	2,000	2,000	1,000	1,000	3,000	9,000		1000 pounds iron over.
Added	100	500	800	800	400	400	1,100	3,500		
Added		150								
Pounds of each...	4,175	9,350	29,300	21,300	18,900	14,900	44,100	132,500	4,000	
	Time.	Time.	Time.	Time.	Time.	Time.	Time.	Time.	Time.	
Blast put on........ Iron flowed.........	12.30 12.38	1.00	3.00		4.00		4.30		4.50	
Blast to iron	M. 8	Oz. 14	Oz. 16	Oz.	Oz. 17	Oz.	Oz. 14	Oz.	Oz. 12	
Dropped bottom	5.15					Pressure of blast in ounces.				
Blast to drop of bottom.	H. M. 4.53									

BYRAM & COMPANY,
IRON WORKS.

435 and 437 Guoin Street.
46 and 48 Wight Street.

DETROIT, MICH.

FUEL USED and IRON MELTED at the Foundry of

IN THE COLLIAU CUPOLA FURNACE.				CHARGES.	
..................Size.				FUEL.	IRON.
Diameter Inside of Lining........54 ins.				Bed 1700	1. 5000
Pressure of Blast....................oz.				2. 250	2. 2000
				3. 250	3. 2000
Lighting,	-	-	12 15 o'clock.	4. 250	4. 2000
Loading Commenced,		1	"	5. 250	5. 2000
Blasting,	-	2	"	6. 250	6. 2000
Closed Tap Hole,		....	"	7. 250	7. 2000
First Iron Taken,		2 16	"	8. 250	8. 2000
				9. 250	9. 2000
Loading Finished,		....	"	10. 250	10. 2000
Blasting Stopped,		....	"	11. 250	11. 2000
Dropped Bottom,		....	"	12. 250	12. 2000

REMARKS:------

Dated at ------ 189----

No.--------

THE CUPOLA ACCOUNTS.

DAILY REPORT OF FOUNDRY DEPARTMENT.
LEBANON STOVE WORKS.

_____ 189___

Charges.	Lbs. of Coal.	Lbs. of Coke.	Lbs. of Shells.	Kind of Pig.	Kind of Pig.	Kind of Pig.	Kind of Pig.	Kind of Pig.	Gates and Scrap.	Lbs. per Charge.
1st										
2d										
3d										
4th										
5th										
6th										
7th										
8th										
9th										
10th										
11th										
12th										
Total										

	Hrs.	Min.
Time of Lighting,		
Time Blast Put On,		
Time First Shank,		
Time Bottom Dropped,		
Length of Heat,		
Condition of Iron,		
Ratio of Iron to Fuel,		

REMARKS: _____

_____ *Foreman.*

THE CUPOLA FURNACE.

[Page 218: Blank melting sheet form of Syracuse Stove Works, printed sideways. Column headings include: Date; No.; Charge; Coal; Coke; Stove Plate; Scrap; Sprues; Pressure of Blast in Ounces (Time, oz.); Blast and Coke Melted (1 Lb. Coal and Coke Melted, Lbs. Iron); Short of Charge (Total); Stove Plate; Sprues; Pounds of Each Charge and Total; Recorded by; Remarks. Row labels: Added; "; "; "; "; "; Lbs. of Each; Blast on Iron; Dropped Bottom; Blast to Drop of Bottom; Blast Put On; Iron Flowed; Condition of Metal; Castings (Sprues, etc.); Lbs.]

REPORT OF CASTINGS IN _____ SHOP.

Date, _____ 189_

Cupola Heat Consisted of		Coal Used	Lbs.	Good Castings	Lbs.	Cost of Moulding	Lbs.
Pig	Lbs.	Coke Used	Lbs.	Bad Castings	Lbs.	" for Special Cores	Lbs.
Pig	Lbs.	Flux Used	Lbs.	Checked Castings	Lbs.	" General "	Lbs.
Pig	Lbs.			Returns from Bottom, &c.		Cupola Expenses	Lbs.
Pig	Lbs.	Ratio Fuel to Iron	To	Loss	Lbs.	Crane Runner	Lbs.
Scrap	Lbs.	Blast on	p. m.	Total		Gang Foreman	Lbs.
Iron Melted	Lbs.	Bottom Dropped	p. m.	Total Cost of		Furnace Man	Lbs.
Iron Borrowed	Lbs.			Good Castings per 100 Lbs.		Sand Mixer	
						Total Cost of	
						Labor on Castings	

Shop Order.	Moulder's No.	No. of Helpers.	No. of Pieces.	Description of Work.	Pattern No.	Weights in Detail.	Castings.			Total No. of Hours.	Wages per Hour.	Price per Piece.	No. of Cores.	Amount of Sand in Pounds.	Amount of Hay Rope in Pounds.	Fuel Used in Baking.	Wages on Cores.	Amount of Loam in Pounds.	Number of Brick Destroyed.	Cost of Loam Moulds and Cores.	Cost of Gravel and Hay Rope Cores.	Cost of Moulding.	Total Cost.	Remarks.
							Good.	Bad.	Checked.															

THE CUPOLA FURNACE.

CUPOLA SLATE FOR CHARGING AND CUPOLA REPORT.

Coke.	Lbs.	Iron.	Iron.	Iron.	Iron.	Iron.	Sprues.	Scrap.	Total.
Bed									
Charges—1st									
2d									
3d									
4th									
5th									
6th									
7th									
8th									
9th									
10th									
11th									
12th									
13th									
Total									

The blanks for these reports and records of them are furnished to the foundry foreman or melter, and preserved in different ways. In some foundries they are furnished in separate sheets, and when filled out and returned are kept in files provided for the purpose. In other foundries they are made out in book form and filled in by the foreman or foundry clerk. Such reports can be kept by a foundry foreman when provided with a small office for doing such work; but when there is no office, as is frequently the case, a report book kept by the foreman soon becomes so soiled that it is useless for reference, and report blanks are generally furnished in separate sheets and either filed or transferred to the report book by the foundry clerk. When only a record of fuel used and iron melted is kept, the report is generally made on a slate upon which lines are scratched similar to those in a printed report, and name and amount of various grades of iron and fuel filled in with the slate pencil. The fuel to be used and amounts of various irons to be melted in each charge are placed upon the slate by the foreman and given to the melter to charge the cupola by, and after the heat is melted the slate is sent to the foundry office to be copied into the cupola account book. This latter is the oldest way of making out these reports.

A cupola account is of no value if not correctly kept, and it should be the aim of every foundry foreman to see that the report he makes of fuel consumed and iron melted is correct, and not, as is frequently done, endeavor to make a good showing for himself, of melting a large per cent. of iron with a small per cent. of fuel, and permit his melter to shovel in extra fuel to make iron sufficiently hot to run the work. Foundrymen can readily ascertain the amount of fuel consumed by comparing the amount reported with the amount purchased. False reports only reduce the foreman in the estimation of his employers, and are frequently the cause of his losing his position.

CHAPTER XI.

PIG MOULD FOR OVER IRON.

IN foundries in which the iron is all poured from hand ladles, there is frequently a small amount of iron left in a ladle that is not sufficient to pour a mold, and cannot be used except when the iron is very hot and the moulder catches in immediately after pouring.

Moulders will not take the time to carry this iron back to the pig bed at the cupola, and it is generally poured upon the floor in the gangway or into the sand heaps; and a great deal of light scrap is in this way made in large foundries, that requires much time and labor to collect and even when carefully collected with much loss in the sand and gangway dirt. To obviate this wastage of iron and labor, many foundries have adopted the cast iron pig mould shown in Fig. 49, and placed one of them in the gangway at the head of each floor, or at convenient distances apart in the gangways for the moulders. All the over iron is poured into these moulds and is collected in a pig of convenient size for handling and melting, greatly reducing the loss of iron and cost of removing.

FIG. 49.

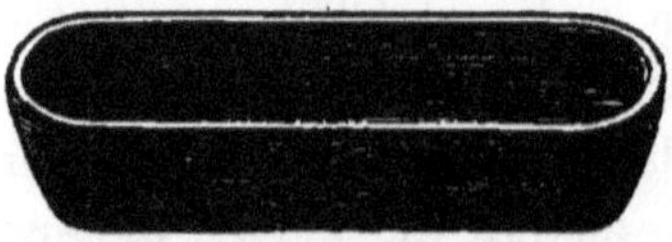

PIG MOULD FOR OVER IRON.

CHAPTER XII.

WHAT A CUPOLA WILL MELT.

THE cupola furnace was originally designed for melting cast iron for foundry castings, and at the present time is principally employed for that purpose, but it is now also used in the melting of almost all of the various grades of manufactured iron and steel, and many other metals.

It is extensively employed in the melting of pig iron in the manufacture of Bessemer steel, and in the melting of iron for castings to be converted into steel and malleable castings after they are cast. It is also used in melting steel for steel castings, but as it makes an uncertain grade of steel is only employed for the more common grade of castings.

It is also employed in melting tin plate scrap, sheet iron, wrought iron and steel wire, gas pipe, bar iron, horse shoes and all the various grades of malleable wrought and steel scrap, found in a promiscuous pile of light scrap and used in the manufacture of sash, elevator and other weights, and melts them readily, producing a very hot fluid metal, and when properly managed is the very best furnace for this purpose.

It is to some extent used in the smelting of copper ores and the melting of copper, in the manufacture of brass, and also in the melting of brass for large castings; but in melting brass, the alloy is oxidized to so great an extent that an inferior quality of brass is produced to that obtained from crucibles.

Lead is frequently melted in cupolas. It melts more slowly than would naturally be expected, and it is very difficult to retain it in a cupola in the molten state, as it is almost impossible to put in a front through which it will not leak, and the ladle is generally heated and the tap hole left open.

The quantity of cast iron that can be melted in a cupola per hour depends upon the diameter and height of cupola, and at the present time varies from one hundred pounds to hundreds of tons. The number of hours a cupola will melt iron freely when properly managed, is only limited by the length of time the lining will last. Cupolas have been run continuously from one o'clock Monday morning until twelve o'clock Saturday noon, melting fourteen tons per hour.

The size and weight of a piece of cast-iron that can be melted in a cupola at one heat, depends upon the size of the cupola.

As a rule, any piece of iron that can be properly charged in a cupola can be melted. In steel-works cupolas, ingot moulds weighing five tons, are melted with ease in the regular charges of the cupola.

At the foundry of the Pratt & Whitney Co., Hartford, Conn., a large charging opening is placed in the cupola for the purpose of charging large pieces of iron to be melted, and almost any piece can be melted in one heat that can be placed in the cupola.

At the foundry of the Lobdell Car Wheel Co., Wilmington, Del., an oblong cupola with charging door placed at the ends was constructed shortly after the War of the Rebellion to melt cannon and other heavy government scrap, and large cannon weighing many tons were melted in this cupola without previously breaking them up.

CHAPTER XIII.

MELTING TIN PLATE SCRAP IN A CUPOLA.

TIN plate scrap is melted in the ordinary foundry cupola the same as cast iron scrap, but more fuel is required to melt it. The best results are obtained with 1 pound of coke to from 3 to 4 pounds of scrap and a mild or light blast. Various ways of preparing the scrap for charging, such as hammering or pressing it into ingots and forming it into compact balls, have been tried; but as good results are obtained by charging it in bulk, and it is generally added in this way. The charges are made of about the same weight as charges of iron in a cupola of similar size, but more fuel is added. The scrap when first put in the cupola is very bulky and takes up a good deal of room, but when heated it settles down into a compact mass, and takes up very little more space than a charge of cast iron scrap. Tin plate scrap settles rapidly, but melts slower than cast iron scrap or pig.

Numerous attempts have been made to recover the tin deposited upon the iron by heating the scrap in various ways to a temperature at which tin melts, but the coating of tin is so light it will not flow from the iron. All such attempts to recover it have proved failures. The iron, or rather steel, which is coated with tin is a very soft and tough material, but when melted the tin alloys with it, and the metal produced is very hard and brittle. The molten metal from this scrap has very little life, chills rapidly in the spout, ladles or molds, must be at a white heat when drawn from the cupola, and must be poured as quickly as possible. When not melted extremely hot the metal expands or swells in cooling to so great an extent as to tear a sand mold to pieces or break an iron mold where it cannot escape. When the metal is melted very hot this expansion does not take place to so great an extent, and a sand or iron mold may be used for any work into which it is to be cast.

The molten metal is more susceptible to the effect of moisture than iron, and is instantly thrown out of a mold when sand is worked too wet and cannot be made to lay in it. The sand must, therefore, be worked as dry as possible. The metal is very hard and brittle, and only fit for sash and other weights, and even these when light and long must be handled with care to avoid breaking. The weights when rough cannot be chipped or filed smooth, and sash weights made of this metal are generally sold at a less price than iron weights; for when rough they wear out very quickly the wooden box in which they are hung, and builders dislike to use them. A foundryman who recently had a contract from the Government for a number of weights of several tons each, to be used for holding buoys in the ocean, made them from tin plate scrap. When cast they were so rough that he remarked it was a good thing they were to be sunk in the mud under the ocean, for they were not fit to be seen.

In a number of experiments we made in melting this scrap, we found we could produce a gray metal from it about as hard as No. 3 pig iron, by melting it with a large per cent. of fuel and a very light blast. But the metal was very rotten and had little if any more strength than when white. We tried a number of experiments to increase its strength, but in none of them did we succeed to any extent. Melting it very hot and running it into pigs and remelting the pig improved the strength in some degree; but this was expensive, and the results did not justify the expense. We also made a number of tests to learn the amount of metal lost in melting this scrap, and found with a light or proper amount of blast to do good melting there was practically no loss. With a strong blast the loss was heavier, and in one heat, with a very heavy blast, we lost 10 per cent. of the metal charged. The metal from this heat was a little stronger and also a little harder, which was probably due to oxidation of the tin and iron by the strong blast before melting. In melting old roofing tin, rusted scrap and old cans, the loss in melting varied from 10 to 25 per cent,, which was probably due to rust, paint and solder used in putting the work together.

Tin acts as a flux when melted with iron, and renders it more fusible. Scrap from which the tin has been removed by acids to recover the tin or by the process employed in the manufacture of chloride of tin, is more difficult to melt in a cupola than when covered with tin, and more fuel and time are required to melt it, but a better grade of iron is produced from it. Scrap of this sort should be melted soon after the tin is removed from it, for it rusts very quickly, and when rusted to any extent produces nothing but slag when melted.

Scrap sheet iron is more difficult to melt than tinned scrap and is seldom melted in a cupola, for better prices are paid for it by rolling mills than foundrymen can afford to offer.

Galvanized sheet-iron scrap cannot be melted at all in a cupola in large quantities, for the zinc used in galvanizing it, acting like the zinc solution used in the Babcock fire extinguishers, cools the fire in the cupola to a marked degree. When melting tinned scrap any galvanized scrap that has been mixed with it must be carefully picked out, for even in small quantities it lowers the heat in a cupola to such an extent that the metal from the tinned scrap cannot be used, and must be poured into the pig bed if it runs from the cupola at all. There are a number of ways of doctoring the metal from tin-plate scrap when it melts or flows badly, by the use of gas and oil, retort carbon, etc., but they do not improve the quality of the metal to any extent, and it is very doubtful if they increase its melting or flowing properties.

A cupola of any suitable size can be employed for melting tin-plate scrap and an entire heat of the scrap may be melted alone, or it may be mixed with cast iron scrap or pig, and melted, or again, it may be melted alone directly after a heat of iron. It is a common practice in many small foundries to melt this scrap in the cupola for sash and other weights directly after melting a heat of iron for soft castings. An extra heavy charge of fuel is placed upon the last charge of iron to check the melting for a few minutes by preventing the scrap settling into the melting zone, and the soft iron is all melted and drawn off before the scrap begins to come down. In melting long heats of this scrap it is necessary to flux the cupola with limestone or shells in sufficient quantities to produce a fluid slag. The flux should be put in on the first charge of scrap in very small cupolas and on the second or third charge in large cupolas, and on each charge throughout the heat afterward. The slag hole should be placed at the lowest point consistent with the amount of molten metal to be collected in the cupola at one time, and opened as soon as the first charge of scrap, upon which flux is placed, has melted. The slag hole may be opened and closed from time to time, but it is better not to make the hole too large, and leave it open throughout the heat. The flow of slag then regulates itself and there is no danger of it running into the tuyeres. In melting a few hundredweight of this scrap in a cupola, after melting a small heat of iron, it is not necessary to charge flux in sufficient quantities to produce a fluid slag to be tapped, unless the cupola is very small and shows signs of bunging up. In this case flux must be charged with the iron, and slag tapped early in the heat, to keep the cupola in condition to melt the scrap after the iron is melted.

When constructing a cupola expressly for melting tin-plate scrap the charging door or opening should be placed about 6 inches above the scaffold floor, so the scrap may be dumped in from a barrow and save handling it a second time with forks. The charging door should be much larger than in a cupola of the same diameter for melting iron and should be not less than 3 or 4 feet square in any case, and for cupolas of very large inside diameter the opening should be equal to one-half or three-fifths the diameter of the shell, and 4 or 5 feet high. The height of the door above the bottom depends upon the diameter of the cupola. In large cupolas it should be placed 18 or 20 feet above the bottom and in smaller cupolas as high as possible without danger of the stock hanging up in the cupola before settling into the melting zone. The lining material must be carefully selected, for a poor fire brick will not last at the melting zone through one long heat; in fact, none of the fire brick lasts very long at this point and it is generally necessary to put in a few new ones after each heat. High silicon brick is said to last better than any other brick, but some of the native stone linings which we have described last longer in melting this scrap than any of the fire brick, and they are generally used for lining cupolas for this work. The cost of melting tin-plate scrap in a cupola is from $1 to $2 per ton more than the cost of melting iron. The amount of profit in melting this scrap for weights, &c., depends, like all other foundry business, upon the location and size of the plant and the management of the business; but at the present time, even under favorable circumstances, the profits are small.

CHAPTER XIV.

COST OF MELTING.

THERE is probably less known about the actual cost of melting iron in cupolas for foundry work than about any other branch of the foundry business. But few foundrymen make any attempt at keeping a cupola or melting account. Many of those who do, keep it in such a way that they not only fail to learn the cost of melting, but are misled by the account to suppose their melting costs them a great deal less per ton than it really does. In the majority of foundries the melting is left entirely in the hands of the melter, who as a rule has no system for doing the work, and has no control over his assistants or interest in having them do a fair day's work. In many of the foundries we visit, twice the number of men are employed as cupolamen as are employed in melting the same amount of iron in other foundries, where the facilities for handling the stock are almost the same, and the expense of lining and daubing material is frequently double with one melter what it is with another in the same sized cupola with the same sized heats.

In many foundries the fuel is not weighed, but is measured in baskets, or the number of shovels counted and the weight estimated. When the fuel is measured in baskets, the baskets always stretch and enlarge, and an old basket frequently holds one-third more than a new one; from 10 to 20 pounds more can easily be piled on the top of a basket after it is filled. Foundrymen who charge their fuel by the basket always use more fuel than they estimate they are using; when the shovels are counted, each shovel may be made to weigh more than is estimated, and a few extra shovelfuls are always thrown in, for fear some were not full. When too much fuel is used in a cupola there is not only a wastage of fuel, but there is slow melting, increased destruction of the lining, and an increased wear and tear of the blast machinery. For these reasons every pound of fuel that goes into the cupola should be accurately weighed. Even when the fuel is supposed to be accurately weighed, there should be some check on the melter, for he will shovel in extra fuel if not watched.

At a foundry we recently visited in New Jersey an accurate account of the melting had been kept for a year; at the end of the year the president of the company had figured up the amount of fuel consumed in the cupola and compared it with the amount purchased, and found they were short 260 tons. At another foundry, where the melter always reported melting 7 pounds of iron to 1 pound of anthracite coal, they ran short 300 tons in a year. This kind of work should be prevented by checking up the melter's report and comparing it with each car-load of fuel consumed.

A cupola book should be provided, with blank spaces for recording the weight of coal or coke in the bed and charges, and the weight of each brand of iron, No. 1, 2 or 3 and scrap, showing the exact mixture of each charge and heat. A note should also be made of the quality of iron produced from the mixture. Such a record is of great value in making mixtures and charging a cupola, if it is properly kept.

The cost of melting per ton is figured in a number of different ways, but to be of any practical value the entire cost of melting should be figured on as follows:

Interest on cost of cupola plant and depreciation in value of same.

Fire brick for relining and repairs.

Fire clay, loam and sand for cupola and ladles.

Repairs to cupola, blast pipe, elevator, scaffold, runway, blower, &c.

Belts, oil, &c., for blower.

One-fourth the entire cost of engine.

Tools, wheelbarrows, buckets, hose, shovels, forks, rakes, hoes, sledges, picks, bars, trowels, bod sticks, tap bars, &c.

Wood for lighting up and drying ladles.

Coal or coke consumed in melting.

Labor employed in removing the dump, making up cupola, milling dump and gates, collecting gates, scrap and bad casting from foundry, placing iron and fuel on scaffold, charging, breaking and piling iron in yard, breaking up bad castings, daubing ladles, &c.

When the cost of all these items has been learned, and the amount divided by the number of tons melted, it will be found that the cost of melting is about $2 per net ton of iron in the ladles. In foundries with all the modern improvements for handling the stock the cost is a little less than $2 per ton, and in foundries with none of the improvements for handling the stock and no system in melting, the cost per ton is as high as $3. When there is doubt as to the accuracy of weights in charging, the weights should be compared with the fuel purchased and castings sold, and the cost of melting may be figured on the weight of castings sold in the place of the amount of iron melted. To make a cupola report of value, the fuel, labor and tool accounts should be kept separate, and an effort made to reduce the expense of each account.

CHAPTER XV.

EXAMPLES OF BAD MELTING.

MUCH has been written and published on melting by foundrymen and foundry foremen, who invariably give an account of rapid or economical melting done in their foundries; and it is seldom, if ever, that they publish accounts of poor melting or poor heats melted by bad management of their cupolas, or in their attempts to reach that perfection in melting of which they write. In giving points on melting for the benefit of others, it is as essential that causes of poor melting should be known that they may be avoided, as it is that those essential to good melting should be known that they may be practiced, and we therefore present a few instances of poor melting that have come under our observation in foundries we have visited, or in which we have been called upon to render assistance to overcome troubles in melting which were both annoying and expensive. In these instances we only give examples of what may occur in any foundry, and has occurred in many of them, where foundrymen are wholly dependent on their melters.

In 1878 we were engaged in making some experiments in melting with oil at the stove foundry of Perry & Co., Sing Sing, N. Y., at that time the largest stove works in the country. They were melting from 50 to 60 tons per day in four cupolas entirely with convict labor, and the results in melting were very unsatisfactory. Mr. Andrew Dickey, one of the firm and manager of the works, came to us one day after some very bad heats and asked us to take charge of their cupolas, set our own wages, and carry on our experiments at the same time. We took charge of their cupolas the following day and soon had their melting going along smoothly, but we did not like the job, and suggested to Mr. Dickey that we should teach a man to melt who could take our place when we were ready to leave, and this he consented to do. A man was selected who proved an apt scholar, and we soon had him instructed in all the details of melting, and when we left he took full charge of the cupolas.

Two years later we received a despatch from Perry & Co., stating that they wished to see us as soon as possible at their Sing Sing Works. Upon our arrival there late in the afternoon, Mr. Dickey informed us they were having trouble with all their cupolas, and it had been impossible of late to get a good heat out of any of them, and wished us to see what was the trouble. We found the same man in charge whom we had two years previously taught to melt, and inquired of him what the trouble was. He said he did not know, that he had fully followed our instructions and had no trouble in melting until within the last few weeks; during this time the cupolas had been melting very badly. He had increased and decreased the fuel in the bed and charges, increased it in one part of the heat and decreased it in another, varied the amount of iron on the bed and in the charges, but had been unable to locate the trouble. We asked him to describe how the cupolas melted, and he said they melted the first few tons, which was about the first two charges, fast and hot; after that the melting gradually grew slower until near the end of the heat, when melting almost ceased; the cupolas were so bunged up every heat that they could scarcely be dumped, and it was only after a great deal of labor with bars that a hole could be gotten through, so that they would cool off by the next morning. The iron was of an uneven temperature, frequently too dull for pouring and in some parts of the heat white hard, although nothing but soft iron had been charged. He thought the trouble must be in the blast—that old "no blast" story that foundrymen hear so often, when melters do not know how to manage a cupola and have to lay the blame on something. We informed him that the trouble could not be

EXAMPLES OF BAD MELTING 235

Fig. 50.

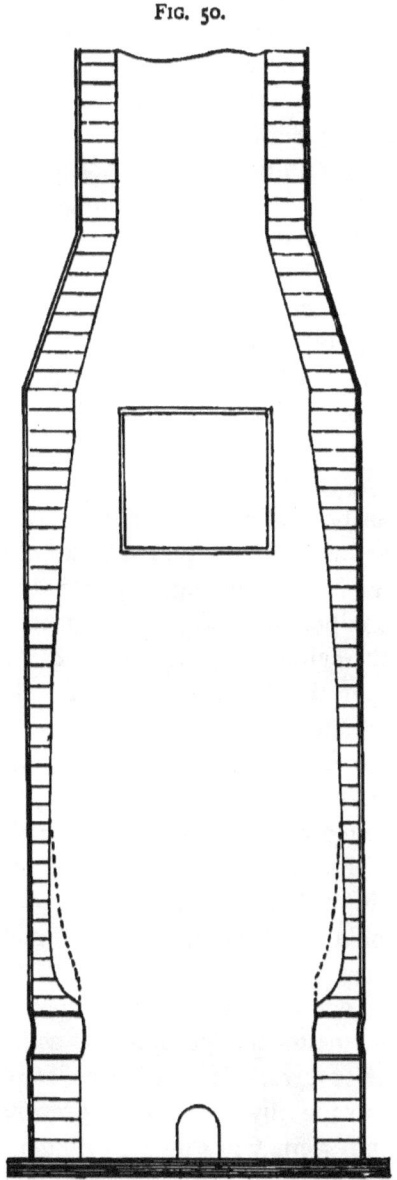

SECTIONAL VIEW LINING OUT OF SHAPE. NO. 1.

in the blast, or the cupolas would not have melted the first two charges fast and hot; that the trouble was the stock logged in, settled or settled unevenly after melting the first two charges, which was the cause of the uneven melting in the latter part of the heat, and he must have permitted the linings to get into a shape that produced this condition in the cupolas. He did not think this possible, for he had followed our directions for shaping a lining, but admitted that he frequently found pieces of unmelted pig and scrap in the cinder above the tuyeres when chipping out, which confirmed our theory, and we looked no further for the cause of poor melting.

The following morning the cupolas were almost closed up with cinder slag and iron, and after a great deal of labor in breaking down and chipping out we found the linings in the shapes shown in Figs. 50 to 53.

Cupola No. 1 had not been lined for a long time, and the lining was burned away until it was very thin all the way up. This did not prevent the cupola melting, but should have made it melt faster; for as a cupola is enlarged in diameter by burning out of the lining its melting capacity increases; but in this case the melter had permitted the lining to become hollow around the cupola just above the tuyeres. When the stock settled, that on the outer edges logged in this hollow, became chilled and threw the blast to the centre of the cupola. After a few tons had been melted the chilled stock over the tuyeres increased rapidly until the melting was restricted to an opening in the centre, which gradually closed up with the fan blast, and the longer the cupola was run the slower it melted, until melting ceased altogether.

In No. 2 the lining was not burned away to so great an extent as in No. 1, but the melter had permitted it as in No. 1 to become hollow over the tuyeres. He had been troubled with molten iron running into the tuyeres, and to prevent it doing so had built the lining out from 3 to 4 inches with daubing over each tuyere. This cupola like the others was 60 inches in diameter with six oval tuyeres each 4 by 12 inches laid flat. Over

EXAMPLES OF BAD MELTING.

FIG. 51.

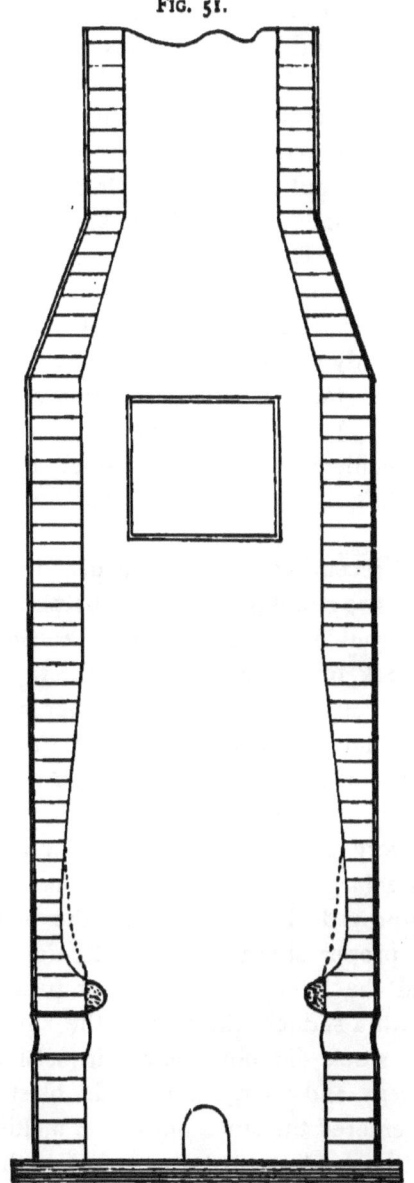

SECTIONAL VIEW LINING OUT OF SHAPE. NO. 2.

each of these tuyeres was a projecting hump 3 to 4 inches thick and 16 to 18 inches long; add to the thickness of these humps a hollow in the lining of 4 to 6 inches and a shelf from 8 to 10 inches wide was formed over each tuyere upon which the stock could not help lodging, and could not be melted after lodging. When the cupola was first put in blast it melted very well, but after the stock began to lodge gradually, melted more slowly until it finally bunged up. The convict who had charge of this cupola informed me that every day, when chipping out, he found pieces of pig iron and unburned coke lodged over the tuyeres, and molten iron frequently ran into the tuyeres when melting. To prevent this, he had gradually built the lining out over the tuyeres (from day to day), until the shape we have described was reached; but it neither prevented the stock lodging nor the molten iron flowing into the tuyeres, but increased the trouble.

No. 3 (Fig. 52) had recently been newly lined, and melted differently from the other two cupolas. It was in a better shape over the tuyeres, and the trouble in melting was not caused by the hanging up of the stock from lodgment over the tuyeres, but by the escape of blast around the lining. The cupola had been lined with 9 inch brick and its diameter greatly reduced by the heavy lining, and as a result the cupola melted more slowly than with the old lining. To make it melt faster, the melter had chipped it out very close every day and permitted the lining to burn out to enlarge the cupola at the melting point. This would have improved the melting had the belly in the lining been given a proper shape; but no attempt had been made to shape it, and the lining was burnt out to a depth of from 4 to 6 inches with a sudden offset from the small to the large diameter. The stock did not expand in settling to fill this sudden enlargement, and a large part of the blast escaped into the belly and re-entered the stock above the melting zone. This naturally threw the heat against the lining at the top of the belly and cut it out very rapidly, and would have ruined the lining in a week's time had the cupola been permitted to

EXAMPLES OF BAD MELTING.

FIG. 52.

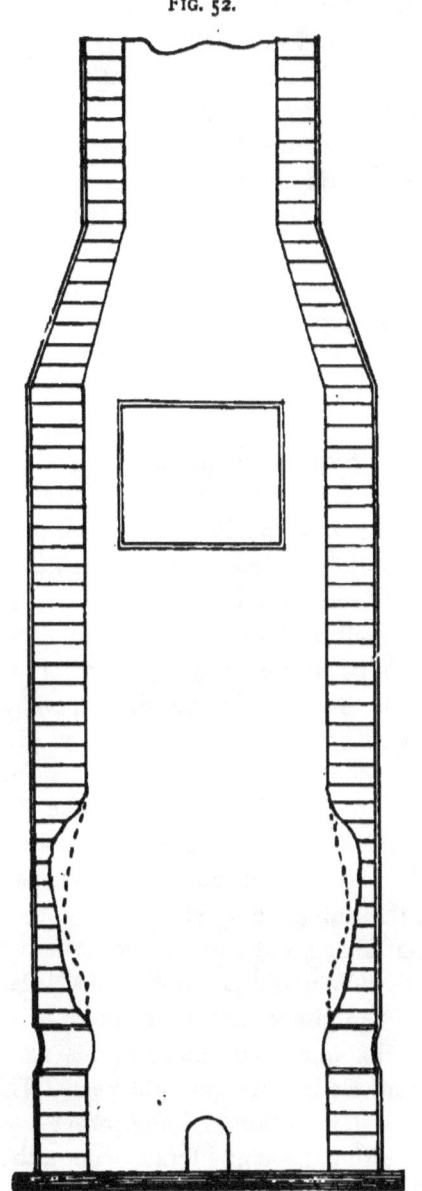

SECTIONAL VIEW LINING OUT OF SHAPE. NO. 3.

continue to work in this way. The belly in the lining was filled with stock when charging, and the melting was very good until the stock settled and the blast began to escape in the manner described, when it rapidly grew slower until it stopped altogether, and this cupola which had been relined to make it melt better was the poorest melting one of the lot.

In Fig. 53 the lining had been permitted to belly out over the tuyeres at a very low point and a shelf formed, upon which the stock lodged by building the lining out over the tuyeres, but the humps over the tuyeres were not so long as those in Fig. 51, and the stock had settled between the tuyeres to a greater extent than over them. This uneven settling of the stock had thrown the heat against the lining at different points and burnt it out in holes all the way up to the charging door.

Here were four cupolas, all of the same diameter, having the same number of tuyeres, with the lining of each one in a different shape, but all having the same objectionable feature—a hollow in the lining over the tuyeres, which was the real cause of bad melting. We had all the humps over the tuyeres chipped off and the linings daubed up perfectly straight for six inches above the top of the tuyeres, all around the cupola, and filled in the lining above with split brick and daubing, giving each cupola the shape indicated by the dotted lines. The cupolas were then charged as they were before the trouble began, and each one melted hot, even iron, throughout the heat and dumped clean. As soon as the man we had taught to melt saw us shape up a small section of the lining, he said: "Why, you told me to keep the linings in that shape and showed me how to do it two years ago." We said: "Why did you not do it?" He said he had forgotten it, and when the cupolas began to work badly, did not know what to do, and in fact had lost his head and let every melter under him do as they thought best. This is frequently the case with good melters. They forget points that they have learned in melting, have no literature upon the subject from which to refresh their memories, or melters to consult who are competent to advise, and gradually drift into a routine of

EXAMPLES OF BAD MELTING.

Fig. 53.

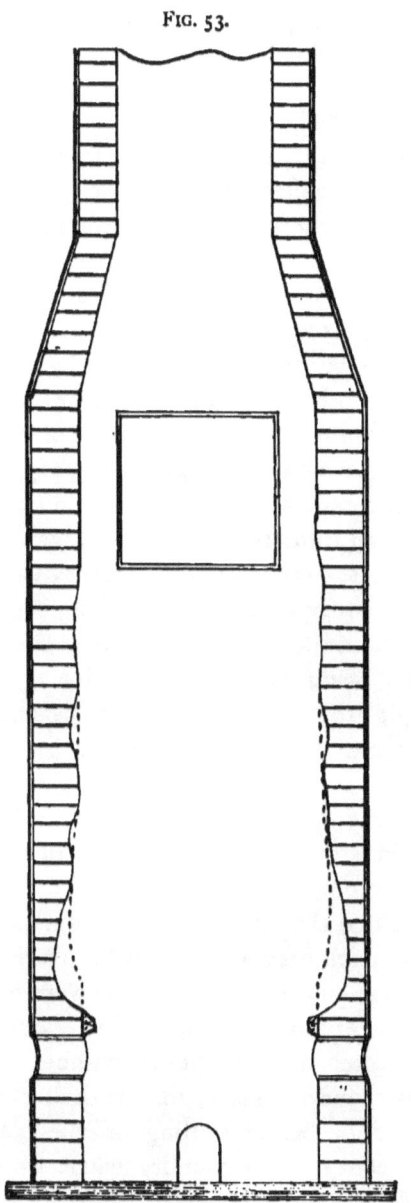

SECTIONAL VIEW OF LINING OUT OF SHAPE. NO. 4.

work, and when anything goes wrong with the melting do not know how to overcome the difficulty.

BAD MELTING AT A WEST TROY STOVE WORKS.

In 1882, we visited the foundry of Daniel E. Paris & Co., West Troy, N. Y., and while waiting for Mr. Paris, looked over the cupola. We found the lining in a condition indicating very poor melting and knew they were having some trouble with their iron. When Mr. Paris returned and learned who we were, he informed us that their foundry had recently burned down and they had moved into the present one, which had for some time before been idle. The boiler and engine were small and they were having some trouble in melting for want of power to drive a Sturtevant blower, which when run at a proper speed was large enough for the cupola. They were also endeavoring to melt up a lot of scrap from their recent fire, and had also procured some of the best brands of No. 1 Pennsylvania irons and Scotch pig to melt with it, but were having some hard castings. He wished to know if we could suggest anything to help them out until they could put in a new engine and boiler, and find some softer pig iron to work up the scraps, and he took us out to look over the works to see what change could be suggested.

We looked over the blower and machinery, which were only those employed in stove mounting, and then went into the engine and boiler room, where we found a good-sized engine and boiler and decided that they were large enough to run the blower and all the machinery in the works at the same time. The engineer at once informed us that they were too small and he could not run any of the mounting machinery when the blast was on, or pump water into the boiler, without reducing the speed of the blower, and he had to fill the boiler and stop the engine for half an hour before putting on the blast to get up steam. We then went into the foundry, where we found a well arranged eupola of fifty-four inches diameter inside the lining, and learned that they were melting about eight tons of iron

EXAMPLES OF BAD MELTING. 243

each heat; that from four to four and a half hours were required to run off the heat, and they were melting seven pounds of iron to the pound of anthracite coal. The iron melted so slowly that it was difficult to catch four hand-ladles full to pour off a *four up* before the first ladle-full was too dull to run the work, and the iron was sometimes so hard that the plate cracked when taken out of the sand or when knocking off the gates.

We then went upon the scaffold, where we found the coal when charged was not weighed, but measured in a basket and dumped from the basket into the cupola. We afterwards weighed a basket of coal filled as the melter generally filled it, and found it weighed almost twice as much as the melter stated, and with the extra weight of coal in the basket and the extra shovelfuls the melter said he threw in to fill up holes, we concluded that they were melting about three pounds of iron with one of coal, in place of seven to one as claimed by the melter. No slate was used in charging, sprews and gates were not weighed, but the weight estimated by counting the shovelfuls, and pig was weighed by counting the pieces, estimating four pieces to the hundred weight. The greater part of the coal, when dumped into the cupola from the basket, fell directly under the charging door, where it remained; and the greater part of the iron naturally went to the opposite side of the cupola, and this uneven charging naturally produced uneven melting.

We pointed out to Mr. Paris that his cupola lining was not glazed in front of the charging door, but was rough and jagged, as linings generally are in cupola stacks, which is an indication that too great a quantity of fuel is being consumed in melting and that by using less coal better melting would be done. He thought seven to one was very good melting and knew none of the foundries in Troy were doing any better, and did not think iron could be melted sufficiently hot for their work with a greater ratio of iron to fuel than was being consumed in their cupola. But he was getting very poor results in melting, and after considerable talk he concluded to let us try a heat the following day with less fuel.

The following morning when we went round to have the cupola prepared for a heat, we found the matter of less fuel had been talked over by the entire foundry force and by them condemned. They argued that dull iron had been melted with the quantity of fuel used, and could not be poured at all if less fuel were used. It is a curious fact that moulders working piece work and losing work every day from dull iron will object to a stranger, or any man whatever but the melter making any change in the management of the cupola, or as they term it experimenting with the cupola. While getting the cupola ready for a heat, the moulders came to us at the cupola or in the yard, one after another, and asked us all kinds of questions about melting, and Mr. Paris came also and asked us if we were sure we could melt iron hot enough for their work with less fuel than they were using, also if we had ever done so before; and we found that we would have to be very careful what we said or did, or we would not be permitted to run off a heat.

The melter was an old hand, who had melted iron in a number of the foundries in Troy and was considered good. He was very much opposed to having us do any better melting in the cupola than he had done without a new engine and boiler, which he declared must be put in before anything better could be done. He knew all about it, and to teach this man to melt with less fuel would only be a waste of time, for he would probably in less than a week drift back into the same old rut if not closely watched, and would condemn our way of managing a cupola. So we told Mr. Paris we could teach his foreman to melt in a few days so that he could oversee the work and teach a man to do it in case his melter was sick or quit, and that it would be much better for them than for us to show their melter how to work with less fuel. After consulting the foreman it was decided that we should teach the foreman, and he went on the scaffold with us. He had the cupola made

EXAMPLES OF BAD MELTING.

up as we directed, sent to a store and purchased a new slate and arranged a system of mixing and charging the iron so that it would produce an even grade when melted, having had the scales dug out of a pile of rubbish in a corner and cleaned up, and the iron and fuel placed conveniently for charging.

After everything had been arranged for the heat we had a little time to spare, and made it a point to see some of the leading moulders and explained to them that we had shaped the lining so that the cupola would melt faster and with a little less fuel than they had been using, and make hot iron. We also saw the engineer and informed him that we would charge the cupola in a way that it would demand less blast, and if he filled his boiler and had a good head of steam on just before putting on the blast, he could run all the machinery required for mounting when the blast was on. These explanations seemed to satisfy everybody, and the foreman was so enthusiastic in learning to melt that we had no further fear of being run out of the works, and were looked upon as the man who understood his business until the heat was all charged into the cupola, when the melter went into the foundry and said to the moulders: "Be jabers yees will not pour off to-day boys, for that cupola will not make hot enough iron for yees with all the coal I was after putting in, and that man has left out half of the coal I put up for the heat. Yees may as well go home and save your moulds for to-morrow's heat; for yees will not run your work to-day."

From that time until the blast went on, we were looked at shyly by all the moulders except two, who had seen us melt in other foundries; but the foreman and these two assured them that we understood our business and they would have a good heat, which probably saved us from being driven out, for there was a tough lot of stove-moulders in Troy in those days, who considered their rights sacred and that no punishment was too great for any man who encroached upon them.

When the blast was put on, the moulders gathered round the

cupola and watched every tap until the iron came down so hot and fast that the first turn could not handle it, and the second turn was called up, and they were all kept on the run until the end of the heat. Getting iron so fast and hot was something the moulders had never been used to in that foundry, and a number of them wished to know if we were trying to kill them all by giving them the iron so fast. But all were delighted with getting hot iron to pour off their work and getting through so early; and as we went along the gangways to see how the castings were turning out, a number of them asked us to wait until they were shaken out and have a glass of ale with them, which was the great drink of the Troy moulders. Had we waited for them we probably would not have reached our hotel that evening, for almost all of them dropped into a nearby saloon after they were through with their day's work, and we should have been asked to drink with every one of them.

In this heat we had used considerably more coal than we considered necessary, as we were not familiar with the working of the cupola and desired to be on the safe side and make hot iron, even though the melting was a little slow, which was the case. Two hours were required for the heat, but even this length of time was fully two hours better than they had been doing, and all the machinery required for mounting was run during the heat without stopping the engine for half an hour to get up steam before putting on the blast.

On the following day we reduced the coal a little more, and on the third day reduced it until we were melting six and a half pounds of iron to one of coal, and the heat was melted in one hour and thirty minutes. This was as fast as the moulders could handle the iron; and as we did not consider it safe to melt iron for stove plate with less fuel, although we could have done so, and they did not desire it melted any faster, we made no further attempt to save fuel or reduce time of melting.

The foreman learned very rapidly, and at the end of three days was fully competent to oversee the work, and they had no further trouble in melting or with hard iron, and were able to

EXAMPLES OF BAD MELTING.

melt up all the scrap from their recent fire with the brands of pig iron they had on hand, and it was not found necessary to put in a larger engine and boiler to get a sufficient blast, after they had learned how to manage the cupola.

The cause of bad melting in this foundry was plainly indicated to an experienced melter at first glance by the lining in front of and around the charging door, namely, too great a quantity of fuel in the cupola and too small a volume of blast for that fuel. So large a quantity of fuel was charged for a bed that the iron placed upon it did not come within the melting zone, and could not be melted until the surplus fuel burned away and permitted it to settle into the zone. Each charge of fuel to replenish the bed was too heavy, and the greater part of it had to be consumed before the iron placed upon it was permitted to enter the melting zone, and the slow melting was due to the time required in consuming the surplus fuel before the melting could take place. The hard iron in parts of the heat was due to uneven charging, which permitted the scrap at times to be melted by itself and drawn from the cupola without being mixed with melted pig, and the entire mass of iron was hardened by being subjected for a long time to a high degree of heat before it was permitted to enter the melting zone and be melted.

The speed of the blower had been increased to fully double the number of revolutions per minute given in the directions for running it, to increase the volume of blast; but the volume of blast had been decreased in place of being increased, as was supposed it had been by the increase of speed, and the cupola received less blast.

We had no means of definitely determining to what extent it was decreased, but from the appearance of the blast in the cupola at different stages of the heat, before and after decreasing the speed of the blower, we concluded that the volume of blast was increased fully one-half, by running the engine at its normal speed and reducing the speed of the blower to the number of revolutions given in the directions for running it.

This is one of the cases where the cupola air-gauge in common use would have been of value, for it would have indicated a high pressure of blast before the speed of the engine was increased, and located the trouble at the cupola in place of at the engine.

WARMING UP A CUPOLA.

In 1881 we visited the plant of the Providence Locomotive Works, Providence, R. I. The superintendent, Mr. Durgon, we believe was his name, wished to know if we were the Kirk that wrote "The Founding of Metals." We informed him that we were, and he replied that we might know all about a cupola, but our directions there given for constructing a cupola were no good, for he had constructed a cupola on that plan and it was a complete failure. It would not make hot iron, or melt half the amount per hour stated, or melt the heat before bridging over and bunging up. We informed him that if he had constructed the cupola exactly on the plan given it would do the work stated it would do. He invited us to go into the foundry and look the cupola over, and if it was not right he would make it right. We accepted the invitation and looked the cupola, blower and pipes all over, and could find no fault with them. The cupola was in blast at the time and we watched it melt for an hour, and it certainly was a complete failure. The iron from the beginning to the end of the heat was dull, the melting slow, and the castings dirty and much harder than they should have been with the quality of iron melted.

We knew that the trouble lay in the management of the cupola, and decided to go round the next day and see the melter make it up for a heat. This the superintendent decided to let us do, although he thought he had the best melter in New England and the trouble could not be in the management of the cupola. On the following day we were on hand early and found the cupola badly bridged and bunged up. The melter soon had it chipped out and daubed up in good shape, and we saw that the trouble was not in the shape of the lining. He then put in a very nice sand bottom from which

EXAMPLES OF BAD MELTING.   249

there could be no trouble in melting. He next put in shavings and a large quantity of wood, which he burned to dry the daubing. After this had been dried he added more wood and a good bed of hard coal which he burned up to warm the cupola for melting, and he certainly did give it a good warming, for when the doors were opened for charging the lining was heated to a white heat from the bottom to the stack. He then added a little more coal to level up the bed, and began charging.

As soon as we saw the extent to which the lining had been heated and the bed burned, we knew that the cause of the poor melting lay in the bed. In warming the cupola up for melting, the life had all been burned out of the coal and but little of it left to melt with. The cupola was filled with ashes below the tuyeres, and even if iron was melted hot it would be chilled in its descent through these ashes to the bottom of the cupola. The fuel thrown in just before charging was flaked off, broken and burned up by the intense heat almost before the iron could be charged, and had it not been that an extra high bed was put in before warming up, not a pound of iron would have been melted.

We had frequently seen beds burned too much, but had never seen one burned to the extent of this one, or a cupola heated so hot before charging, and we stayed on the scaffold during the filling of the cupola with stock to see if the intense heat in the cupola had any effect upon the stock that would improve the melting in any way. The first charge seemed to be heated to a considerable extent by the hot lining and bed, and prepared for melting. After this charge was put in, the cupola cooled off very rapidly, and before it was filled there was scarcely any perceptible heat at the charging door, and the stock could not have been heated to any extent above the first or second charge, by warming of the cupola. When the cupola had been filled the blast was put on, and the iron melted exactly as we had seen it do the day before, dull and slow. The cupola had been properly made up; plenty of fuel had been put in to make hot iron; charges of fuel and iron were of about the right proportion, and had been properly placed in charging, and there could be no doubt that the trouble in melting lay in the bed, as before stated.

The following day the superintendant put the melter on the other cupola and gave us full charge of the one constructed on our plan. We had it made up in about the same way as the melter did; put in our shavings, wood and all the bed, but a few shovelfuls to level up with before lighting up. After lighting up we waited until the heavy smoke was burned off and the fire began to show through the top of the bed. We then leveled up the bed and began charging. The only change we made in charging was to reduce the fuel in the bed about one fourth, and that in the charges a little. When the blast was put on iron came down in about ten minutes, melted fast and hot throughout the heat, and the same amount of iron was melted in one half the time it had been the previous day. This convinced the superintendant that the cupola was all right, for it did all we claimed it would do and a little more, and it convinced us that there was nothing to be gained in melting by warming up a cupola before charging.

BAD MELTING, CAUSED BY WOOD AND COAL.

In one of the leading novelty foundries in Philadelphia that we visited some years ago they were employing two cupolas, one 40 inches and the other 30 inches inside diameter, to melt 8 tons of iron, and it was very difficult to melt that amount in these cupolas. We knew that something was wrong and went upon the scaffold to look into the cupolas and found the melter just putting in the wood for lighting up, He had put in quite a lot of finely split wood, and had another barrow ready to add. After this was in, he went down and got three more barrows of cord-wood sawed in two and added this and then some long wood, and when he had it all in, the cupola was filled to the bottom of the charging door. He then filled the cupola with coal to the top of the charging door, putting in the largest lumps he could find. We asked him why he put in so large a quantity of wood, and he said it was necessary to light the coal; and we presume it was, for some of the pieces of coal were as large as he could lift and place in the cupola, and it would require considerable heat to start a fire with such large coal; and he said they could not melt with any smaller coal. We tried to convince him that the cupola would melt better with less wood and smaller coal, but this was impossible, for he was an old melter and knew all about it.

Either one of these cupolas would have melted the amount of iron they were getting in the two, and in less time, had they been properly managed; but this was not done and the firm afterwards put in two Colliau cupolas to do the work. The cause of poor melting in these cupolas was too great a quantity of hard wood, which took a long time to burn out and in burning out the bed was burned to so great an extent that the cupola was filled with wood ashes and coal ashes before melting began. The large lumps of coal also contributed to the poor melting by making an open fire through which the blast escaped freely without producing a hot fire, such as would have been produced by smaller coal.

POOR MELTING IN A CINCINNATI CUPOLA.

In Fig. 54 is seen a sectional elevation showing the condition of a small cupola we saw in Cincinnati, Ohio, a few years ago. This cupola would not melt, the founder said, and could not be made to melt. He had put in a new fan, and now his melter wanted a blower, and said the cupola would not melt without a forced blast. We examined the cupola, and suggested to the founder that he needed a new melter worse than a new blower.

The cupola had not for a long time been properly chipped out, and a belt of cinder and slag varying in thickness from four to six inches had been permitted to adhere to the lining around the cupola above the melting point, and another belt of cinder and slag projected from the lining. Between these two projecting belts the lining had burned away, making a deep hollow at

252   THE CUPOLA FURNACE.

the melting point. Entirely too much fuel had been consumed

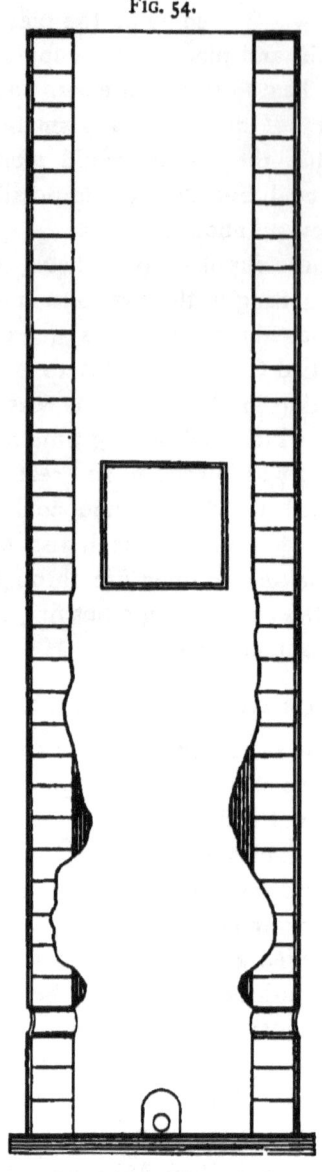

FIG. 54.

ILLUSTRATION OF BAD MELTING.

EXAMPLES OF BAD MELTING. 253

in melting or the belt of cinder and slag could not have formed above the melting point.

We had all the projecting humps chipped off and the hollows filled in with fire-brick and daubing, so as to give the lining an even taper. The cupola was then properly charged, and there was no trouble in melting iron hot and fast.

UNEVEN BURNING OF THE BED.

We were once compelled to dump a cupola at the foundry of Perry & Co., from the carelessness of the melter in placing the shavings and wood in the cupola in such a way that they did not light up the fuel evenly, and in putting on the blast when the bed was only burned up on one side. We had not noticed it, and he thought the blast would make it burn up on the other side. This it did not do, and after the cupola had been in blast a short time, it had to be dumped.

The careless way in which shavings and wood are often thrown into a cupola from the charging door, frequently causes an uneven burning of the bed and bad melting. We had a number of poor heats in our own foundry, due to this kind of carelessness, before discovering the cause of them.

We might relate many more examples of poor melting in various foundries, but these will probably suffice, as the causes of poor melting when a cupola is properly constructed will generally be found in the shape of the lining, burning of the bed, or quantity of fuel used in melting; examples of which are here given.

CHAPTER XVI.

MELTERS.

THERE is no man about a foundry for whom we have more respect than a practical and scientific melter. He is generally a self-made man and has learned the art of melting himself. He is a man of intelligence, who, perhaps, has been a melter's helper and a close observer of the work, and when given charge of a cupola, has followed in his footsteps or improved on the methods of his predecessors. He may have been a man who was given a few instructions in melting when he first began, and has become an expert through his own efforts. He is respected by the foreman and moulders, and well-paid by his employer. There is no man about a foundry for whom we have more pity than a poor melter, for he seldom melts two heats alike, and is cursed by the piece moulders who have lost their work through bad iron. Gibed by the day moulders, lectured by the foreman, looked black at by his employer, poorly paid, and respected by no one about the foundry, his lot is a hard one.

A poor melter is not always to blame for doing poor work, for he may have been a foundry laborer who was put to work as a melter, and never given proper instruction in the management of a cupola. Again, a good melter may be made a poor one from being interfered with by others who do not understand melting. Foundrymen in conversing with each other learn that they are melting ten pounds of iron to the pound of fuel. The foundryman not being a practical man, does not inquire the size of the heat or cupola in which it is melted, the conditions under which it is melted, or the kind of work the iron is for. He does not stop to think that the other foundryman may be lying to him, or is deceived by his melter and does not know how many pounds of iron he is melting to the pound of fuel. But he goes to his foundry and insists that iron must be melted at a ratio of 10 to 1. The conditions in his foundry may be totally different from those of the other one, and iron may not be melted at a ratio of 10 to 1 in the other foundry. The melter, if he is a practical man, knows this, or finds it out the first heat, and to hold his job shovels in extra fuel, unbeknown to any one, and if he is watched, does not get it in evenly or at the proper time, and the result is uneven melting and dull iron. Foundrymen do not always furnish their melters with proper tools for chipping out and making up the cupola, a suitable material for repairing and keeping up the lining, a proper flux for glazing the lining and making the cupola melt and chip out free, and a man who would be a good melter if given a chance, is frequently made a poor one by being hampered in his work for want of tools and material to work with. He is blamed for poor melting when it is really not his fault. Good melters frequently get into a rut or certain way of doing their work, for want of text-books and other literature on melting to read and study, or association with men of their calling, and become very poor melters. As a lawyer who does not read law-books that are up to the times and associate with his colleagues, becomes a pettifogger, so does a doctor who does not study his text-books and medical literature, diagnoses all cases as one of two or three diseases, has one or two prescriptions which he prescribes for all cases. The man of learning, or a man who knows it all, when left to himself for years gets to know nothing; and so it is with melters when left to themselves. They forget many things they are not called upon to practice every day, and in time get into a rut or routine from which they unconsciously gradually degenerate if the mind is not refreshed by reading or contact with other melters. It should be the aim of every melter to converse with other melters upon cupola matters at every opportunity, and to read and study all literature upon the subject, whether good or bad; for, if good, he may learn something new, and, if bad, it stimulates the mind to reason why it is not good, and how it can be improved upon. It recalls to mind facts in his own experience which have long been forgotten, and he learns something, at all events. It is to the interest of every foundryman who depends upon his melter for results to keep him posted upon all that is new in the business, and he should furnish him all the new literature on the subject that comes into his office or is published.

CHAPTER XVII.

EXPLOSION OF MOLTEN IRON.

MOLTEN iron is a very explosive body, and under certain conditions explodes with as loud a report and as much violence as gunpowder. Under other conditions it is not at all explosive, but the conditions under which it explodes must be fully understood and avoided by melters and moulders to prevent dangerous accidents.

A stream of iron flows from a tap-hole and spout smoothly if the front and spout lining have been properly dried. When wet the iron explodes as it emerges from the tap hole and is thrown in small particles some distance from the cupola. The instant a stream of iron strikes a wet spout it explodes and the entire stream is thrown from the spout in all directions with great force. In a damp spout the iron boils and small particles may be thrown off, but the explosion is not so violent as from a wet spout.

A wet bod causes molten iron to explode the instant it comes in contact with the stream, and it is impossible to close a tap hole with it. A bod containing a little too much moisture causes a less violent explosion and a tap hole may be closed with it, but in closing it, the iron explodes and is frequently thrown from the tap hole with great force past the sides of the bod before it is pressed into the hole. When the bod is in place in the hole one or more small explosions frequently take place, and the bod-stick must be firmly held against the bod to prevent it being blown out. The kick or thump felt against the end of a bod-stick when pressing a bod into place is due to these explosions, and not to the pressure of molten iron in the cupola, as is generally supposed. Bod material should be no wetter than moulding sand properly tempered for moulding.

When the iron is very hard, a stream of very hot iron throws off a great many sparks from a dry spout. These sparks are caused by an explosion of the iron due to the combination of oxygen with the combined carbon of the iron, and the sparks are the oxide of iron. They contain very little heat, and melters or moulders do not hesitate to enter showers of these sparks to stop in or catch the stream of iron. The sparks from explosions caused by dampness are of an entirely different character, and burn the flesh or clothing wherever they strike.

A wet, cold or rusted tapping bar thrust into a stream of iron in the tap hole or spout, causes the iron to explode. Tap bars should, therefore, always be heated before they are put into the stream of iron.

When iron falls from a spout upon a hard floor, it spatters and flies in small particles to a considerable distance from the place it first strikes, and it is dangerous to go near the spout as long as the stream is falling upon the floor.

When iron falls from a spout upon a wet, muddy floor, it explodes instantly, and small particles of molten iron may thus be thrown a hundred feet from the cupola. If the stream continues to run upon the floor, one explosion follows another in rapid succession, or a pool of molten iron is formed, which boils and explodes every few minutes, as long as there is any moisture in the floor and the iron remains liquid. The floor under a spout should always be made of loose dry sand, with a hole in it to catch any iron that falls from the spout.

The floor under a cupola should always be dry, and when paved with brick or stone, should be covered with an inch or two of dry sand before dumping, to prevent fluid iron or slag in the bottom of the cupola spattering or exploding when dumped.

Molten iron explodes violently when a piece of cold, wet or rusted iron is thrust suddenly into it, as the writer has reason to know from practical experience, when working at stove moulding in the winter of 1866 and 1867. Knowing that a rusty or wet skimmer made iron explode, we always took the precaution of putting our skimmers into the foundry heating stove and heating them to a red heat before catching iron. One day we had taken the precaution, heating a skimmer to a red heat and putting it in a convenient place for use. A small boy who was around the foundry and sometimes skimmed our iron before pouring, saw the red-hot skimmer, and took it out and put it in the snow, while we were catching a ladle of iron. As soon as we set the ladle on the floor he ran in with the skimmer dripping wet, and before we could prevent him, thrust it into the molten iron. The iron exploded instantly and was thrown all over us as we leaned over the ladle, burning us so severely that we were not able to be out of the house for several weeks, and we still carry many scars from those burns. The iron was thrown with great violence, and passed through our clothing and a thick felt hat, like shot from a gun. The exploded iron passed over the boy's head and he was burned slightly, but never was seen about the foundry again, and probably never became an iron moulder.

Molten iron when poured into a damp or rusted chill-mould or a wet sand-mould, explodes and is thrown from the mould, and escaping from a mould upon a wet floor or into the bottom of a wet pit, explodes. In the foundry of Wm. McGilvery & Company, Sharon, Pa., a deep pit for casting rolls on end was put in the foundry floor and lined with boiler plate. The first roll cast in this pit was one eleven feet long, weighing about five tons, moulded in a flask constructed in ring sections and clamped together. The mould was not properly made and clamped, and when almost filled with molten iron gave way near the bottom and permitted the iron to escape into the pit, the bottom of which was covered with wet sand or mud. The iron at once exploded and forced its way up through ten feet of sand that had been rammed about the mould in the pit, and was thrown up to the foundry roof at a height of forty feet. The molten iron continued to explode until fully four tons were thrown from the pit in small particles, and the foundry burned to the ground.

Molten iron explodes when poured into mud or brought in contact with wet rusted scrap, but does not explode when poured into deep or clean water. At a small foundry that stood near the Pittsburg & Erie canal, in Sharon, Pa., many years ago a wager was made by two moulders that molten iron could not be poured into the water of the canal without exploding. A ladle of iron was accordingly taken to the canal and poured into the water without any explosion taking place. A few days later an apprentice boy who had witnessed this experiment undertook to pour some into water in an old salt kettle that sat in the yard near the foundry and contained rusted scrap and mud under the water. An explosion at once took place that almost wrecked the foundry. The water in this case was not of sufficient depth to destroy the explosive property of the molten metal before it came in contact with the rusted scrap and mud at the bottom of the kettle.

Moulders frequently pour the little iron they have left over, after pouring off their day's work, into a bucket of water to heat the water for washing in cold weather. This was a common practice of the moulders in the foundry of James Marsh, Lewisburg, Pa., until one day iron was poured into a bucket of water in which clay wash had been mixed and contained mud at the bottom. It exploded instantly with so great a violence that all the windows were blown out of the foundry, and this stopped the heating of water for washing, in that way, at that foundry.

At another foundry, iron poured into clear water in a rusted cast-iron pot exploded, doing great damage.

At the foundry of North Bros., Philadelphia, Pa., during the flood in the Schuylkill river June, 1895, the cupola was prepared for a heat and the blast put on; but before the heat could be poured off water soaked into the cupola pit and had to be bailed out to prevent the pit being filled. The heat was all poured before water came upon the moulding floors, but the bottom of the cupola pit was soaking wet, and the melter, in his eagerness to leave the foundry before it was flooded, dropped the bottom without drawing off the molten iron remaining in the cupola. The instant the molten iron and slag dumped from the cupola came in contact with the wet floor of the pit, a violent explosion took place, scattering molten iron, slag and fuel in all directions and blowing all the windows out of the foundry. Had the melter taken the precaution to have drawn off all the molten iron before dumping, and thrown a few shovelfuls of dry sand under the cupola to receive the first slag to fall upon the bottom, this explosion would not have taken place.

At the foundry of The Skinner Engine Co., Erie, Pa., a violent explosion took place in their cupola which almost entirely wrecked it. At the time of this explosion, a lot of small steam cylinders were being melted in the cupola, and in some of these cylinders the ports of the steam-chest had been closed by rust, leaving the steam-chest filled with water, from which it could not escape. The foreman, David Smith, had given the melter orders to see that each of these cylinders was broken before being put into the cupola, but this order had by the melter been disregarded, and the explosion was attributed to the water confined in one of the cylinders being converted into steam and exploding with such violence as to wreck the cupola.

At the foundry of The Buffalo School Furniture Co., Buffalo, N. Y., an explosion took place in 1895 in their sixty-inch cupola, about seven minutes after the blast was put on for a heat, which blew the heavy cast iron door from the tuyere box, on each side of the cupola; and also blew out the front and broke the heavy cast-iron bottom doors. A number of men who chanced to be near the cupola were severely burned, but fortunately none were killed. This explosion was attributed to a number of causes, one which was the formation of gas in the cupola before the blast was put on, which was exploded by the addition of oxygen from the blast. But this could hardly have been the cause, for the blast had been on fully seven minutes before the explosion occurred, and had this been the cause the explosion would have taken place almost as soon as the blast was put on. Another cause given for the explosion was that dynamite had been placed in the cupola concealed in some pieces of scrap-iron. This may have been the case, or some other explosive body may have been concealed in the scrap; but it is just as probable that it was due to steam generated from water confined in some piece of the scrap, by rusting of the opening through which it was admitted to the casting; as in the case at the foundry of The Skinner Engine Co.

A damp ladle causes iron to boil, and if the daubing is very thick may cause it to explode. A wet daubing or water in a ladle explodes the iron the instant it touches it. Wet or rusted scrap iron placed in a ladle to chill the molten iron, causes the iron if tapped upon it, or if thrown into a ladle of iron, to explode. Such an explosion may be prevented by heating the scrap to a red heat just before using it to chill the iron.

CHAPTER XVIII.

SPARK CATCHING DEVICES FOR CUPOLAS.

FOUNDRYMEN, whose plants are located in closely built up neighborhoods, are very much annoyed by sparks thrown out of their cupolas lighting upon the roofs of adjoining buildings and setting them on fire. In some cases they have on this account been compelled to move their plants from towns and cities to the suburbs. Many plans have been devised and tried for arresting these sparks; one of the oldest and most efficient of which is the design shown in Fig. 55. This arrangement was devised when the old-fashioned cupolas with brick stack were in vogue, and was generally put up in such cases where cupola sparks were very objectionable. It consisted in constructing the stack upon an iron plate supported by iron columns, on a level with the top of the cupola. The end of this plate extended over the top of the cupola, with an opening in the plate equal to the inside diameter of the cupola, and on the plate was put a short stack, in which was placed the charging door, the top of which was arched over toward the main stack, with which it connected on the side.

Any sparks that arose from the cupola were thrown into the bottom of the main stack by the arch in the direction indicated by the arrows and were removed when cold, as often as the bottom of the stack filled up to such an extent as to interfere with the arrest of the sparks.

This arrangement was very effective in arresting sparks, but was not found to be a very convenient one for attaching to our modern cupolas, and numerous other plans have since been devised and used.

264   THE CUPOLA FURNACE.

FIG. 55.

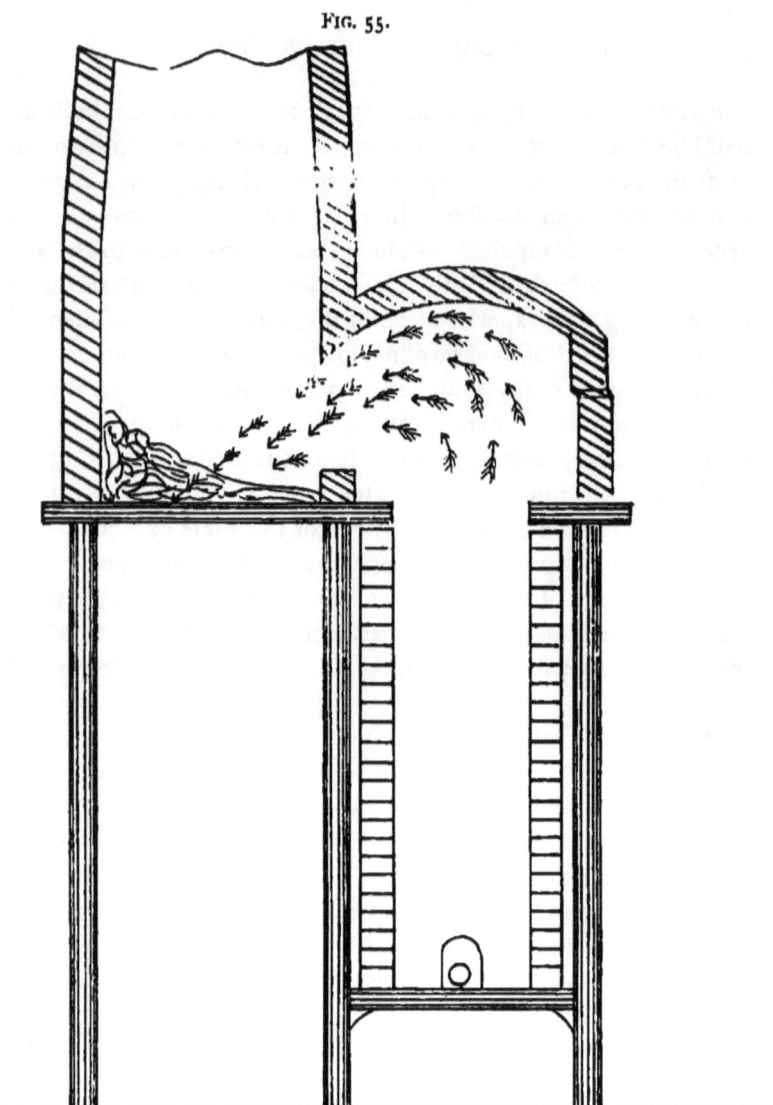

SPARK CATCHER IN OLD STYLE CUPOLA.

SPARK CATCHING DEVICE FOR MODERN CUPOLAS.

In Fig 56 is seen a more modern spark-arrester than the one just described. In this device, the casing is cut in two at the

SPARK CATCHING DEVICES FOR CUPOLAS.

FIG. 56.

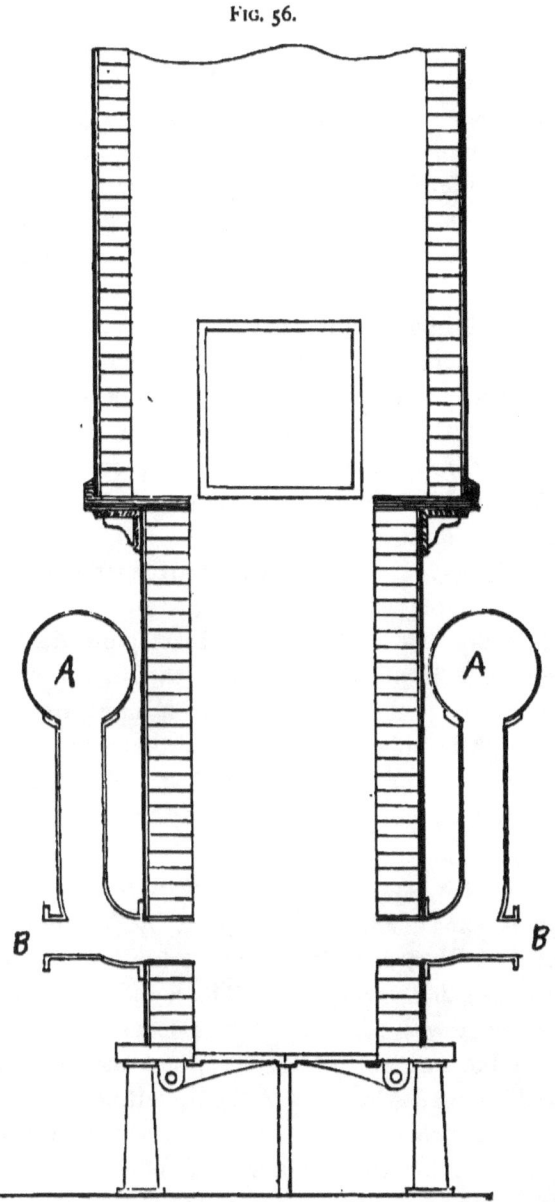

SPARK CATCHING DEVICE IN MODERN CUPOLA.

bottom of the charging door and an iron plate or ring placed upon the top of the cupola casing, where it is supported by the casing and cast-iron brackets riveted or bolted to it on the outside. The inside of the plate or ring generally covers the top of the cupola lining to protect it when charging the stock, and the outside extends over the cupola casing from six to twelve inches. On this plate the stack casing, which is of larger diameter than the cupola casing, is placed and lined with a thin lining. The spark-arresting device consists in making the stack larger than the cupola so that the blast loses its force when it emerges from the cupola, and enters the stack, and the sparks carried out of the cupola fall back into it before reaching the top of the stack. The extent to which the stack should be enlarged to be effective in arresting sparks depends upon the height of it; low stacks requiring to be of a larger diameter than high ones.

In this illustration is shown a very neat arrangement for supplying blast to a cupola when a belt air chamber riveted to the cupola shell is not used. The main blast pipe AA which encircles the cupola is placed up out of the way, in catching iron or removing large ladles. The branch pipes are cast in one piece and tightly bolted or riveted to the main pipe and cupola casing, to prevent the escape of blast. The peep holes BB are cast in the pipe, and close with a tight-fitting swing cap and latch.

RETURN FLUE CAPULA SPARK CATCHER.

In Fig. 57 is shown a device designed by John O'Keefe, Superintendent of Perry & Co's Stove Works, Albany, N. Y., for catching sparks and saving fuel. The foundry of the firm in which this device was constructed, was located on Hudson St., in a closely built up part of the city, and they were very much annoyed by sparks from their cupola setting fire to roofs of buildings in the vicinity, and it became necessary to prevent sparks escaping or move their foundry. A number of devices, such as hoods, etc., were tried, but none of these proved effective, and a return flue was constructed. The arch or dome A was

SPARK CATCHING DEVICES FOR CUPOLAS.

Fig. 57.

RETURN FLUE CAPULA SPARK CATCHER.

thrown across the cupola stack above the door, and the flue B led out of the cupola just below the dome and down to the foundry floor, from which point it returned to the stack above the dome. When the cupola was in blast, waste heat from the cupola struck the dome and was thrown back upon the stock in the cupola, or was forced down through the flue B and returned to the cupola stack through the flue C above the dome. When the cupola was put in blast it was found that so large an amount of heat and gas escaped from the door that the cupola could not be charged when in blast, and it became necessary to make a small opening through the dome to permit part of it to escape. Had the cupola been of a size to admit of all the stock being charged before the blast was put on and the door closed, during the heat, there is no doubt considerable fuel might have been saved, and faster melting done. But as it was, no fuel was saved, and there was no perceptible change in the time required to melt a heat. The device was effective in preventing the escape of sparks and small pieces of fuel from the stack, for they were all thrown back into the cupola or deposited in the bottom of the flue, from which they were removed through the opening D at the bottom of the flue, as frequently as found necessary.

OTHER SPARK CATCHING DEVICES.

Another device for arresting sparks is to place a half circle fire-brick arch opened at both ends on the top of the stack, making its total length and breadth equal to the outside diameter of the stack. This plan arrests the sparks in their upward course and some of them fall back into the cupola, but many are carried out at the ends of the arch by the blast and fall upon the foundry roof, and on windy days may be carried to adjoining roofs.

Iron caps or hoods are also placed one or more feet above the top of cupola stacks to arrest sparks; but they, like the arch, only arrest the sparks in their upward flight and throw many of them down upon the foundry or scaffold roof.

Another plan for preventing the eseape of sparks is to sus-

SPARK CATCHING DEVICES FOR CUPOLAS.

pend an iron disk of a few inches smaller diameter than the stack in the stack near the top. The sparks strike this disk and are thrown back into the cupola. But this device cannot be used in contracted stacks with a strong blast, and in large ones the cohesive properties of the iron are soon destroyed by the heat and gases of the cupola, and if not frequently replaced there is danger of it breaking from the jar in chipping out the cupola, and falling upon the melter.

THE BEST SPARK CATCHING DEVICE.

The cause of sparks being thrown from a cupola is the strong blast forced into the cupola at the tuyeres, which carries small pieces of fuel out at the top of the stack during the heat, and large pieces near the end of a heat, when the stock is low in the cupola and the blast passes through it more freely. The lifting power of the blast is increased by confining it in a contracted stack, and good-sized pieces of fuel may be thrown several feet above the top of a small stack; but the instant the blast escapes from the top of the stack it expands and its lifting power is lost, and sparks or pieces of fuel fall by their own weight and may in their descent be carried to some distance by a strong wind.

To prevent them being carried out of the stack, it is only necessary to provide sufficient room in the stack for the blast to expand, after escaping from the cupola, and lose its lifting force, when the sparks will fall back in the cupola and be consumed. This may be done by constructing the stack casing of the same diameter as the cupola casing, and lining it with a thin lining of four-inch fire-brick supported by angle iron, so that the cupola lining may be removed or repaired without disturbing the stack lining. Cupolas constructed in this way, when the stack is of proper height, do not throw out sparks. When it is not desirable to have a very high stack, the enlarged stack shown in Fig. 56 may be used. The first cost of a stack of this kind is a little greater than that of a contracted one, but when properly constructed and lined, will last the life of a cupola. In fact we never knew one, if properly lined when constructed, requiring to be relined or repaired, and the saving effect by preventing damage to roofs, lumber, flasks, etc., from sparks will soon pay for the extra cost of construction. The objection usually made by foundrymen to large stacks is that they do not give sufficient draught for lighting up. This may be the case when the top of the stack is only a few feet above the charging door, but when given a proper height for arresting sparks there is always sufficient draught for lighting up. There are many cupolas constructed upon this plan in use at the present time, and they give better satisfaction than those with contracted stack.

CHAPTER XIX.

HOT BLAST CUPOLAS.

A NUMBER of plans have in this country been at different periods devised for utilizing the heat escaping from the top of a cupola when in blast, for heating the blast before entering the cupola at the tuyeres. The best arranged cupolas of this kind that we have seen are those shown in Fig. 58. This pair of cupolas was made at Albany, N. Y., by the firm of Jagger, Treadwell & Perry. With a view of saving fuel and improving the quality of iron for light work, the two cupolas DD of thirty and forty-five inches diameter, respectively, inside the lining, and eight feet high were constructed, and were made of boiler plate; the bottom and top plates between which the cupolas were placed were supported by four iron columns, and on the top plate were fitted the brick arches BB, which connected the cupolas with the brick ovens EE. In the rear of each cupola, between the ovens, was placed the high stack A. Each oven was filled with cast-iron pipe CC, through which the blast passed before entering the cupolas. When in blast, the escaping heat from the cupolas passed downward through the ovens as indicated by the arrows, and entered the stack A from the bottom of the ovens. The pipes were by the escaping heat carried up to a red heat, and the blast in passing through these coils of pipe was heated to a sufficient degree before entering the cupolas to melt lead. This plan was a success so far as heating the blast was concerned, but the blast could not be carried up to the above degree until the cupolas had been in blast for some time. Hence very little fuel was saved, for no economy in fuel could be effected until the blast was heated, and the cupolas had to be fully charged with fuel for the first

THE CUPOLA FURNACE.

FIG. 58.

HOT BLAST CUPOLAS.

half of the heat. No perceptible improvement was made in the quality of the iron by the heating of the blast, and the greatest objection to these cupolas was the difficulty of keeping the coils of pipe intact. The heating of the pipe to a red heat every time the cupolas were put in blast and permitting them to cool before the next heat, in a short time destroyed the cohesive properties of the iron, and the pipe frequently broke after or during a heat and permitted the blast to escape into the oven. These breaks became so frequent and annoying after the pipe had been in use for a short time, and were so expensive to repair, that the slight saving effected in fuel did not justify a continued use of the hot blast, and it was abandoned. The cupolas were for a long time used without the hot blast, and the ovens proved excellent spark catchers. No sparks were ever thrown from the top of the high stack, and the ovens had frequently to be cleaned to remove them.

At the stove foundry of Ransom & Co., Albany, N. Y., a cupola was constructed with a large stack, and coils of pipe for heating the blast were placed in the stack directly over the cupola. The blast when passed through these pipes was heated to a high degree after the cupola had been in blast for a short time, but the pipes in this case broke after repeated heating and cooling, as in the ovens of the Jagger, Treadwell and Perry cupolas, and after the killing of a melter, by a piece of pipe falling upon him from the stack while picking out the cupola, the pipes were all removed from the stack and heating of the blast was discontinued. Several attempts have been made to take the escaping heat direct from the top of a cupola and return it into the cupola through the tuyeres; but in all cases this plan has, for lack of means to force the hot air into the cupola, proven a failure.

Exhaust pipes have been connected with the stack of a cupola and the inlets of the blower placed near the cupola, and hot air drawn from the stack by the blower and returned to the cupola through the tuyeres. This arrangement supplied a hot blast to the cupola with no expense for heating the blast, and was in the early part of a heat in which it was tried, a success, when only a small amount of heat escaped from the cupola and the air drawn from the stack was heated only to a limited extent. But, as the melting progressed and the stock settled low in the cupola, the air drawn from the stack was heaten to so high a a degree as to heat and destroy a blower through which it was passed in being returned to the cupola. Could hot air have been taken from a cupola stack and returned to the cupola through the tuyeres without passing it through a blower, it would, no doubt, have effected a great saving in fuel in the days of low cupolas, when a large amount of the heat from fuel direct was not utilized in melting. But this could not be done, and after a number of experiments to secure a hot blast in this way, the plan was given up as a failure.

The blast for a cupola can be heated in a hot-blast oven similar to those some years ago used in heating the blast for furnaces, and which was done by furnaces specially constructed for the purpose, and not with gas taken from the furnaces as at the present time. But these ovens would be required to be kept continually hot to prevent breakage of the pipes by repeated heating and cooling. The saving of fuel effected in melting with a hot blast obtained in this manner, would not be sufficient to pay for the expense of heating the blast for a cupola that is only in blast for a few hours each day; and it is doubtful if the saving effected would justify the heating of the blast, if a cupola was kept constantly in blast, or the hot blast changed from one cupola to another as soon as the heat was melted.

WASTE HEAT FROM A CUPOLA.

A number of plans for utilizing the heat escaping from a cupola, besides using it for heating the blast, have been devised; such as utilizing it for heating the iron before charging it into the cupola, drying cores, ladles, etc. All these experiments were made years ago, when from six to ten feet was considered to be the proper height for a cupola, and fully one-half of the heat escaped from the top; but it was not until the height of cupolas was increased that a practical means of utilizing all the heat of the fuel in melting was found. In a high cupola all the heat escaping from the melting zone is utilized to heat the stock in the cupola and prepare it for melting before the stock settles into the melting zone. The height that a cupola should be made in order to utilize all the heat depends upon its diameter, volume of blast, and the way in which the stock is charged. Cupolas of twelve to twenty inches in diameter must be made low, so that the stock in case it hangs up in the cupola may be dislodged with the bar, and all the heat cannot be utilized in these small cupolas except when a very small volume of blast is used. In this latter case the melting is slow, and it is more economical to permit part of the heat to escape, and do fast melting with a strong blast. Cupolas of large diameter may be made of a sufficient height to utilize all the heat, no matter how great the volume of blast or how openly the stock is charged. Cupolas of large diameter now in use in many foundries are from fifteen to twenty feet high, and those in the Carnegie Steel Works, Homestead, Pa., are thirty feet high. In these cupolas whole bars of pig iron are charged, and all the stock is dumped into the cupola from barrows, and no pains taken to pack it close to prevent the escape of heat. Yet no heat escapes from the top of the cupola when filled with stock, and it has not been found necessary to line the iron stacks with brick to prevent them being heated by heat escaping from the cupolas.

In low cupolas heat may to a large extent be prevented from escaping by breaking the pig and scrap into small pieces, and when charging packing it close. More time is then required for the heat to work its way through the stock in escaping from the melting zone, and a greater amount of it is utilized in heating the stock and preparing it for melting before it settles into the melting zone.

CHAPTER XX.

TAKING OFF THE BLAST DURING A HEAT—BANKING A CUPOLA—BLAST PIPES, BLAST GATES.

EXPLOSIONS IN BLAST PIPES, BLAST GAUGES, BLAST IN MELTING.

THE length of time the blast can be taken off a cupola after it has been in blast long enough to melt iron, and put on again and good melting done, depends upon the condition of the stock in the cupola at the time it has been stopped.

The blast may be taken off a cupola that has only been in blast for a short time, is in good melting condition and filled with stock, for many hours if the melted iron and slag are all drawn off and the tuyeres carefully closed to exclude the air and prevent melting and chilling after the blast has been stopped. We have known a cupola in this condition in case of a break-down in the blowing machinery to be held from four o'clock in the afternoon until eight o'clock the following morning, and good melting done when the blast was again put on.

In this case, the tuyeres were packed with new molding sand rammed in solid to completely exclude the air, and the molten iron all drawn off, after the tuyeres had been closed for a short time and the tap hole closed with a bod. Before putting on the blast in the morning, the tuyeres were permitted to remain open for a short time, to allow any gas that may have collected in the cupola during the stoppage to escape and avoid an explosion, which might have occurred had a large volume of blast been forced into the cupola when filled with gas.

Cupolas that have been in blast for some time and from which the blast is removed toward the end of the heat when the cupola is comparatively empty, or in bad shape for melting, cannot be held for any great length of time, even if the tuyeres are at once closed and every precaution taken to prevent chilling and clogging. This is due to the gradual settling of a semi-fluid slag and cinder above the tuyeres, and the closing up of small openings in it through which the blast was distributed to the stock; and in case of accident to the blower it is better to dump the cupola at once than to attempt to hold it for any length of time.

Cupolas, in which all the iron charged has been melted and drawn off, may be held over night, if the cupola has been properly fluxed, the slag drawn off, and a fresh charge of coke put in, with a liberal charge of limestone on top of it to liquefy any slag that may over night have chilled in the cupola. Small cupolas are frequently managed in this way; the tuyeres are closed and the tap hole permitted to remain open to admit sufficient air to ignite the fresh coke.

In the morning after the cupola has been filled with stock and the blast put on, the limestone on the bed is the first to melt, and if in sufficient quantity makes a fluid slag that settles to the bottom, freeing the cupola of any clogging that may have taken place during the stoppage.

BANKING A CUPOLA.

Since writing the foregoing paper we have received the following practical illustration of keeping a cupola in good condition for melting for many hours after it had been charged and the blast put on, from Mr. Knœppel, Foundry Superintendent, Buffalo Forge Co., Buffalo, N. Y. In this case melting had not begun before the pulley broke and the blast was taken off, but the same results would have been obtained from banking the cupola in this way if melting had begun and the cupola been in blast for a short time.

"Banking a cupola is something that does not come in the usual course of foundry practice, but there are times when the knowledge of how it is to be done would be a source of profit, as well as loss of time being averted. By request having been induced to allow this letter to appear in your valuable publication on 'Cupola Practice;' hence will try[and give you the details as near as I can from memory, although I wrote an article on this subject in the 'American Machinist,' December 10, 1891, which I am now unable to get.

"In the latter part of October, 1891, just as we were about to put on the blast in our foundry cupola and the fan making a few revolutions, the main pulley broke, running the main shaft to the fan or blower of our cupola. After considerable trouble, loss of time and delay in trying to get a new pulley, which was of wood pattern, we finally succeeded in getting one of the proper size, and had it put on the shaft; but the belt being a little tight, and also anxious to get off the heat, in slipping the belt on the pulley, it was cut in such a shape that it became useless for that day. By this time it was beyond our regular hour for quitting. At first there seemed no way out of the dilemma but to drop the bottom. The thought of re-handling the hot material and fuel, the extra labor attached therewith, suggested the idea of holding up the charges until next morning, when repairs would be completed. After a few moments' consultation, proceeded as follows: Let me say first that the cupolas was lighted at 1:45 p. m. and at 6 p. m. began the operation of banking the cupola, having had four hours and fifteen minutes' time for burning the stock, and being charged with eleven tons of metal. The cupola was of the Colliau type 60" shell lined to 44" at bottom and 48" at melting zone, having six lower tuyeres, 7"x9", upper tuyeres being closed. Height of tuyeres from bottom when made up 18", blast pressure 10 oz., revolutions of blower about 2100, manufactured by the Buffalo Forge Co., and known as No. 10, the adjustable bed type. The cupola bed was made up of 600 lbs. Lehigh lump coal and 800 lbs. Connellsville coke, the succeeding charges 50 lbs. of coal and 150 lbs. coke, coal being an important factor in this heat on account of its lasting qualities. We first cleaned and cleared all of the tuyeres, packed each one with new coke, and then filled and rammed them tight with floor moulding sand to prevent any draft getting through them, and had the top of charges covered with fine coal and coke dust, and tightened that also to stop the draft in that direction. The object in using coal dust was this: should any get through into the charges, it would not cause much trouble. After all was completed, gave orders to the cupola men to be on hand at 6 a. m. next morning, clean out the tuyeres and top of cupola, and ordered the men to be ready for pouring off at 7 a. m. The next morning all were on time. I had the tuyeres poked with bars, so that the blast might have easy access to center of cupola, and started the blast at 7 : 15, bottom being dropped at 8 : 45 ; total time from time of lighting cupola until bottom dropped, was nineteen hours. At first the iron was long in coming down and first 500 lbs. somewhat dull, but made provision for that and put it into dies, which turned out to be very good. The balance of the heat was hot enough for any kind of casting—our line being light and heavy, and had to be planed, bored and otherwise finished with some stove repair casting in with this heat engine casting, cylinder and a class of work that requires fluid metal. I am confident that if this method is carefully followed, it can be done at all times, but would not advise it in small cupolas, less than 36" inside measurement; and should the melt be in progress, it could not be successfully done at all. Should I be placed in a similar position, would resort to the same means with more confidence and certainty of success.

"Yours respectfully,
JOHN C. KNŒPPEL,
Foundry Supt. Buffalo Forge Co., Buffalo, N. Y.

BLAST PIPES.

In constructing a cupola, one of the most important points to be considered is the construction and arrangement of blast pipes and their connection with the cupola, for the best constructed cupola may be a complete failure through bad arrangement of pipes and air-chambers.

Not many years ago it was a common practice of foundry-men to place blast pipes underground. The main pipe was generally made square and constructed of boards or planks spiked together, no care being taken to make air-tight joints, and the escape of blast was prevented by ramming sand or clay around the pipe when put in place. Connections from the main pipe to the cupola were made by means of vertical cast-iron pipes to each tuyere, as shown in Figs. 31 and 32. The main pipes were generally constructed with square elbows and ends, and the tuyere pipes were placed over an opening in the top of a branch of the main pipe on each side of the cupola. The square turns and ends of the pipe greatly reduced the force of the blast, and the capacity of the pipe was frequently reduced by water leaking into it or a partial collapse of the pipe, and the volume of blast delivered to a cupola was very uncertain even when the pipes were new, and could not be depended upon at all when the pipes became old and rotten. Iron pipes arranged in this way were also a source of continual annoyance and uncertainty from water or iron and slag from the tuyeres getting into them and reducing their capacity for conveying blast. This way of arranging cupola pipes has generally been abandoned, and they are now commonly placed overhead or up where they are least liable to injury and may be readily examined to see that there is no leakage of blast from a pipe.

Blast pipes may be made of wood, tin plate, sheet iron, cast iron, or galvanized iron. Wooden pipes shrink and expand with changes of weather and moisture in the atmosphere, and it is almost impossible to prevent the escape of blast from such pipes. Tin and sheet iron pipes, when placed in a foundry, are very rapidly rusted and destroyed by steam and gases escaping from moulds and the cupola, if not thoroughly painted outside and in. Cast iron pipes are heavy, difficult to support in place, liable to break when not properly supported, or leak at the joints, and the best for foundry use are those made of galvanized iron. In constructing pipes of this material, an iron of a proper gauge for the size of pipe should be selected, and their shape should, whenever possible, be round, for round pipes are more easily constructed and have the largest effective area with a given perimeter of any known figure. Pipes should be made in lengths convenient for handling, say 8 or 10 ft., having joints lapped nearly 2 inches in direction of the air current. Joints should be riveted about every 4 inches to hold them securely together and prevent sagging of the pipe between supports, and to insure their being tight they should be soldered all the way around. Section ends should be placed over supports and laps of from 3 to 4 inches made at each joint and also soldered. The end of the main pipe when not connected direct with an air chamber on the cupola should be divided into two or more branches of equal capacity for connection with the tuyeres or air belt, and rounded curves or elbows used in changing the direction of pipes. A pipe should never terminate abruptly, and branches should not be taken out of the side for supplying the cupola, as is frequently done. The area of main pipes and also branch pipes should be increased as the distance from the blower to the cupola is increased; and as a guide for increasing their diameter in proportion to the length of pipe, we do not think we can do better than give our readers the excellent table prepared by the Buffalo Forge Co., Buffalo, N. Y., as follows.

DIAMETER OF BLAST PIPES.

It will be seen, by reference to the following table, that the diameter of pipe for transmitting or carrying air from one point to another, changes with the length or distance which the air is carried from the blower to the furnace, or other point of delivery.

As air moves through pipes, a portion of its force is retarded by the friction of its particles along the sides of the pipe, and the loss of pressure from this source increases directly as the length of the pipe, and as the square of the velocity of the moving air.

This fact has long been known, and many experimenters and engineers, by close observation and long-continued experiments, have established formulas by which the loss of pressure and the additional amount of power required to force air or gases through pipes of any length and diameter may be computed.

As these formulas are commonly expressed in algebraic notation, not in general use, we have thought it desirable to arrange a table showing at a glance all the necessary proportionate increase in diameter and length of blast pipes and conical mouth-pieces, in keeping up the pressure to the point of delivery. It is often the case, where a *blower is condemned as being insufficient*, the cause of its failure is that the pipe connections are too small for their lengths, coupled with a large number of short bends, without regard to making the pipe tight, which is a necessity.

The table, diameter of pipes, given below, showing the necessary increase in the size of pipes in proportion to the lengths, is what we call a practical one, and experience has proved the necessity for it.

TABLE SHOWING THE NECESSARY INCREASE IN DIAMETER FOR THE DIFFERENT LENGTHS.

LENGTH OF PIPE.	30 FT.	60 FT.	90 FT.	120 FT.	150 FT.	180 FT.	210 FT.	240 FT.	270 FT.	300 FT.
Diameter of Blower Outlet in Inches.	Diameter of Pipe should be	Diameter of Pipe should be	Diameter of Pipe should be	Diameter of Pipe should be	Diameter of Pipe should be	Diameter of Pipe should be	Diameter of Pipe should be	Diameter of Pipe should be	Diameter of Pipe should be	Diameter of Pipe should be
3	3¼	3⅝	4	4¼	4½	4¾	5	5⅛	5⅜	5½
3½	3¾	4⅛	4½	4⅞	5	5¼	5½	5⅝	5⅞	6⅛
4	4⅜	4¾	5⅛	5⅜	5¾	6	6¼	6½	6¾	7
4½	5	5⅜	5¾	6	6⅜	6¾	7	7¼	7½	7⅞
5	5½	6	6⅜	6¾	7⅛	7½	7¾	8⅛	8⅜	8¾
6	6½	7	7⅝	8	8½	9	9⅜	9¾	10⅛	10½
7	7⅝	8¼	8⅞	9⅜	10	10⅜	10⅞	11⅞	11¾	12⅛
8	8¾	9½	10⅛	10¾	11⅜	11⅞	12⅜	12⅜	13⅜	13⅞
9	10	10¾	11½	12⅛	12¾	13⅜	14	14½	15⅛	15⅝
10	11	11⅞	12¾	13½	14¼	14⅞	15½	16⅛	16¾	17⅜
11	12	13	13⅞	14¾	15⅝	16⅜	17⅛	17¾	18½	19⅛
12	13⅛	14¼	15¼	16⅛	17	17⅞	18⅝	19⅜	20⅛	20⅞
13	14¼	15⅜	16½	17½	18⅜	19¼	20⅛	21	21¾	22⅝
14	15⅜	16⅝	17¾	18⅞	19¾	20¾	21¾	22⅝	23½	24¼
15	16½	17¾	19	20⅛	21¼	22¼	23¼	24¼	25⅛	26
16	17½	19	20⅜	21½	22⅝	23¾	24¾	25⅞	26⅞	27¾
17	17⅝	20⅛	21½	22¾	24	25¼	26⅜	27½	28½	29½
18	19¾	21⅜	22¾	24¼	25½	26¾	27⅞	29⅛	30⅛	31¼
19	20⅞	22½	24	25½	27	28¼	29½	30¾	31⅞	33
20	22	23⅝	25⅜	27⅞	28⅜	29¾	31	32¼	33½	34¾
21	23	24⅞	26⅝	28¼	29¾	31⅛	32½	33⅞	35¼	36⅜
22	24⅛	26⅛	27⅞	29½	31⅛	32⅝	34⅛	35½	36⅞	38⅛
23	25¼	27¼	29⅛	30⅞	32½	34⅛	35⅝	37⅛	38½	39⅞
24	26½	28½	30⅜	32¼	34	35⅝	37¼	38¾	40¼	41⅝
Length of pipe	30 ft.	60 ft.	90 ft.	120 ft.	150 ft.	180 ft.	210 ft.	240 ft.	270 ft.	300 ft.
Length of mouth-piece	9 in.	15 in.	21 in.	27 in.	33 in.	36 in.	42 in.	48 in.	54 in.	60 in.

The connection of blast pipes with cupolas is also a matter to which entirely too little attention is given, and is frequently the cause of poor melting when cupola is otherwise properly constructed. As stated elsewhere, tuyeres should be large enough to admit blast to a cupola freely, and to obtain good results in melting it must be fully and evenly distributed to the tuyeres. When blast is delivered direct to tuyeres through branch pipes, the branches should be taken off the main pipe in as near a direct line with the current of the blast in the mainpipe as possible, and its course to the tuyeres should be changed by long curves or round elbows in the pipes, to prevent the velocity of the air being checked and blast thrown back in the pipe. The combined area of all the branch pipes should be equal to the area of the main pipe and not less as is frequently the case, owing to a mistake being made through the erroneous idea that a multiple of the diameter of two or more small pipes is equal to the area of one large one of their combined diameters. If this were the case two five-inch pipes would have an area equal to one ten-inch pipe, which is not so, as will be seen by the table on p. 285, which may be of value to foundrymen in arranging their blast pipes.

DIAMETER AND AREA OF PIPES.

Diameter.	Area.	Diameter.	Area.	Diameter.	Area.
2	3.141	$8\frac{1}{4}$	53.456	$14\frac{1}{2}$	165.13
$2\frac{1}{4}$	3.976	$8\frac{1}{2}$	56.745	$14\frac{3}{4}$	170.85
$2\frac{1}{2}$	4.908	$8\frac{3}{4}$	60.132	15	176.71
$2\frac{3}{4}$	5.939	9	63.617	$15\frac{1}{4}$	182.65
3	7.068	$9\frac{1}{4}$	67.200	$15\frac{1}{2}$	188.69
$3\frac{1}{4}$	8.295	$9\frac{1}{2}$	70.882	$15\frac{3}{4}$	194.82
$3\frac{1}{2}$	9.621	$9\frac{3}{4}$	74.662	16	201.06
$3\frac{3}{4}$	11.044	10	78.539	$16\frac{1}{4}$	207.39
4	12.566	$10\frac{1}{4}$	82.516	$16\frac{1}{2}$	213.82
$4\frac{1}{4}$	14.186	$10\frac{1}{2}$	86.590	$16\frac{3}{4}$	220.35
$4\frac{1}{2}$	15.904	$10\frac{3}{4}$	90.762	17	226.98
$4\frac{3}{4}$	17.720	11	95.033	$17\frac{1}{4}$	233.70
5	19.635	$11\frac{1}{4}$	99.402	$17\frac{1}{2}$	240.52
$5\frac{1}{4}$	21.647	$11\frac{1}{2}$	103.86	$17\frac{3}{4}$	247.45
$5\frac{1}{2}$	23.758	$11\frac{3}{4}$	108.43	18	254.46
$5\frac{3}{4}$	25.967	12	113.09	$18\frac{1}{4}$	261.58
6	28.274	$12\frac{1}{4}$	117.85	$18\frac{1}{2}$	268.80
$6\frac{1}{4}$	30.679	$12\frac{1}{2}$	122.71	$18\frac{3}{4}$	276.11
$6\frac{1}{2}$	33.183	$12\frac{3}{4}$	127.67	19	283.52
$6\frac{3}{4}$	35.784	13	132.73	$19\frac{1}{4}$	291.03
7	38.484	$13\frac{1}{4}$	137.88	$19\frac{1}{2}$	298.64
$7\frac{1}{4}$	41.282	$13\frac{1}{2}$	143.13	$19\frac{3}{4}$	306.35
$7\frac{1}{2}$	44.178	$13\frac{3}{4}$	148.48	20	314.16
$7\frac{3}{4}$	47.173	14	153.93		
8	50.265	$14\frac{1}{4}$	159.48		

THE CUPOLA FURNACE.

In connecting blast pipes direct with tuyeres, either by long branch pipes from the main pipe or short ones from a belt air chamber not attached to cupola shell, care should be taken to have as few joints or connections in the pipes as possible, and every joint should be made in such a way that the jar made in chipping out and charging the cupola will not cause the joints to leak after they have been in use a short time. In leading pipes out of an air chamber they cannot always be placed in line with the current of the blast, and must be filled from pressure of blast in the air chamber, but the connecting pipes may be shaped to guide the blast smoothly from the air chamber to its destination.

In Fig. 56 is shown as perfect a connection of air chambers of this kind as can be made. In this illustration the belt pipe $A\ A$ is placed up out of the way and of danger of being injured when making up or working the cupola, and the branch pipes to each tuyere are straight and smooth inside and the pipe is given a curve at the bottom to throw the blast into the tuyere without having the force of its current impaired, and the tuyeres are of a size to admit the full volume of blast from the pipe. Only two joints are required in connecting the air chamber with the cupola, and these are made in such a way that they may be securely bolted or riveted, and all leakage prevented.

In contrast with the neat arrangement of pipes on this cupola is shown the other extreme of poor arrangement in illustration Figure 59. This is a section of a "perfect cupola" illustrated and described in *The Iron Age* some years ago, and while other parts of the cupola may have been perfect, this part was certainly very imperfect. The air chamber and its connecting pipes are made of cast iron. The connecting pipes are cast in three pieces, necessitating the making of four joints. The air box is cast in two pieces, requiring another joint; and a peep-hole and an opening for escape of slag and iron running into the tuyeres, is placed in the pipe, making in all seven joints and openings in each connection to be made and kept air-tight. The jar in working the cupola, together with the small explosions of gas that frequently take place in cupolas and pipes,

FIG. 59.

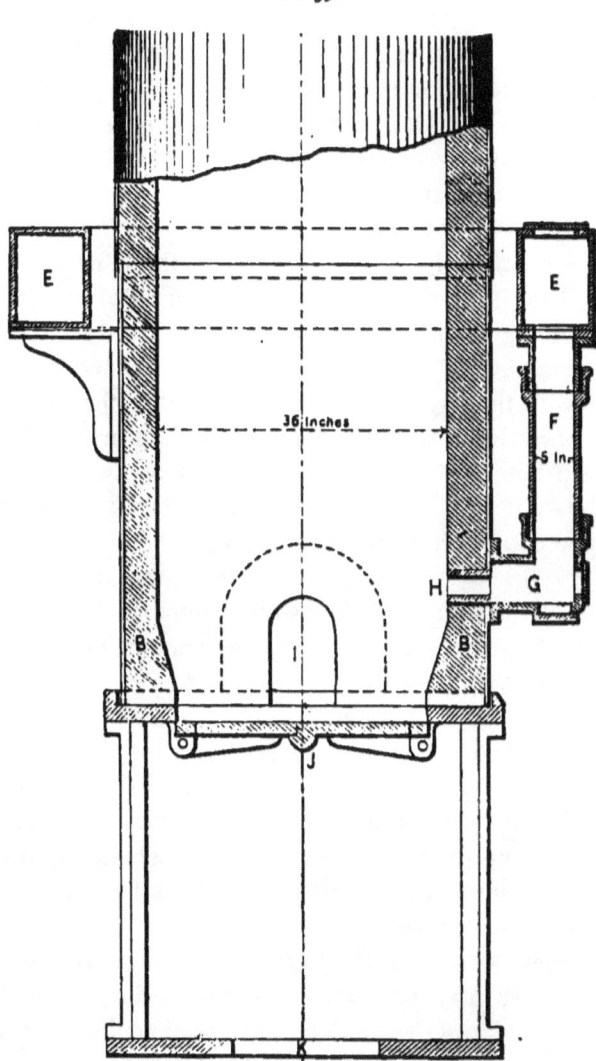

POOR ARRANGEMENT OF BLAST PIPES.

would naturally tend to loosen many of these joints, and a large amount of blast would be lost through leakage of joints. The many joints make more or less roughness in the pipes, thus impeding the blast. The turn in the pipe for connection with the tuyere is square and the course of the current of air is abruptly changed, and the tuyere is entirely too small to admit the full volume of blast from the pipe to the cupola, and only by a heavy pressure of blast could the air be forced into the cupola in sufficient quantities to do good melting.

In·Fig. 57 is shown another way of connecting a belt air-chamber with the tuyeres. In this case the pipe is made of galvanized iron, and the tuyere boxes are made of cast-iron and are large, giving abundant room for changing the direction of the blast current. Only two joints are made in connecting the air-chamber with the cupola; beside these joints, the end of the tuyere box is closed with a large door, the full size of the box, and a peep-hole is placed in the door, making two more openings to be kept air-tight. Many cupolas are in use having their blast connections arranged in this way, and while the arrangement is very good, it is not perfect, and a great deal of blast is lost through leakage of joints—the principal loss occurring around the large door and at the joint connecting the galvanized iron pipe with the cast-iron tuyere box.

The very best way of connecting blast pipes with cupola tuyeres is by means of a belt air-chamber riveted to the cupola casting, as shown in Figs. 39, 43 and 45, or by an inside air-chamber, as shown in Figs. 31 and 46. In either case the air-chamber is riveted to the cupola shell and the joint made perfectly air-tight, and in case of jar to the cupola, the air-chamber being part of the cupola, oscillates with it, and the jar in chipping out and charging does not loosen the joint and cause leakage of blast. The blast pipes may also be securely riveted or bolted to the air-chamber and a perfectly tight joint made. In constructing cupolas in this way, care should be taken to make the air-chamber of a sufficient size to admit of a free circulation of blast and supply all the tuyeres with an adequate amount for good melting. When the air-chamber is small, the blast pipe

TAKING OFF THE BLAST DURING A HEAT. 289

should be connected with it on each side of the cupola, and on the side or top as found most convenient. When the chamber is large and there is an abundance of room for the escape of blast from the pipe, one pipe is sufficient and it may be connected on the side or top. When attached on the side it should be placed in line with the circle of the cupola as shown in Fig. 48, to cause the current of blast to circulate around the cupola and facilitate its escape from the pipe. When

FIG. 60.

BLOWER PLACED NEAR CUPOLA.

the current of blast is thrown directly against the cupola casing or bottom of the chamber in a narrow air-chamber, the mouth of the pipe should be enlarged, to facilitate the escape of blast into the chamber; for cupolas of this construction may be made a complete failure by failing to provide a sufficient space at the end of the pipe for escape of blast into the air-chamber, when the chamber is of a sufficient size to supply the cupola. Connections with the inner air-chambers of limited capacity should

19

be made on each side by means of an air or tuyere box placed outside as shown in Fig. 6, and the pipe connected on top to equalize the volume of blast supplied to each tuyere.

Long blast pipes often cause poor melting, from the volume of blast delivered to a cupola, being reduced by friction in the pipes, and in all cases the blower should be placed as near the cupola as possible. In Fig. 60 is shown a very neat arrangement in placing a blower near a cupola and at the seme time having it up out of the way of removing molten iron or the dump from the cupola, and the space under it may be utilized for storing ladles, etc. In this illustration is also shown a very perfect manner of connecting the main pipe with an air chamber. The pipe is divided into two branches of equal size in line with the current of blast from the blower, and connected with the air chamber on each side by curved pipes arranged in such a way as not to check the current of air as it passes through the pipe.

BLAST GATES.

These devices are especially designed for opening and closing blast pipes, such as are employed for conveying air between blowers and cupolas. There are several different designs of blast gates, but the one shown in Fig. 61 is the one most commonly used by foundrymen. They are manufactured and kept in stock by all the leading manufacturers of blowers, and cost from one dollar upwards, according to size of blast pipe.

The employment of the blast gate places the volume of blast delivered to a cupola under control of the melter, which feature is frequently very important in the management of cupolas in melting iron for special work, or in case of accident or delay in pouring. In foundries in which the facilities for handling molten metal are limited and melting must at times be retarded, to facilitate its removal from the cupola as fast as melted, and in foundries where the amount of iron required to be melted per hour is limited by the number of molds or chills employed, from which castings are removed and the molds refilled, it is very important that the blast should be under control of the melter. In such foundries the cupolas are generally of small diameter and frequently kept in blast for a number of hours at a time, and it is often desired to increase the volume of blast to liven up the iron and decrease it, to reduce the amount melted in a given time.

The blast gate places the blast under control of the melter and enables him to increase or diminish its volume as deemed

FIG. 61.

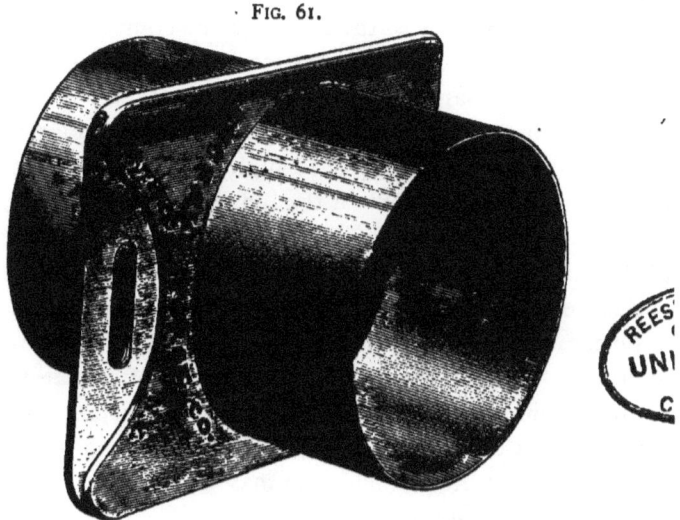

BLAST GATE.

necessary to obtain the best results in melting. They are often of value in regular cupola practice to reduce the volume of blast and retard melting for a few minutes while pouring a large piece of work, in foundries where the facilities for handling large quantities of molten iron are limited, and the speed of blower cannot be reduced without reducing the speed of machinery in other parts of the works or stopping the blower entirely, which is not good practise after a cupola has been in blast for some time.

The gate is also a safeguard against gas explosions, which often occur from the accumulation of gas in pipes during the temporary stoppage of the blower. The gate should always be placed in the pipe near the cupola, and closed before stopping the blower and not opened until it is again started up.

EXPLOSIONS IN BLAST PIPES.

Violent explosions frequently take place in cupola blast pipes, tearing them asunder from end to end. These explosions are due to the escape of gas from the cupola into the pipes during a temporary stoppage of the blower in the course of a heat. The explosion is caused by the gas being ignited when the pipe becomes over-charged, or the instant the blower is started and the gas is forced back into the cupola. Such explosions generally take place in pipes placed high or arranged in such a way as to have a draught toward the blower. But they may occur in any pipe if the cupola is well-filled when a stoppage takes place and the blower is stopped for a great length of time.

Such explosions may be prevented by closing the blast-gate if placed near the cupola, or by opening the tuyere doors in front of each tuyere and admitting air freely to the pipe. Such precaution should always be taken the instant the blast is stopped, as a pipe may be exploded after only a few minutes' stoppage of the blower, and men may be injured or the blower destroyed by the explosion.

BLAST GAUGES.

A number of air or blast gauges have been designed and placed upon the market for determining the pressure of blast in cupola blast pipes and air-chambers. These gauges are of a variety of design, and are known as *steel spring*, *water* and *mercury gauges*. They are connected with a blast pipe or air-chamber by means of a short piece of gas-pipe or a piece of small rubber hose, through which the air is admitted to the gauge. The pressure of blast is indicated by a face dial and hand on the spring gauge, and the graduated glass tube of the water and mercury gauges, pressure being shown up to two pounds, in fractions of an ounce. These gauges, when in good order, indicate very accurately the pressure of blast on a cupola, and when tuyeres and pipes are properly arranged, show to some extent the resistance offered to the free escape of blast from the pipe and the condition of the cupola in melting. But they do not indicate the number of cubic feet of air that pass into a cupola in any given length of time, and a gauge may show a pressure of six or eight ounces when scarcely a cubic foot of air is passing into a cupola per minute.

With a pressure blower these gauges show a gradual increase of pressure in the pipe when a cupola is clogging up, and may enable a foundryman to prevent bursting of the pipe; but with a none-positive blower they show nothing that is of any value to a foundryman in melting, so far as we have been able to learn. The volume of blast is what does the work in a cupola, and not the pressure; and a high pressure of blast does not always indicate a large volume of blast, but rather the reverse, for little if any pressure can be shown on a gauge when blast escapes freely from a pipe.

We have seen two cupolas of the same diameter, one melting with a two-ounce pressure of blast and the other with a six-ounce pressure, and the cupola with the low pressure doing the best melting. This was simply because with the low pressure the air was escaping from the pipe into the cupola and with the high pressure it was not, and the high pressure was wholly due to the smallness of the tuyeres which prevented the free escape of blast from the pipe into the cupola.

A definite number of cubic feet of air has been determined by accurate experiments to be required to melt a ton of iron in a cupola, and an air-gauge to be of any value in melting must indicate the number of cubic feet of air that actually enter a cupola at the tuyeres. We have at the present time no such gauge, and in the absence of such a gauge the foundryman's best guide as to the number of cubic feet of air supplied to his cupola is the tables furnished by all manufacturers of standard blowers, giving the number of revolutions at which their blowers should be run, and the number of cubic feet of air delivered at each revolution. From these tables a foundryman may figure out the exact number of cubic feet of air his cupola receives, provided there is no leakage of air from pipes or tuyeres and the tuyeres are of a size that will permit the air to enter the cupola freely.

BLAST IN MELTING.

A cupola furnace requires a large volume of air to produce a thorough and rapid combustion of fuel in the melting of iron or other metals in the furnace. Numerous means have been devised for supplying the required amount of air, among them the draught of a high chimney or stack, and the creating of a vacuum in the cupola by means of a steam jet, placed in a contracted outlet of a cupola as shown in Figs. 28 and 29. These means of supplying air are a success in cupolas of small diameter and limited height, but even in these cupolas the volume of air that can be drawn in is not sufficient to produce rapid melting, and it is doubtful if iron could be melted at all in a cupola of large diameter and of a proper height to do economical melting, by either of these means of supplying air. Owing to the peculiar construction of a cupola furnace and the manner of melting, the free passage of air through it is restricted by the iron and fuel required; and rapid melting can only be done when air for the combustion of the fuel is supplied in a large volume, which can only be by a forced blast.

A number of machines have been devised for supplying this blast, among the earliest of which were the leather bellows, trompe or water blast, chain blast, cogniardelle or water-cylinder blast, cylinder or piston blower. These have, as a rule, given away to the more modern fan blower and rotary positive blast blower, a number of which will be described later on.

The relative merits of a positive aud non-positive blast, is a very much disputed question. It is claimed by many, that with a positive blast a definite amount of air is supplied to a cupola per minute or per hour, while with a non-positive blower or fan there is no certainty as to the amount of air the cupola will receive. This is very true, for a cupola certainly does not receive the same amount of air from a fan blower when the tuyeres and cupola are beginning to clog as it does from a positive blower when there is no slipping of the belts. But is it advisable to supply a cupola with as large a volume of blast when in this condition as when working open and free? Does not the large volume of blast have a chilling effect upon the semi-fluid mass of cinder and slag, and tend to promote clogging about the tuyeres while keeping it open above the tuyeres; while blast from a non-positive blower would perculate through small openings in the mass, and be more effective than a large volume of blast from a positive blower forming large openings in it through which it escaped into the cupola?

These are questions we have frequently tried to solve by actual test; but it is so difficult to find two cupolas of the same dimensions melting the same sized heats for the same class of work, one with a positive and the other with a non-positive blast, that we have never been able to test the matter. We have melted iron with nearly all the blowers now in use and with a number of the old-style ones, and think there is more in the management of a cupola than there is in a positive or non-positive blast. Good melting may be done with either of them, when the cupola is properly managed, and it cannot be done with either of them when the cupola is not properly managed. Until the management of cupolas in every-day practice is reduced to more of a system than at present, it will be impossible to determine any practical advantage in favor of either blower over the other. So far as we are concerned, we have no preference in blowers, but make it a rule to charge a cupola more openly when melting with a non-positive blast, for the reason that stock may be packed so closely in a high cupola, that the volume of blast that is permitted to enter at the tuyeres may not be reduced by preventing its escape through the stock.

The amount of air that is required for combustion of the fuel in melting a ton of iron has been determined by accurate experiments to be about 30,000 cubic feet, in a properly constructed cupola in which the air was all utilized in combustion of the fuel. This amount of air if reduced to a solid would weigh about 24,000 lbs., or more than the combined weight of the iron and fuel required to melt it. In a cupola melting ten tons per hour, 300,000 cubic feet of air must be delivered to the cupola per hour to do the work. It will thus be seen, that a very large volume of blast is required in the melting of 10 tons of iron. To deliver this amount of air to a cupola from a blower that is capable of producing it in the shape of a blast, the blast pipes must be arranged in such a way that the velocity of the air is not impeded by the pipes; and the tuyeres must be of a size to admit the air freely to the cupola. This is not always the case, for we have seen many cupolas in which the combined tuyere area was not more than one-half that of the blower outlet The object in making the tuyere area so small was to put the air into the cupola with a force that would drive it to the center of the stock. This was the theory of melting in the old cupolas with small tuyeres, but this is wrong, for air cannot be driven through fuel in front of a tuyere, as an iron bar could be forced through it, even with a positive blast; and when the air strikes the fuel it cannot pass through it, but escapes through the crevices between the pieces of fuel. These crevices may change its direction entirely, and the same force that drives it into the cupola impels it in the direction taken, which will be the readiest means of escape, and is more liable to be up along the lining than toward the center of the cupola. For, as a rule, stock does not pack so close near the lining as toward the center, and the means taken to prevent the escape of blast around the lining is the very thing that causes it to escape in that way. Since blast cannot be driven through fuel to the center of a cupola and can only escape from the tuyeres through the crevices between the pieces of fuel, the only way to force it to the center of a cupola is to supply a sufficient volume of blast to fill all of the crevices between the pieces of fuel. This can only be done by discarding the small tuyeres and using a tuyere that will admit blast freely to a cupola.

TAKING OFF THE BLAST DURING A HEAT. 297

In placing tuyeres in a cupola, it must be remembered that the outlet area of a tuyere is governed by the number of crevices between the pieces of fuel in front of the tuyere through which the blast may escape from the tuyere. With small tuyeres a large piece of fuel may settle in front of the tuyere in such a way that the tuyere outlet is not equal to one one-hundreth part of the tuyere area, in which case the tuyere is rendered useless, and may remain useless throughout the heat. For these reasons small tuyeres should never be placed in a cupola. For small cupolas we should recommend the triangular tuyere, Fig. 14, for the reason that it tends to prevent bridging, and its shape is such that it is less liable to be closed by a large piece of fuel than a round tuyere of equal area. The vertical slot tuyeres, Figs. 10 and 11, are also for the same reason good tuyeres for small cupolas.

For large cupolas we think the expanding tuyere, Fig. 3, is the best, and if we were constructing a large cupola we should use this tuyere in preference to any other, and make the outlet at least double the size of the inlet, and should place the tuyeres so close together that the outlets would not be more than six or eight inches apart. This would practically give a sheet blast, and distribute air evenly to the stock all around the cupola. The width of the tuyere can be made to correspond with the diameter of cupola, and may be from three to six inches, and should be of a size that will permit blast freely to enter the cupola. Parties who have been melting with small tuyeres and put in large ones upon this plan, must change their bed and charges to suit the tuyeres, for this arrangement of tuyeres would probably be a complete failure in a cupola charged in the same way as when not more than one-fourth of the blast supplied by the blower entered the cupola.

The largest cupolas in which air can be forced to the center from side tuyeres with good resnlts would appear from actual test to be from four and a half to five feet. Larger cupolas than this have been constructed, and are now in use, but they do not melt so rapidly in proportion to their size as those of a smaller diameter. To illustrate this, we might cite the Jumbo Cupola of Abendroth Bros., Port Chester, N. Y., already described, in which the the diameter at the tuyeres is 54 inches, and above the bosh 72 inches, in which 15 tons of iron have been melted per hour for stove-plate and other light castings.

The Carnegie Steel Works, Homestead, Pa., have cupolas of seven and one-half feet diameter at the tuyeres and ten feet diameter above the bosh, in which the best melting per hour is only fourteen tons. The area of this cupola at the tuyeres is almost three times that of Abendroth's cupola, yet the amount of iron melted per hour is actually less than that of the smaller cupola. Tuyeres have been arranged in different ways in this large cupola, and from one to four rows used, yet the melting was not in proportion to the size of cupola. This would seem to indicate that the cupola was not properly supplied with blast near the centre, and the melting done in the center was caused principally by the heat around it; which is probably the case, for the cupola is kept in blast night and day, for six days, and melting must take place in the centre, or the cupola would chill up.

There are many cupolas of sixty inches diameter at the tuyeres in use in which good melting is done, but this would seem to be the limit at which good melting takes place in a cupola supplied with blast from side tuyeres, for above this diameter the rapidity of melting does not increase in proportion to the increase in size of cupola.

There has been considerable experimenting done during the past two or three years with a center blast tuyere for admitting blast to the center of a cupola through the bottom. We have had no practical experience with this kind of tuyere for the last twenty-five years, when we placed one in a small cupola with side tuyeres and found no advantage in it; probably for the reason that a sufficient quantity of air for an even combustion of the fuel was supplied to the centre of the cupola from the side tuyeres.

During the past few years, we have visited a number of foundries in which the center blast was being tried, but in every case the tuyere was out of order or not in use at the time of our visit. The great objection to this tuyere seems to be its liability to be filled with iron or slag and rendered useless. Should this objectionable feature be overcome by such practical foundrymen as Mr. West or Mr. Johnson, who are experimenting with centre blast, it would certainly be a decided advantage in melting in cupolas of large diameter, in connection with side tuyeres. In cupolas of small diameter with side tuyeres, we do not think a center blast would increase the melting capacity of a cupola, for the reason that air can be forced to the center of a small cupola from side tuyeres, when properly arranged and of a proper size.

With a center blast alone, it is claimed that considerable saving is effected in lining and fuel. It is reasonable to suppose that a saving in lining might be effected by a centre blast; for the most intense heat that is created by the blast is transferred from near the lining to the center of the cupola, and the tendency to bridge is greatly reduced. As to the saving of fuel, there never was a new tuyere that did not "save fuel," and there have been hundreds of them, but consumption of cupola-fuel is still too large.

CHAPTER XXI.

BLOWERS.

PLACING A BLOWER.

A BLOWER should always be placed at as near a point to a cupola as is consistent with the arrangement of the foundry-plant, and it should be laid upon a good, solid foundation, and securely bolted to prevent jarring, as there is nothing that wrecks a blower so quickly as a continual jar when running at high speed. In Fig. 60 is shown a convenient way of placing a blower near a cupola, aud at the same time having it out of the way. But when so placed, the blower should be laid upon a solid frame-work of heavy timber, and securely bolted down to prevent jarring when running. It should also be boxed in to prevent air being drawn in from the foundry, and have an opening provided for supplying air from the outside, for air drawn from a foundry when casting and shaking out are taking place is filled with dust and steam, which are very injurious to a blower and pipes.

A blower should never for the same reason be placed in a cupola-room or a scratch room in which castings are cleaned; for it is impossible to exclude dust from the bearings when so placed, and when a bearing once begins to cut, it makes room for a greater amount of dust, and cuts out very rapidly in blowers run at high speed. Dust and steam also corrode and destroy blast wheels which are inside the blower and out of sight, and a blast wheel may be almost entirely destroyed and not discovered until it is found the cupola is receiving no blast. To prevent a blower from being destroyed in this way, and insure a proper volume of blast for a cupola, the blower should be placed in a clean, dry room and supplied with pure air from the outside.

If it cannot be so placed near a cupola, it had better be placed at some distance, in which case the blast pipe must be enlarged in proportion to its length, as described elsewhere. When a blower is placed in a closed room, windows should be opened to admit air when it is running, and when the air about the room is filled with dust, a pipe or box for supplying pure air should be run off to some distance from the blower and the room kept tightly closed.

FAN BLOWERS.

BUFFALO STEEL PRESSURE BLOWER.[*]

The manufacturers make claim for their blower as follows;
In Fig. 62 is shown the latest improved construction form of

FIG. 62.

STEEL PRESSURE BLOWER.

the Buffalo Steel Pressure Blower, for cupola furnaces and forge fires. A distinguishing feature of this blower, common to

[*] Manufactured by Buffalo Forge Co., Buffalo, N. Y.

those of no other manufacture of the same type, is the solid case, the peripheral portion of the shell being cast in one solid piece, to which the center plates are accurately fitted, metal to metal. It will thus be seen that the objectionable and slovenly "putty joint" is entirely dispensed with. Ready access to the interior of the blower, without entirely taking it apart, is also thus afforded. With blowers of other manufacture, the "putty joint" feature of the shell or casing is an indispensable adjunct, although it is a construction point which is, at the best, something to be avoided in an efficient machine.

The Buffalo Steel Pressure Blower is designed and constructed especially for high pressure duty, such as supplying blast for cupolas, furnaces, forge fires, sand blast machines, for any work requiring forcing of air long distances, as in connection with pneumatic tube delivery system. It is adapted for all uses where a high pressure or strong blast of air is required. The journals are long and heavy, in the standard ratio of length to diameter of six to one, and embody a greater amount of wearing surface than those upon the blower of any other construction. Attention is directed to the patented journals and oiling devices employed on this blower, which are unique features. The bearings are readily adjustable, and any wear can be taken up, which is an important point attending the durability and quiet running of a perfect machine.

The Buffalo Steel Pressure Blower possesses the fewest number of parts of any like machine; in fact, the blower is practically one piece, so that under any service the bearings invariably are in perfect alignment, vertically and laterally, with the rest of the machine. In the items of durability, smooth running and economy of power, it is thus rendered far superior to any blower with the so-called universal journal bearing which is commonly employed.

In every point of construction, the greatest pains have been taken to simplify all parts and at the same time to give them the greatest strength. To adjust, repair and keep in order a Buffalo Blower is a very small matter and readily understood by a machinist of average ability.

For obtaining the best results from a blower of given size, when used for melting iron in foundry cupolas, much depends upon the proper lay-out of the blast piping between the blower and the cupola, and also upon the proper proportionment, arrangement and design of the cupola tuyeres, Several forms of cupolas are now upon the market, economical in the use of fuel and fast melting, which are the points most sought for in cupola construction. It is a common but erroneous idea that a blower large for the work will give better results, in a given diameter of cupola, than a smaller one. In the tables which accompany the blower, we give the proper sizes of blower for different diameters of cupolas; but it must be borne in mind, that if the tuyerage is not of sufficient area, or if the blower has to be located at some distance from the work to be accomplished, these points enter for consideration. Frequently, foundrymen, when experiencing difficulty in obtaining satisfactory melts, throw the whole cause of the trouble upon the blower, when the fault does not lie at this point. It is safe to say that failures are due more largely to the mismanagement of a cupola and improper application of the blower, than to any other cause.

The Buffalo Steel Pressure Blower is especially adapted for foundry cupolas, and is guaranteed to produce stronger blast with less expense for power, than any other.

BLOWER ON ADJUSTABLE BED, AND ON BED COMBINED WITH COUNTERSHAFT.

Unless considerable care is taken in putting up countershafts, and some special attention is given to keep them in perfect alignment, trouble is often experienced, especially in keeping the belts on the larger sizes of blowers, on account of the great speed at which they have to run to produce high pressures. To overcome such features, this house designed the adjustable bed, and the adjustable bed combined with countershaft arrangements, which is illustrated in Fig. 63. The blower on adjustable bed, alone, without the countershaft, is very convenient for taking up the slack in belts while the fan is in motion and driven by belt from main line.

In Fig. 63 is shown the latest construction form of Buffalo Steel Pressure Blower on adjustable bed with combined countershaft. Its use will be found to result in a decided saving in the wear and tear upon belts, which, in a short time, more than justifies the extra initial expense of the arrangement. The cost will be found little in excess of ordinary method, and a few turns of the nut on the end of the adjusting screw, which is clearly shown directly under the outlet of the blower, after first unloosening the holding-down bolts, which should afterward be re-tightened,

FIG. 63.

BLOWER AND COUNTERSHAFT.

accomplish, in a very few moments, what, previous to the introduction of this apparatus, has caused considerable delay and annoyance. It will readily be seen that the usual frequent relacing of belts, to make them sufficiently tight to avoid slipping, is hereby entirely obviated.

Positive alignment of the countershaft with the shaft of the blower by this arrangement causes the belt to track evenly, run smoothly and avoid the usual wear by their striking against the hanger or side of the blower. As will be readily appreciated, the tightening screw gives the same uniform tension to both belts, and this may be regulated at will of operatior. A telescopic mouth-piece, as is shown by the cut, is placed upon each blower purchased in this form, which enables the machine to be moved upon its bed without any disarrangemnnt of the blast piping.

Especial attention is called to the fact that the arrangement of blower on adjustable bed combined with countershaft, as illustrated in Fig. 63, occupies the smallest amount of space consumed by any apparatus of this kind manufactured in the world. Ordinary tight and loose pulleys are placed upon the countershaft from which the power is transmitted to the countershaft of this apparatus. When this feature is not desirable, which is often the case where power is transmitted from the main line without the intervention of a countershaft, the adjustable bed countershaft may be furnished with the blower, so that it will extend at the right or left, as desired, and the tight and loose pulleys are then placed thereon; we then have a right or left hand apparatus. The space between the two pulleys which drive the blower is not wide enough to permit of the introduction of tight and loose pulleys.

BLOWER ON ADJUSTABLE BED, COMBINED WITH DOUBLE UPRIGHT ENGINE.

We would call attention to the blower in the adjustable bed form and also in the combination with countershaft. The further combination as secured in the introduction of a double upright enclosed engine for supplying the power, affords the very highest economy and convenience. This arrangement gives positive control over the tension of belts, ensures the greatest rigidity, ease in adjustment, perfect alignment, and when it is desirable, an immediate change in the speed of the blower. The latter is a very desirable feature, especially in cupola work, because in hot weather it requires an increased volume of air to melt the same quantity of iron over that of cold weather. It will readily be seen that this arrangement possesses marked advantages over blowers with power by belt transmission, as they may be run whenever desired, and are independent of other sources of power.

THE CUPOLA FURNACE.

The design of engine, together with the workmanship and material employed, is identically the same as upon the regular Buffalo Double Upright Enclosed Engine. This design of engine is peculiarly fitted for driving steel pressure or cupola blowers. In foundries or forge shops, much dust and dirt are present in the atmosphere, but the running parts of the engine are thoroughly protected therefrom. As will be seen by reference to Fig. 64, this engine is furnished with a common oil chamber

FIG. 64.

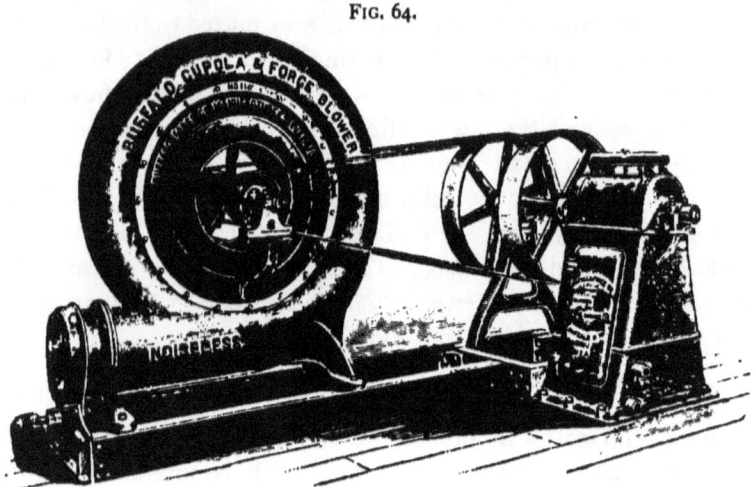

BLOWER AND UPRIGHT ENGINE.

on top of frame, from which oil tubes of different sizes, according to the function each is to perform, lead to every reciprocating part. Continuous running is possible without the repeated opening and closing of the door in engine. The engine, which is built in a variety of sizes for the different blowers, being especially designed and adapted for high rotative speed, possesses short stroke, and the reciprocating parts are perfectly balanced.

BUFFALO ELECTRIC BLOWER BUILT IN "B" AND STEEL PRESSURE TYPES.

The "B" Volume Blower, illustrated in Fig. 65 is built with electric motors of approved type as a part of the fan, and connected directly to the fan shaft. Electric fans afford greater

BLOWERS. 307

convenience even than direct attached engine fans. They are unrivaled in their adaptability to all classes of work, and to all locations. To start and stop is simply a matter of moving a switch or pushing a button, according to the arrangement. No engines or belts are required, and they are always ready for immediate use.

One special feature of their great convenience, to which particular attention should be called, is the fact that the fans can be set up in any position without affecting the running of the

FIG. 65.

ELECTRIC BLOWER.

motor. This so adapts the fans that they may be located to discharge or exhaust from any desired direction, which entails the least complication of pipe connections. The "B" volume type of blower and exhauster, when built as an electric fan, can readily be furnished in the different styles of discharges described for this design.

All types of fan built by this house can be readily fitted and furnished with direct attached electric motors, though, in the case of very large steel plate fans, it is usually more desirable to

308   THE CUPOLA FURNACE.

employ an independent motor, conveniently located, and then belt to the fan. All the fans supplied are of standard high grade, but are somewhat especially designed to receive the motors. That the highest efficiency may be secured, electric motors of approved design and special construction are built for the propulsion of the different varieties of fans. They are also capable of continuous use with only ordinary attention. For ventilating work, these fans may be employed in a multitude of positions where the introduction of an engine and boiler required to derive the power for driving other varieties of fans would be impossible. All that is required is a wire connection with a power circuit, and the fan is ready for immediate operation. Electric fans may be driven at a high speed, therefore they are of large capacity. The combination of electric motor and fan, with proper care and under ordinary conditions of use, is noiseless in operation and is the acme of convenience.

The Buffalo Steel Pressure Blower is frequently furnished with electric motors attached direct to the shaft. It is desirable, especially in the larger sizes, to arrange the combination of steel pressure blower and motor substantially as shown in Fig. 65, substituting the motor for the engine. By properly proportioning the pulleys on countershafts, any pressure required for ordinary duty can be given while the motor is making its regular speed.

BUFFALO BLOWER FOR CUPOLA FURNACES IN IRON FOUNDRIES.

In the following table are given two different speeds and pressures for each sized blower, and the quantity of iron that may be melted per hour with each. In all cases, we recommend using the lowest pressure of blast that will do a given work. Run up to the speed given for that pressure, and regulate the quantity of air by the blast gate. The proportion of tuyerage should be at least one-ninth of the area of cupola in square inches, with not less than four tuyeres at equal distances around cupola, so as to equalize the blast throughout. With tuyeres one-twentieth of area of cupola, it will require double the power to melt the same quantity of iron, and the blast will not be so evenly distributed. Variations in temperature affect the working of cupolas very materially, Hot weather requires an increase in volume of air to melt same quantity of iron as in cold weather.

TABLE OF SPEEDS AND CAPACITIES AS APPLIED TO CUPOLAS.

Number of Blower.	Square Inches Blast.	Diameter Inside of Cupola, in Inches.	Pressure in Ounces.	Speed—No. of Revs. per Minute.	Melting Capacity in Lbs. per Hour.	Cubic Feet of Air Required per Minute.	Pressure in Ounces.	Speed—No. of Revs. per Minute.	Melting Capacity in Lbs. per Hour.	Cubic Feet of Air Required per Minute.
4	4	20	8	4732	1545	666	9	5030	1647	717
5	6	25	8	4209	2321	773	10	4726	2600	867
6	8	30	8	3660	3093	951	10	4108	3671	1067
7	14	35	8	3244	4218	1486	10	3642	4777	1668
8	18	40	8	2948	5425	2199	10	3310	6082	2469
9	26	45	10	2785	7818	3203	12	3260	8598	3523
10	36	55	10	2195	11295	4938	12	2413	12378	5431
11	45	65	12	1952	16955	7707	14	2116	18357	8358
11½	55	72	12	1647	22607	10276	14	1797	25176	11144
12	75	84	12	1647	25836	11744	14	1797	28019	12736

SMITH'S DIXIE FAN BLOWER.

This blower is constructed with a view to deliver a large volume of air under moderate pressure with the least possible expenditure of power. It is the nearest to noiseless in operation of any fan blower made, and is of the simplest and strongest construction, the latest design, and is the lightest running fan in the world. It has steel shafts, wheels and casing, and is thoroughly tested and fully warranted. The construction of the case or shell of this blower is entirely different from anything heretofore made, and owing to its adjustable hanger bracket and feet on all four sides, it is adapted for any possible location or position. The illustrations, Figs. 66 and 67, show how the blower may be changed from bottom to top horizontal discharge, by simply turning the bearing brackets on the side of blower cases. It can also be changed to bottom or top vertical discharge as well, in less than five minutes' time, by simply loosening four bolts on either side of the blower case and turning the bracket one-fourth round either way. Tighten up again and the blower is changed and ready for operation as you want it. This blower is adapted to cupola furnaces aud forges, and for all purposes where a strong blast of air is required, or a large volume of air such as is needed in the melting of iron in foundry cupolas, for which purpose a large number of these

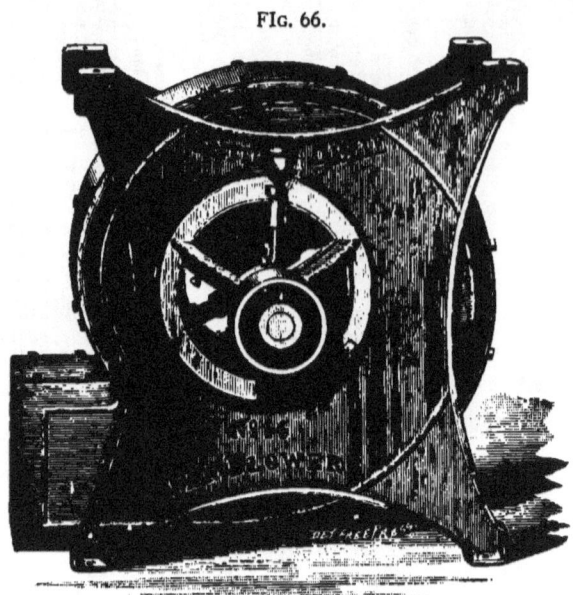

FIG. 66.

Bottom Horizontal Discharge.
SMITH'S DIXIE FAN BLOWER.

blowers are now in use. The proper arrangement of pipes in connecting a blower with a cupola is a matter of great importance, if the full volume of blast from the blower is to be delivered to the cupola. The friction of air through long or crooked pipes, which are much too small for the distance the air is to be conducted, or the pipes having one or more short, abrupt angles between blower and cupola, is often the cause of much annoyance and dissatisfaction, and frequently blowers of all makes are condemned as worthless, when the piping alone is at fault. A blower delivering air two hundred and fifty feet from the blower through an eight-inch diameter of pipe, the area of which is the same as the combined area of the tuyeres in cupola, will not deliver over two-thirds the pressure at the

Fig. 67.

Top Horizontal Discharge.
SMITH'S DIXIE PAN BLOWER.

cupola that there is at the blower. Under like conditions a twelve-inch diameter of pipe would deliver 15-16 of the pressure at the blower to the cupola. Built by the American Blower Co., Detroit, Michigan.

FORCED BLAST PRESSURE BLOWERS.

THE MACKENZIE BLOWER.

In Fig. 68 is shown a section of the Mackenzie Positive or Pressure Blower, which is probably the first rotary positive blower introduced in this country, and is certainly the first one to come into general use for foundry cupolas. This blower was designed by the late P. W. Mackenzie and introduced in connection with the Mackenzie cupola, and was a decided improvement upon the rotary fan blower then in common use. The blower, although an old one, is said by those who have used it to be a good one, and a large number are at the present time in use in foundries in various parts of the country. A description and claims for the blower are furnished by its present

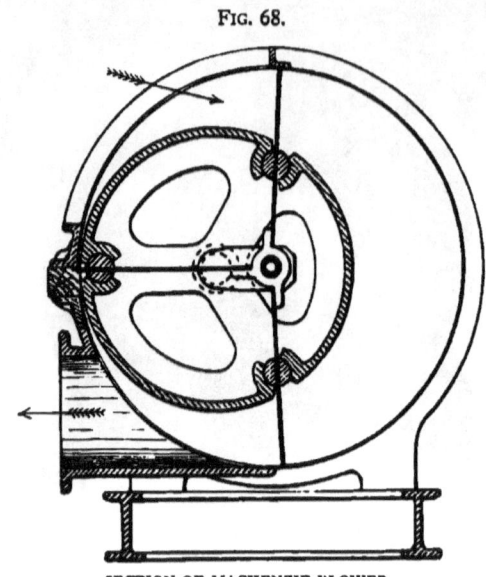

FIG. 68.

SECTION OF MACKENZIE BLOWER.

manufacturers, Isbell Porter Co., 46 Bridge St., Newark, N. J., as follows:

It is a well-known fact that a trustworthy blast, thoroughly penetrating the charge, is of the utmost importance in the economical working of a cupola, saving in many instances twenty to thirty per cent. of coal. The Mackenzie blower is a positive or pressure blower, that is, it delivers a definite quantity of air for each revolution, regardless of the condition of the cupola. This, of course, is essential to the proper melting of iron. It requires less speed, has the least possible friction of parts, and consequently uses less power than any other blower made. The late P. W. Mackenzie in experiments with blowers found that no positive blower required more than six-tenths of the power required for the best fan blowers, when the pressure exceeded four-tenths of a pound per square inch.

This blower is practically noiseless in operation, and its durability is unequaled.

We build eight sizes:

No.	Dia. of Shell.	Capacity per 100 revolutions.	Size of Outlet Opening.	Floor Space occupied.	Price.
3	22"	1300 cu. ft.	9½" x 13¼"	58" x 36"	
4	32	1800 "	10 x 18½	71 x "	
5	48	3200 "	12½ x 19¼	70 x 54	
6	"	4500 "	13 x 25	82 x "	
7	"	5600 "	" x 36½	94 x "	
8	60	7500 "	19½ x 24	96 x 75	
9	"	10000 "	" x 26	108 x "	
10	"	12500 "	" x 27½	120 x "	

No. 3 will supply blast for No. 1 and No. 2 MACKENZIE CUPOLA with 2 to 3 H. P.
" 4 " " No. 3 " " " 3 " 4 "
" 5 " " No. 4 and No. 5 " " " 4 " 5 "
" 6 " " No. 6 " No. 8 " " " 6 " 7 "

It will melt faster and with less power in straight cupolas than any other blower in use.

No. 3 will supply blast for Cupolas   to   30" dia. with 3 to 3½ H. P.
" 4 " " " 30" " 36" " " 3½ " 5 "
" 5 " " " 36 " 48 " " 5 " 7 "
" 6 " " " 48 " 60 " " 7 " 8 "

The construction and operation of the machine will be readily understood from the cut. The blades are attached to fan boxes, which revolve on a fixed center shaft. Motion is imparted to them by means of a cylinder to which are attached the driving pulleys. Half-rolls in the cylinder act as guides for the blades, allowing them to work smoothly in and out as the cylinder revolves. At each revolution the entire space back of the cylinder between two blades is filled and emptied three times.

DIRECTIONS FOR SETTING UP BLOWER.

Set the machine upon a level and substantial foundation, in a room free from dust. The main pipe should be equal in capacity to the combined capacity of the tuyere pipes.

For blowers Nos. 3 and 4, the diameter of main pipe should be thirteen to fifteen inches, and for Nos. 5 and 6, the diameter of main pipe should be from sixteen to eighteen inches; all connections must be permanently air-tight, and all curves made easy. The blades should be oiled freely for a few days, then they will show plainly where oil should be used. The shaft upon which the fan boxes revolve is hollow, and the opening to the oil passage in shaft will be seen outside the hanger and on top of shaft. Fill the shaft with oil when the machine is started, and supply a small portion occasionally when running. Keep the passages open so that the oil will find its way readily to the bearings. Use good oil, give the machine proper care, and it will last for years without repairs.

THE GREEN PATENTED POSITIVE PRESSURE BLOWER.

This is a blower of a new design recently placed upon the market by the Wilbraham-Baker Blower Co., Philadelphia, Pa., to take the place of the Baker blower, for many years manufactured by them. The new blower is said to be a great improvement upon the Baker blower, which is one of the best in use for foundry cupolas. Claims for the Green blower are made by the manufacturers as follows:

This blower is designed to occupy the minimum space, displace the largest volume for the space occupied, exhaust and deliver in a practically even volume, have the least weight combined with ample strength, be entirely void of complications, complicated shapes, sliding parts or sliding motion, the least liability to get out of order, the least weight to revolve, be entirely free of internal friction in case of wear of the journals, and do the work with the minimum power.

The working parts are two perfectly balanced impellers, each of which is a single strong casting, well ribbed inside and firmly fastened to a steel shaft of ample dimensions, extending the full length of blower, the shaft being flattened where it passes through the body of impeller.

The journal bearings are bushed with phosphor bronze and are detachable from blower, being bolted and dowel pinned to

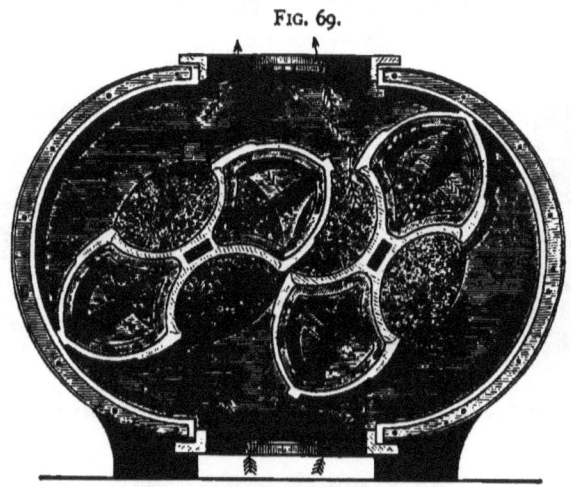

FIG. 69.

SECTIONAL VIEW.

the head plate, easily removed and returned to their original central position.

The blower is geared at both ends, the gearing being of ample proportions, cut in the most accurate manner, and enclosed in an oil-tight cover, free from dust and dirt, and continuously in oil. The case of blower is well-proportioned, strongly ribbed and firmly bolted together.

The head plates, in addition to being well-ribbed, are further strengthened by having the hoods or extensions, into which the circular ends of the impellers project, a part of head-plate casting.

The circular parts of casing and the pipe plates are also ribbed, and the pipe plates are fitted in between and bolted to the head plates and circular casing.

The finished surfaces of the impellers are two circles, which roll together without friction, forming an even and continuous practical contact; the point of contact being always on the pitch line of the gears and traveling at the same speed at all points of the revolution.

The gear wheels are keyed to the shafts close to ends of journal bearings, forming collars at each end of blower, prevent-

FIG. 70.

COMPLETE IMPELLER.

A single casting with Steel Forged Shaft in position. Two such pieces compose the interior working parts of Blowers and Exhausters.

ing the impellers from rubbing endwise against the interior sides of head plates.

These provisions insure an entire absence of internal friction at all times, and are a positive preventive of possible accidents.

The following Fig. (71) shows Standard Green Blower with discharge outlet on either side. Fig. 69 a sectional view, and Fig. 70 the complete impeller.

The inlet and outlet flanges are tapped for tap or screw bolts and provided with loose flanges for attaching light sheet-iron pipe.

Directions for setting up.—Set the blower perfectly level and solid. Brick or stone is best for a foundation; timber is liable to rot and allow blower to get out of level. See that oil holes are clean before attaching oil cups, and set cups to feed properly. Before attaching pipes see that nothing has fallen into the blower. Fasten a coarse wire screen on end of inlet pipe. Wipe the gear wheels and gear casing perfectly clean before attaching the casing, and cover the joint between parts of gear casing with red or white lead. Put a supply of good heavy oil inside the gear covers, say a pint in each small, and a quart in each large cover, and draw off this oil and replace with fresh oil

FIG. 71.

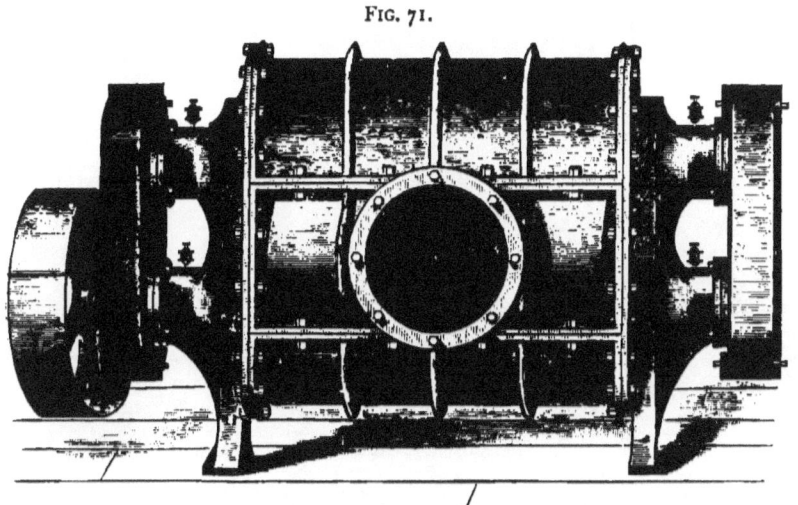

STANDARD GREEN BLOWER.

about once each month. Use a good fluid oil on journals and gear-wheels. No lubricant required inside of blower.

Efficiency of blower.—The blower is not guaranteed to accomplish any given duty; the blower simply furnishes the air at its discharge outlet; the result obtained depending upon the disposition of the air after it leaves the blower. Tight iron pipes must be used so that all the air delivered by the blower will reach the desired point. For overground, galvanized iron pipes, riveted and soldered, are good, and for underground, cast iron is the best. Cast iron blast gates are recommended. Light wrought iron or brass gates are liable to leak and impair the efficiency of the blower. A blast gate having the gate pass entirely through the frame is the best.

Power.—For estimating the approximate amount of power required to displace a given amount of air at a given pressure, it is customary to add 25 per cent. to the net result obtained by using the following rule. Rule.—Multiply the number of cubic feet delivered per minute by the pressure in ounces per square inch (at the blower) and the product by .003; divide the last amount by 11.

STANDARD FOUNDRY BLOWERS DRIVEN BY PULLEY.

DIMENSIONS IN INCHES.

Number.	Displacement per Revolution.	Total Length.	Side Outlet.		Top or Bottom Outlet.		Pulley.		Inlet and Outlet.		Average Weight.
			Height of Head Plate.	Width of Head Plate.	Height of Head Plate.	Width of Head Plate.	Diameter.	Face.	Dia. of Opening.	Dia. of Flange.	
No. ¼	¾ cu. ft.	41"	24"	15"	15"	22"	16"	5"	6"	9"	650 lbs.
" ½	1½ "	52"	31"	23"	23"	30"	20"	5"	7"	10"	1350 "
" 1	3 "	58"	37"	27"	27"	35"	24"	6"	10"	13"	1800 "
" 2	5½ "	66"	43"	29"	30"	41"	30"	6"	11¾"	14¾"	2600 "
" 3	9 "	78"	47"	34"	35"	46"	36"	7½"	14"	18"	4000 "
" 4	15 "	85"	58"	42"	42"	57"	40"	8"	17"	21"	6000 "
" 4½	20 "	96"	58"	42"	42"	57"	44"	8"	17"	21"	6500 "
" 5	25 "	106"	64"	46"	46"	62"	48"	10"	19"	24"	9000 "
" 5½	35 "	112"	72"	50"	50"	70"	48"	12"	21"	25½"	11000 "
" 6	42 "	129"	81"	59"	60"	80"	60"	15"	24"	29"	16500 "
" 7	67 "	156"	93"	66"	67"	92"	72"	18"	30"	37"	28000 "
" 8	112 "										
" 9	200 "										

(Subject to correction.)

THE CUPOLA FURNACE.

SPEED OF FOUNDRY BLOWERS.

No. 1 blower displaces 3 cubic feet per revolution. Suitable for cupola 24 to 28 inches diameter for melting:

    ¾ tons per hour.....................125 revolutions per minute.
    1¼ "   " 210   "   "
    1¾ "   " 290   "   "

No. 2 blower displaces 5½ cubic feet per revolution. Suitable for cupola 24 to 34 inches diameter, for melting:

    1¼ tons per hour.....................115 revolutions per minute.
    2½ "   " 230   "   "
    3   "   " 275   "   "

No. 3 blower displaces 9 cubic feet per revolution. Suitable for cupola 28 to 40 inches diameter, for melting:

    2 tons per hour.....................110 revolutions per minute.
    3 "   " 165   "   "
    4¼ "   " 245   "   "

No. 4 blower displaces 15 cubic feet per revolution. Suitable for cupola 32 to 45 inches diameter, for melting:

    3 tons per hour.....................100 revolutions per minute.
    5 "   " 170   "   "
    6½ "   " 220   "   "

No. 4½ blower (small No. 5), displaces 20 cubic feet per revolution. Suitable for cupola 36 to 50 inches diameter, for melting:

    4 tons per hour.....................100 revolutions per minute.
    6 "   " 150   "   "
    9 "   " 225   "   "

No. 5 blower displaces 25 cubic feet per revolution. Suitable for cupola 42 to 56 inches diameter, for melting:

    5 tons per hour.....................100 revolutions per minute.
    8 "   " 160   "   "
    10 "   " 200   "   "

No. 5½ blower (small No. 6), displaces 35 cubic feet. Suitable for cupola 48 to 64 inches diameter, for melting:

8 tons per hour................115 revolutions per minute.
10 "   " 145   "   "
14 "   " 200   "   "

No. 6 blower displaces 42 cubic feet per revolution. Suitable for cupola 50 to 70 inches diameter, for melting:

9 tons per hour................110 revolutions per minute.
12 "   " 145   "   "
15 "   " 180   "   "

No. 7 blower displaces 67 cubic feet per revolution. Suitable for cupola 66 to 78 inches diameter, or two cupolas 48 to 56 inches diameter, for melting:

14 tons per hour................105 revolutions per minute.
18 "   " 135   "   "
20 "   " 150   "   "

No. 8 blower displaces 112 cubic feet per revolution. Suitable for cupola 74 to 92 inches diameter, or two cupolas 54 to 66 inches diameter, for melting:

20 tons per hour................90 revolutions per minute.
25½ "   " 115   "   "
30   "   " 135   "   "

No. 7½ blower displaces 85 cubic feet per revolution.

No. 9 blower displaces 200 cubic feet per revolution.

Speed.—For blowers running continuously at pressures of about two pounds per square inch, the following maximum speed is recommended:

No. of blower	1	2	3	4	4½	5	5½	6	7	8
Revolutions per minute	250	250	225	200	200	175	175	150	135	100
Displacement per rev. in cubic feet	3	5½	9	15	20	25	35	42	67	112

CONNERSVILLE CYCLOIDAL BLOWER.

The Connersville Positive Pressure Blower, manufactured by the Connersville Blower Co., Connersville, Ind., is one of the latest designs of blower, and has only been manufactured for a few years. A description of it is taken from the excellent circular which is well worth reading by those contemplating the purchase of pressure blowers, and is as follows:

21

322   THE CUPOLA FURNACE.

The cycloidal curves, their nature, peculiarities and possibilities, have always been an attractive study, not only to the theoretically inclined, but more particularly to those interested in the many important applications of these curves in practical mechanics. The especial value of combining the epi- and hypo-cycloids to form the contact surfaces of impellers for rotary blcwers, gas exhausters and pumps has long been recognized, and many attempts have been made to utilize them in that connection, but in vain. While conceded to give the theoretically correct form to a revolver or impeller, it came to be regarded as impossible to produce such surfaces by machinery with suffi-

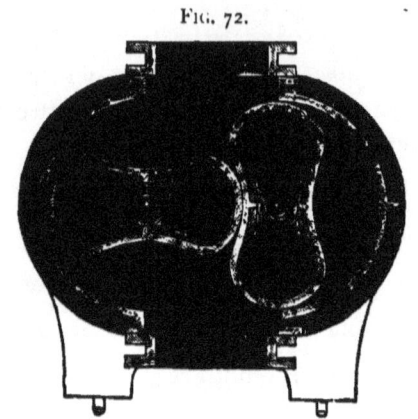

FIG. 72.

SECTIONAL VIEW OF CONNERSVILLE CYCLOIDAL BLOWER.

cient accuracy to admit of their use in practice with any degree of satisfaction. It remained for us to demonstrate that it could be done, and in a highly successful manner as well.

Fig. 72 is an illustration showing a cross section of our new cycloidal blower, and particularly of the revolvers or impellers, their form, relation to each other, and to the surrounding case. A glance only is required to discern the superiority of this method of construction over all others.

The vital part of every machine of this class is the impeller, as on it depend economy of operation and efficiency in results.

That we have the ideal form for an operating part is self-evident. It will be noted that there are two impellers only, and each is planed on cycloidal lines with mathematical accuracy. Now, it is one of the well-known peculiarities of the epi-cycloidal and hypo-cycloidal curves, when worked together as in our machines, that there is a constantly progressive point of contact* between the impellers. As a result of this regular advance of the point of contact, the air is driven steadily forward, producing a smooth discharge that is conducive to the highest economy.

The advantage of this arrangement over the use of arcs of circles to approximate contact curves is very great, as it is a well-demonstrated fact that circular arcs whose centers are not co-incident with the centers of revolution can not keep practical contact through an angle of more than four or five degrees. On the contrary, the contact does not progress continuously, but jumps from one point to another across intervening recesses as the impellers revolve, leaving pockets in which the air is alternately compressed and expanded, producing undesirable pulsations in the blast, a waste of power, and necessitating two points of contact at one time in four positions in each revolution.

Another advantage of the cycloidal form is that, at the point of contact, a convex surface is always opposed to a concave surface; that is, the epi-cycloidal part of one impeller works with the hypo-cycloidal part of the opposite impeller. The consequence of this is to produce a *long contact* or distance through which the driven air must travel to get back between the impellers, instead of the short contact that results when two convex surfaces oppose each other, as is the case in other machines of this character.

Attention has previously been directed to the fact that the

* Wherever the expression "point of contact" is used in this description, it must not be understood to mean that the impellers actually touch at such points, but that it is *the point of nearest* approach. In practice it has been found advisable to allow a very slight clearance rather than have the parts rub together, as thereby friction and wear are entirely eliminated, while on account of the "long contact" referred to, the leakage is insignificant. Our method of planing the impellers enables us to make the clearance very slight.

point of contact between the impellers continuously progresses; indeed, the path it describes is a circle. One result of this continuously-progressive contact, as before mentioned, is a smooth, reliable blast. Another is, as has also been noted, the absence of any pockets or cavities in which air can be gathered, compressed and then discharged back toward the inlet side of the machine, thereby entailing a waste of power and shortening the life of the blower by subjecting the impellers, shaft and gears to a needless shock, strain and wear. Furthermore, the impellers can be in contact only at one point at the same instant—in no position is it possible for them to touch each other at two points at once; hence, there are no shoulders to knock together when the speed is more than nominal.

On account, also, of there being no popping due to the expansion of air when released from the pockets in which it has been caught and compressed, and no pounding of the impellers together, that disagreeable din and vibration usually associated with machines of this class is eliminated, and our blowers run with practically no noise. This is a feature that will commend itself to parties having had experience with other pressure blowers.

Another point contributing to the evenness and uniformity of the discharge is the fact that the extremities of the impellers are curved. Thus, as they sweep past the outlet, there is a gradual equalization of the pressure instead of a sudden shock, such as results from the passage of two sharp edges, which shocks are so detrimental to all working parts, as has been noted.

From what has been stated, we scarcely need to add that the machine is positive in its action. All the air that enters the blower is inclosed by the impellers, forced forward and discharged through the outlet pipe. The leakage is insignificant, and there is no compressed air allowed to escape backward. Hence, all the power applied to the machine is used for the purpose intended—the maintenance of an even blast, and none of it is wasted on needless work.

Furthermore, as the contact between the impellers and the

surrounding case is perfect at all times, the amount of pressure that can be developed and sustained depends solely on the strength of the machine and the power applied.

Each of the two impellers is cast in one piece and well ribbed on the inside to prevent changes in form under varying conditions. It is part of our shop practice to press the shaft into the impeller with a hydrostatic press, finish the journals to standard size, mount the impeller on a planer and *plane its entire surface accurately*. By this means we secure perfect symmetry and exactness with respect to the journal on which it revolves, and, as a consequence, can produce a machine that will run more

FIG. 73.

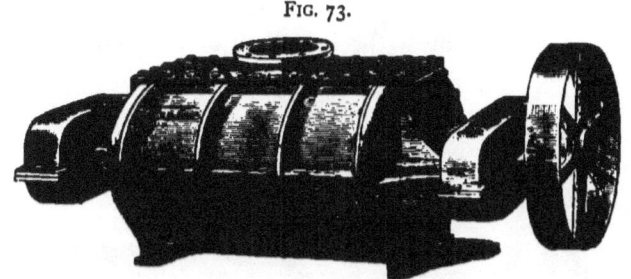

HORIZONTAL BLOWER.

smoothly, and in either direction, at a higher speed and pressure than it has been possible to attain heretofore.

It will be observed that the cycloidal curves produce an impeller with a broad waist. We have availed ourselves of this to use a high-grade steel shaft of about *twice the sectional area of those found in competing machines*. The advantages of this need not be enumerated.

In Fig. 73 we illustrate the styles of blowers that are most largely sold, *i. e.*, those pulley driven. It will be noticed that we use one pulley only. We can, however, when desired, put a pulley on each end, but because of the large shafts, wide-faced gears, and the fact that there is a bearing the entire distance from the gears to the impellers, it is seldom necessary. In any event, we do not recommend two belts very highly, as, owing to the difference in the amount of stretch in the leather, it is usually the case that one transmits most of the power. Indeed, it sometimes occurs that they work against each other.

NUMBERS, CAPACITIES, ETC., OF THE CYCLOIDAL BLOWERS.

Number of Blower	¼	½	1	2	3	4	5	6	7	8
Capacity in cubic feet per revolution	⅔	1½	3	5¼	8	12½	24½	42	67	100
Ordinary speed	400	350	300	275	250	200	175	150	125	100
Diameter of pipe opening	4	6	8	10	12	14	16	20	24	30

By "ordinary speed" we mean what would be about an average of every-day duty. It must be understood, however, that the peculiar form of the impellers of our blowers, in con-

FIG. 74.

VERTICAL BLOWER AND ENGINE ON SAME BED-PLATE.

nection with the other superior points in construction, to which we have called attention, permits of higher speeds than competing machines.

BLOWERS. 327

The speed at which positive pressure blowers are run may be classed as "slow;" therefore, power can be taken direct from the main line of shafting or from a countershaft driven at the same rate.

Fig. 74 shows a blower with an engine to furnish the required power, both on the same bed-plate. By such a combination all shafting, pulleys, gears and belts are dispensed with, as the crank shaft of the engine is coupled direct to a shaft of the blower, thereby effecting a very simple but most efficient driving arrangement. We recommend the installation of such a plant when the blower is to be located at a considerable distance from the line shaft, as it will be found more economical to pipe steam to the engine than to transmit power by shafting or cable. But

FIG. 75.

BLOWER AND ELECTRIC MOTOR.

even where power is convenient there are many good reasons why it will be found much more desirable to operate the blower with its own engine. For instance, it can be run independent of the other machinery, as necessity or convenience may often require, and also permits the speed of the blower to be varied, as there is a demand for an increased or diminished amount of blast, while otherwise this could not be accomplished without a change of pulleys.

In nearly every town there is now a station for electric-lighting purposes, and managers of it are finding that they can extend the earning capacity of their plants and increase their profits by renting power at a time when otherwise their machinery would be practically idle. We have arranged to have our machines operated by electric motors when desired. In Fig. 75 will be found an illustration of a motor geared direct to a blower, both on the same bed plate. When preferred, however, the motor can be located a short distance away, and the power transmitted to the blower pulley by means of a belt. Foundries and other industries needing power only to run their blowers will find it exceedingly advantageous and economical to adopt this plan. Not only will there be a saving in first cost, but the operating expense will be much less.

Furthermore, the motors can have sufficient power to run the rattler and other light machines about the establishment.

GARDEN CITY POSITIVE BLAST BLOWERS.

In Fig. 76 is shown the Garden City Positive Blast Blower manufactured by the Garden City Fan Co., Chicago, Ill., many of which are in use in foundries, and for which claims are made as follows:

The operation of our blower is not on the fan principle, in which pressure is obtained by a high velocity or speed, but when the air enters the case at the inlet and is closed in by the vanes of the blower, it is absolutely confined and must be forced forward until finally released at the outlet, where it must have escape or the blower stop if outlet is closed. There is positively no chance for loss by backward escapement of air, after it once enters the inlet.

In many respects our blower has points of superiority over any positive blower made, and we call your attention to the following points:

1st. It has no gears whatever. No internal parts that require attention, adjustment or lubrication.

2d. It is only two journal bearings that are external to the blower casing. They are self-oiling. Easy of adjustment.

3d. Has no irregular internal surfaces that require contact to produce pressure, and add friction.

4th. Operating parts are always in perfect balance, thus blower may be safely run at a higher speed than any positive blower made, giving a proportionate increase in efficiency and a smaller blower may be used.

5th. A higher pressure can be be obtained than is possible with any other.

FIG. 76.

GARDEN CITY POSITIVE BLAST BLOWER.

6th. The blowers are practically *noiseless* as compared with all other makes.

ROOTS'S ROTARY POSITIVE PRESSURE BLOWER.

The Roots Blower was designed by Mr. Roots, of Connersville, Ind., nearly forty years ago. It is said that it was originally designed for a turbine water wheel, but when the water was let in it was all blown out, and Mr. Roots at once decided it would make a better blower than a water wheel, and after considerable experimenting, perfected it as a blower. Whether this story be true or not we cannot say, but the machine certainly makes a good blower, and hundreds of them have been used to furnish blast for foundry cupolas. A number of marked improvements have from time to time been made in it since it was originally invented, and the impellers, which were originally made of wood placed upon iron shafts and covered with bee's wax or soap to make them air-tight, are now made entirely of iron and accurately fitted. The shape of the blower cases has also been to some extent changed, and they are now constructed vertical and horizontal, as shown in Figs. 77 and 78. They are

FIG. 77.

ROOTS'S VERTICAL PRESSURE BLOWER.

also made with blower and engine on same bed-plate or with blower and electric power motors on same bed-plate. The following claims are made for it by the manufacturers, P. H. & F. M. Roots Co., Connersville, Ind.:

1. It is simpler than any other blower.
2. It is the only positive rotary blower made with impellers constructed on correct principles.
3. It is the best, because it has stood the test of years and is the result of long experience.
4. In case of wear of the journals, the impellers will not come together and break, or consume unnecessary power, as is the case with competing machines.

BLOWERS. 331

5. The principles upon which our blowers are constructed admit of more perfect mechanical proportions than any other.

6. The only perfectly adjustable journal box for this type of machine is used.

7. The gears are wide-faced and run constantly in oil.

8. The gears and journals are thoroughly protected from dust and accident.

9. Our machine blows and exhausts equally well and at the same time, and the motion may be reversed at any time.

10. All the operating parts are accurately balanced.

FIG. 78.

ROOTS'S HORIZONTAL PRESSURE BLOWER.

The principles upon which our blower is constructed are so radically different from any competing machine that we are enabled to adopt proportions that are mechanically perfect, and hence we can speed our machines much faster than any other, with a far greater degree of safety. We are not compelled to cut down the weight of our blower cases, as other manufacturers do, in order to bring the weight of the complete machine within reasonable bounds. The distribution of metal in the shafting, impellers, gears and cases of all our blowers is perfectly proportioned, and it is the only rotary positive blower made so constructed.

CHAPTER XXII.

CUPOLAS AND CUPOLA PRACTICE UP TO DATE.*

THERE are three kinds of furnaces employed in the melting of iron for foundry work. They are known as the pot furnace, reverberatory furnace and the cupola furnace. These furnaces differ from each other in construction and principle of melting, and in the days of poor fuel the employment of the pot furnace or reverberatory furnace in the melting of iron for special work was necessary to the production of good castings. But, with the discovery of veins of coal more suitable for melting and coking and the advancement made in the manufacture of coke, the amount of deterioration to iron by impurities in the fuel has been reduced to a minimum, and the furnace that will melt with the smallest per cent. of fuel has, as a foundry furnace, been almost universally adopted, and the fuel that melts iron most rapidly has almost entirely taken the place of those melting more slowly. Charcoal, the furnace fuel of years ago, is only used in foundries located in isolated districts where other fuel is not obtainable. The use of hard coal in melting is almost entirely restricted to the anthracite coal field, and coke has become the almost universal fuel for foundry work.

In the pot furnace, one ton of coke is consumed in melting a ton of iron. (2240 lbs.) In the reverberatory furnace, from ten to twenty cwt. of coke is required to melt a ton of iron. In the cupola furnace, a ton of iron may be melted with one hundred and seventy-two (172) pounds of coke. It will thus be seen that the cupola melts iron with a smaller per cent. of fuel than either of the other furnaces. To melt iron in a cupola

* Prepared for the first meeting of the American Association of Foundrymen, May 13, 1896, at Philadelphia, Pa.

with the small amount of fuel stated, the cupola must be properly constructed and managed, which is not always the case, and the consumption of fuel as a rule is much greater, but is still not so large as that required in either of the other furnaces. To reduce the amount of fuel required to the smallest possible figure, a cupola must be of a size that will admit of it being run to its fullest capacity in melting a heat. The tuyeres must be placed low to prevent wastage of fuel in the bed, and the charging aperture must be placed high to utilize all the heat of the fuel in heating the stock and preparing it for melting before it enters the melting zone.

The rule for charging a cupola is to place three pounds of iron upon the bed to each pound of fuel placed there, and ten pounds of iron upon the charges of fuel to each pound of fuel in the charge. This rule is not always accurately followed, but it is approximately so, and when the cupola is so large that the entire heat is melted upon the bed in one charge (according to this rule), only three pounds of iron are melted to the pound of fuel. When ten charges are melted in the same cupola, with the same bed, eight pounds of iron are melted to one of fuel; and the greater the number of charges, the less the per cent. of fuel required in melting, and it is only by melting a large number of charges and keeping the cupola in blast for many hours, that the small per cent. of fuel stated as sufficient to melt a ton of iron, can be made to do the work.

Foundrymen, as a rule, cannot have their cupolas in blast all day, and are compelled to use large cupolas to melt in a given time the amount of iron required for their work, while others prefer to melt their iron rapidly; and it is a question for each foundryman to decide for himself, whether it is more economical to use a large cupola and save time, or a small one and save fuel. The height or distance tuyeres should be placed above the sand bottom is from two to six inches, but they are sometimes placed as high as six feet. The general height for heavy work is from eighteen inches to thirty-six inches. The placing of tuyeres at so great a height is productive of the wastage of a large amount of fuel, for the function of the fuel placed below the tuyeres is to support the stock in the cupola, and it takes no other part in melting and is not consumed in melting the longest heats. Its temperature, when a cupola is in blast, is below that of the melting zone, and molten iron in its descent through the fuel to the bottom of the cupola is not superheated, but its temperature is reduced to such an extent that hot iron can only be tapped from a cupola with high tuyeres, when the melting is so rapid that the molten iron passes down through the fuel under the tuyeres in such a large volume and so rapidly, that it is not chilled in its descent.

While the fuel placed under the tuyeres is not consumed in melting, it is heated to so high a degree and ground up to such an extent in the dump that it is rendered worthless as cupola fuel, and every pound of unnecessary fuel placed in a cupola, by using high tuyeres, is a wastage of it.

It is claimed by many foundrymen that it is necessary to have tuyeres placed high to collect and keep iron hot for a large casting. This is one of the fallacies handed down to us by our foundry forefathers, for iron can be drawn from a cupola with low tuyeres so much hotter, that it can be kept hotter in a ladle for any given length of time when properly taken care of than it can be kept in a cupola with high tuyeres. In all cases, tuyeres should only be placed at a sufficient height above the sand bottom to admit of molten irons being mixed for hand-ladle work, and to give sufficient time between taps for the removiag and replacing of large ladles for heavy work.

Low charging doors are another legacy from our ancestors, and in their day the volume of heat escaping from low cupolas was so great that many attempts were made to utilze this waste for heating the blast, and supplying cupolas with a hot blast. Other attempts were made to divert the heat into side flues or chambers, for heating the iron prior to charging; but this feature of the old cupolas has given way more rapidly to modern ideas than the high tuyere, and cupolas in which charging doors were formerly placed six to eight feet above the bottom, now have them placed ten to fifteen feet and even higher; and all the heat that escaped from the low cupola is now utilized in heating stock in the cupola prior to melting.

The highest cupolas in use in this country at the present time are those of the Carnegie Steel Works, Homestead, Pa. The charging apertures in these cupolas are placed thirty feet above the iron bottom, and the heat to so great an extent, is utilized in heating stock prior to melting, that it has not been found necessary to line the iron stacks, as not sufficient heat to heat them escapes from the cupolas even when the stock is low.

A cupola to do economical melting must not only be properly constructed, but properly managed. In every cupola there is a melting zone or belt in which iron is melted. Below or above this zone iron cannot be melted, but it may be on the lower and upper edge of the zone, and in either case a dull iron is the result. The exact location of the melting zone is determined by the volume of blast. A large volume places the zone a few inches higher and a small one brings it a few inches lower. For this reason, cupolas of exactly the same construction frequently have higher or lower melting zones and require a greater or less amount of fuel for a bed. The size and arrangement of tuyeres often increase or diminish the depth of the melting zone, and to obtain the best results in melting the location and depth of the melting zone must be learned, and the weight of the bed and charges varied to suit the zone.

The top of the melting zone may readily be determined by the length of time required to melt iron after the blast is put on. If iron comes down in five or ten minutes, the iron on the bed has been placed within the melting zone. If it does not come down for twenty or thirty minutes after the blast is on, the iron has been placed above the melting zone by too great a quantity of fuel in the bed, and the delay in melting is due to the time required in removing the surplus fuel by burning it away. If iron comes down in five or ten minutes and is dull, and at least three pounds of iron cannot be melted to one of fuel in the bed before iron comes dull, the bed is too low and the iron, when melting began, was not placed at the top of the melting zone. By noticing the melting in this way, the height of the zone can readily be found and the exact amount of fuel required for a bed determined. The depth of the zone can be found by increasing the weight of the first charge of iron until the latter part of the charge comes dull, which indicates that it is being melted too low in the zone, and the weight of the charge should be decreased.

The quantity of fuel required for a bed and the amount of iron that can be melted upon it having been determined, tests are then made to ascertain the amount of fuel required in the charges and the amount of iron that can be melted upon each charge. The amount of fuel required in each charge is the amount that will restore the bed to the same height above the tuyeres at which it was before melting the first charge. This is found by increasing or decreasing the fuel until the melting becomes continuous and there is no variation in the temperature of the iron at the end of each charge. A stoppage or slacking in the melting denotes that the charge of fuel is too heavy and the iron upon it is placed above the melting zone. Dull iron at the latter end of a charge indicates that the charge of iron is too heavy, and so on throughout the heat.

By carefully noticing the melting in every part of the heat, the peculiarities of any cupola in melting may readily be learned, and a large amount of fuel saved.

These rules for melting are not always followed, and in nine cupolas out of ten too great an amount of fuel is consumed. It is a common practice of melters, if not closely watched, to gradually increase the fuel in charging, and when iron comes dull they attribute it to poor fuel or not enough of it, and in either case more fuel is the remedy, and as a rule dull iron and slow melting are the result. Fast melting cannot be done, or hot iron melted, with too large a quantity of fuel in a cupola. The reason for this is, that with an excess of fuel, the iron is placed above the melting zone and the extra fuel must be consumed before the iron can come within the zone, and the result is slow and irregular melting.

Iron held just above the zone for any length of time is heated to so great an extent that when it enters the melting zone it melts rapidly in one mass, and its descent through the melting zone in a molten state is so rapid that it is not superheated in passing through the zone, and drops through to the bottom of the cupola a dull iron. Slow melting is always the result of an excess of fuel, and dull iron is more often the result of an excess than of a deficiency of fuel.

The per cent. of fuel required in melting when a cupola is properly constructed and managed depends entirely upon the length of time required to melt a heat. In short heats of two to three hours eight pounds of iron to one of coke may be melted, and by careful management good hot iron for light work be made. The melting generally done in heats of this length is between six and seven to one. In long heats thirteen to one has been melted, but this ratio is seldom for any great length of time maintained; for the quality of fuel varies, and foundrymen prefer to use a little extra fuel rather than take chances of a bad heat, and in heats requiring from one to six days to melt the average melting is about ten to one.

It should be the aim of every foundryman to reduce his melting to a system. He should first see that the cupola is properly constructed, and then study its working in the manner described. When this has been learned, a slate should be made out and given to the melter, indicating the exact amount of fuel to be placed in the bed and charges, and the exact amount of iron in each charge, as well as the amount of each brand of iron or scrap in it.

As the lining burns out and the cupola diameter increases, the weight of charges of fuel and iron should be increased to correspond with the enlargement of the cupola. An accurate cupola record should be kept and the amount of fuel consumed in melting compared with each carload or lot of coke, to prove that no extra fuel is being used by the melter. When such an account is accurately kept by every foundryman, much less fuel will actually be consumed in melting than at the present day, and at the same time claims of melting ten to one in short heats and fifteen or twenty to one in long heats will no longer be heard.

CHAPTER XXIII.

CUPOLA SCRAPS.

BRIEF PARAGRAPHS ILLUSTRATING IMPORTANT PRINCIPLES.

Make a heat, take a heat, make a cast, make a mould, run a melt, casting, moulding, are all terms used in different sections of the country to indicate the melting of iron in a cupola for foundry work.

When iron runs dull from a cupola, draw all the melted iron off at once and prevent the newly melted iron being chilled by dropping into dull iron in the bottom of the cupola.

When slag flows from a tap-hole with a stream of iron, when the iron is not drawn off too close, it is due to too much pitch in the sand bottom.

The formation of slag in a spout is due to poor material used in making up the spout.

Some foundrymen do not seem to know what hot iron is, for they call all kinds of B. S. hot iron, if it will run out of the ladle.

The cutting out of the spout lining in holes by the stream of molten iron is due to a deficiency of cohesive properties in the lining material when heated to a high temperature.

When a tap-hole closes up with slag and cannot be kept open, the slag is generally produced by the melting of the material used in making up the front and tap-hole. Slag made in the cupola flows from the tap-hole without clogging it.

A little sand or clay-wash added to the front and spout material will generally correct the deficiencies in the material and save the melter a great deal of trouble with his spout and tap-hole.

In a spout with a broad flat bottom the stream takes a different course at every tap, the spout soon becomes clogged with cinder and iron, the molten iron flows in all directions, and the spout looks like a small frog pond with patches of scum. Make the bottom of the spout narrow and concentrate the stream in the center.

If the sand bottom does not drop readily when the doors are dropped, there is too much clay in the bottom material. Mix a little sand and cinder riddled from the dump with it, or some well-burnt moulding sand.

A hard rammed bottom causes iron to boil in a cupola the same as in a hard rammed mould, and is frequently the cause of a bottom cutting through. A bottom should be rammed no harder than a mould.

Wet sand in a bottom not only causes iron to boil, but hardens it. Bottom sand should be no wetter than moulding sand when tempered for moulding.

Exclusively new sand should not be employed in making a bottom. The old bottom with a few shovels of sand riddled from the gangways makes the best bottom material.

Often, a melter "don't know" why the cupola is working badly, because, if he knew, he would be discharged at once for carelessness.

A bad light-up makes a bad heat. The bed must be burned evenly or it will not melt evenly.

If the wood is not all burned up before iron is charged, the wood smokes and the melter can not see where to place the fuel and iron when charging. Never use green wood for lighting up. When green wood is used for lighting up, the bed is frequently burned too much before the wood is burned out, and the cupola is free of smoke.

Don't burn up the bed before charging the iron. When the fuel is well on fire at the tuyeres and the smoke is all burned off, put in the front, close the tuyeres and charge the iron at once.

If anything happens to delay putting on the blast after the fire is lighted, do not let that delay charging the iron, for the bed will last longer with the iron on it than it will with it off. Charge the iron as soon as the bed is ready for charging; close the front and tuyeres and open the charging door to stop the draught, and the cupola may be left to stand for hours and as good a heat be melted as if no delay had occurred.

A melter who burns up his tapping bars so that two have to be welded together to make one almost every heat, don't know how to put in a front or make his bod stuff.

The amount of fuel wasted every year in the United States by the use of high tuyeres in cupolas is sufficient to make a man very rich.

A new cupola always effects a great saving in fuel, but it is often hard to find the fuel (saved) at the end of the year. A little more practical knowledge in managing the old cupola will often enable the foundryman to find the fuel saved and price of the new cupola besides.

Never run a fan in its own wind merely to show a high pressure on the air-gauge.

The volume of blast supplied to a cupola should be regulated by the speed of the blower and not by the size of tuyeres.

That old "no blast" story of the melter has had its day among practical foundrymen.

The air-gauges in use at the present time for showing the pressure of blast on a cupola are an excellent thing to prevent a poor melter from claiming he has no blast and blaming a bad heat on the engineer, for the gauge always shows a higher pressure of blast when the cupola is bunged up from poor management.

High tuyeres in a cupola are an inheritance left us by our forefathers in the foundry business, of which we have never got rid.

The only general improvement made in tuyeres in the past fifty years has been in increasing them to a size that will admit the blast freely to a cupola. The only local improvement has been in placing them lower.

Molten iron should be handled in a ladle and not held in a cupola. Nothing is gained by holding iron in a cupola to keep it hot.

"I will let that go for to-day, and to-morrow I will take more time and fix it right," is a remark frequently made by melters. That kind of work is often the cause of a very bad heat.

Pig-iron melts from the ends, and the shorter it is broken the quicker it will melt.

Tin-plate scrap may be melted in a cupola the same as cast iron. It throws off sparks from the tap hole and spout similar to hard cast iron.

The fins on castings made from tin-plate scrap must be knocked off with the rammer, for the castings are too hard and brittle to be chipped or filed.

The loss of metal in melting tin-plate scrap in a cupola is not so great as in melting iron when melted with a light blast, but the loss may be as great as 25 per cent. when melted with a very strong blast.

The cost of melting iron in a cupola is about two dollars per ton.

The cost of melting tin-plate scrap in a cupola is from three to four dollars per ton.

Galvanized sheet iron scrap, when melted with tin-plate scrap, reduces the temperature of the molten metal to such an extent that it cannot be run into moulds.

Anthracite coal picked from the dump of a cupola will not burn alone in a stove or core oven furnace, and it is very doubtful if it produces any heat when burned with other coal in a cupola.

Lead is too heavy and penetrating when in a fluid state to be retained in a cupola after it has melted. The ladle should be warmed and the tap hole left open when melting this metal in a cupola.

The best lining material for a cupola in which tin-plate scrap is melted is a native mica soap-stone.

The sparks that fly from a stream of hard iron at the tap hole and spout are the oxide of iron. They are short-lived and burn the flesh or clothing very little.

The sparks from a wet tap-hole or spout are molten iron, and burn wherever they strike.

We have probably chipped out, daubed up and melted iron in a greater number of cupolas and in more different styles of cupola than any melter in the United States, and in heats that require from two or three hours to melt, and we have found that 8 pounds of iron to 1 pound of best coke; 7 pounds of iron to 1 pound of best anthracite coal; 6 pounds of iron to 1 pound of hard wood charcoal; 4 pounds of iron to 1 pound of gas-house coke, is very good melting. We have done better than this in test heats, but do not consider it practicable to melt iron for general foundry work with less fuel than stated above.

The best practical results for melting for general foundry work are obtained from 6½ to 7 pounds of iron to 1 pound of coke; 5 to 6 pounds of iron to 1 pound of hard coal; 4 to 5 pounds of iron to 1 pound of hard wood charcoal; 3 lbs. of iron to 1 pound of gas-house coke.

A less per cent. of fuel is required in long heats than in short ones, for, as a rule, three to one is charged on the bed and ten to one on the charges, and the greater the number of charges melted, the less the per cent. of fuel consumed.

Ten pounds of iron to one of coke are melted at the Homestead Steel Works, in cupolas that are kept in blast night and day for six days.

Less fuel is generally required to melt iron in the foundry office than is required to melt it in a cupola.

Use a light blast when melting with charcoal or gas-house coke.

If you go into the foundry when the heat is being melted and find the tap-hole almost closed, the spout all bunged up and the melter picking at the spout with a tap bar and running a rod into the tap-hole a yard or so in his efforts to get the iron out, and remark to him: "You are having some trouble with your cupola to-day," he will say: "Yes, we have some very bad coke to-day, sir; that last car is poor truck;" or, "We are melting some dirty pig or scrap to-day, sir." He never thinks: "We have a very poor melter to-day, sir."

At the first meeting of the American Association of Foundrymen, held in Philadelphia, May 12, 13, 14, 1896, one of the delegates was Mr. C. A. Treat, a good-sized practical foundryman weighing over 300 pounds, and representing the C. A. Treat Mfg. Co., Hannibal, Mo. After the meeting had effected a permanent organization, transacted all its business and was about to adjourn, Mr. Treat arose and in his quiet way remarked: " Gentlemen: Since we have have formed an organization of foundrymen for our mutual benefit, don't you think it would be a good idea for foundrymen to stop lying to each other?" The burst of laughter that followed this remark was loud and long. It would be a great relief to many foundrymen if some foundrymen would take the hint and stop lying about the large amount of iron melted with a small amount of fuel, fast melting, etc.

A few years ago, a foundryman who was about to publish a work on foundry practice, being desirous to obtain some reliable data on cupola practice, had several hundred blanks printed and sent to foundrymen in different parts of the country, with the request that they fill in the amount of fuel placed in the bed and charges, the amount of iron placed on bed and charges, diameter of cupola, height of tuyeres, etc. He was surprised at the reports received in reply. Many of them showed that the men who filled in the blanks either knew nothing at all about a cupola, or, knowing the report was to be published, were desirous of making an excellent showing of cupola work in their foundries, and in many of the reports the cupola was filled with stock in such a way that not a pound of iron could have been melted in a cupola charged as indicated in the formula. In some cases, the amount of fuel placed in the bed was not sufficient to fill to the tuyeres a cupola of the diameter given; in others, the fuel placed in the charges was not sufficient to cover the iron and separate the charges; and it was only after pointing out these mistakes and returning the reports for correction, in some cases two or three times, that they were put in any kind of shape for publication.

Some fifteen years ago, when we took a more active part in

melting than at the present time, and occasionally published an account of heats melted, we were repeatedly criticised in print by some would-be melters, who were melting anywhere from ten to twenty to one, for using too large a quantity of fuel, and some times were invited to come to their foundries and get a few points on melting before publishing another work on the subject. We have never learned of any of our critics on the fuel question becoming prominent in foundry matters or rich in the foundry business, and presume they have all saved their employers such a large amount of fuel in melting that they have been placed upon the retired list with half pay.

The heats published at that time were the best that could be melted in the cupolas described, and the amount of fuel consumed was generally about seven to one with hard coal and eight to one with best Connellsville coke. The foundrymen who at the present time melt heats of the same size in cupolas of the same diameter, with a less per cent. of fuel, are like angels' visits, few and far between.

NOTE.

PAXSON-COLLIAU CUPOLA.

On page 193 are shown illustrations of the Paxson-Colliau Cupola as built by J. W. Paxson Co., Philadelphia, and on page 194 we have stated that this is a *hot blast* type similar to that made by Victor Colliau.

We desire to correct the above to the extent that the Paxson-Colliau Cupola is not claimed to be of the Hot Blast type, and while it has two zones of melting as has the Colliau, there have been made many changes in its construction, bringing it up to date.

The new Paxson-Colliau has a low safety tuyere which discharges any overflow into a cornucopia-shaped trap fitted with a soft metal plug, which is easily melted, and should any hot iron or slag strike it, it is discharged outside through the bottom plate, as shown by Fig. 44, page 193.

This cupola is also fitted with a fine Screen Charging Door, as shown by Fig. 43, same page, requiring no lining, and is claimed to be more acceptable than the old solid cast iron doors, as it does not warp or crack. It may be fitted with a new arrangement to hold up the bottom doors, instead of the old prop, when a small car or truck on wheels or rake can be placed under the cupola to receive the hot drop and carry it into the foundry if desired, which is often done during cold weather to keep the shop warm, or to keep it out of the way, perhaps in the yard, where it can be cooled off by water, and gotten ready for the cinder mills. The cupola then will cool off quickly, allowing any repairs to be made or patching up the burned-out portions.

A new device of a Spark Arrester is also placed over the Paxson-Colliau Cupola, which confines the sparks and dirt to a small area. The lower tuyeres are rectangular and flared, and the upper ones are oval; they are staggered so that there is very little dead plate; the blast reaches every part where it is wanted, being distributed evenly. Therefore the lining is not affected by the action of the blast to the extent that would be expected where upper and lower tuyeres are used and two zones of melting are at work.

While speaking of this it may be mentioned that we have seen the naked hand placed in the Paxson-Colliau furnace at the charging door during the greater part of the heat without injury. Did you ever try this in an ordinary cupola while running a heat? The hand will be pulled back very quickly. This proves the fact that there is enough oxygen admitted through the small upper tuyeres to make a more perfect combustion of the fuel where it is wanted, both below and above the tuyeres.

A new *mercury blast gauge* is supplied with each cupola. It is made of iron and japanned, except the brass scale-plate and glass tube. This is the neatest looking and most common-sense gauge we have yet seen. A further description of this cupola may be had from the builders, J. W. Paxson Co., Philadelphia, Pa.   E. K.

INDEX.

ABENDROTH BROS., Port Chester, N. Y., cupola report of, 214, 215
   Port Chester, N. Y., cupola with three rows of tuyeres used by, 43
   Port Chester, N. Y., large cupola in the foundry of, 198-202, 298
Accounts, cupola, 214-221
   manner of keeping, 214
Adjustable tuyeres, 45, 46
Air, admission of, to the cupola, 30
   -chamber, 14-16
      admission of blast to, 15, 16
      area of, 5, 6
      belt, connecting blast pipes with tuyeres direct from a, 286-290
      construction of, 14, 15
      location of, 14
      openings in the, 16
      perfect manner of connecting the main pipe with an, 290
      round or over-head, objection to, 15
   chambers, air capacity of, 15
      best, 15
      perfect connection of, 286
   cubic feet of, required to melt a ton of iron, 293, 294
   friction of, in pipes, 281, 282
   gauges, 292, 341
   means of supplying, to a cupola, 294
   required for the combustion of fuel, 295, 296
   restriction to the passage of, through the cupola, 294
   rule for estimating the amount of power required to displace a given amount of, at a given pressure, 318

American Blower Co., Detroit, Mich., Smith's Dixie fan blower built by the, 309-311
Angle iron or brackets for the support of the lining of a cupola, 23, 24
Anthracite coal, amount of, required to melt iron, 90

BAD melting, cause of, 247
   caused by wood and coal, 250, 251
   examples of, 233-253
Banking a cupola, 277-279
Bar for cutting away the bod, 93
Bars, tapping, 92, 93
Baskets for measuring fuel, 230
   increase in size of, 212
Bed, the, 77-79
   best depth of, 125
   burned too much, poor melting due to, 249, 250
   burning the, 340
   up the, for warming the cupola, 77
   depth of fuel in the, 78
   effect of too large a quantity of fuel in a, 128
   fuel required for a, 79
   leveling the top of the, 82
   quantity of fuel for a, 336
   raising or lowering a, 78, 79
   uneven burning of the, 253
   up of the, effect of, 77
Belt air chamber, connecting blast pipe with tuyeres direct from a, 286-290
Bessemer steel works, location of tuyeres in cupolas used in, 53
Blakeney cupola, 204, 205
   tuyere, 35, 36
Blast, 85, 86
   admission of, to the air-chamber, 15, 16
   air-chamber for supplying the tuyeres with, 14
   arrangement for supplying, 266
   cause of apparent deficiency of, 91

Blast, direct delivery of, to tuyeres, 284
 furnace, definition of a, 136
  fuel required in, 1
  use of lime-stone in the, 135, 136
 gate, advantage of the, 290, 291
 gates, 290-292
 gauges, 292-294
 heating the, 274
 in melting, 294-299
 indications of, when first put on, 85
 length of time the, can be taken off a cupola, 276, 277
 machines for supplying, 294
 passage of, through heated fuel, 127
 pipe, preventing gas from the cupola from passing into the, 85
 pipes, 279-281
  blast gates, 276-299
  connection of, with cupolas, 284
  diameter of, 281-290
  explosion in, blast gauges, blast in melting, 276, 277
  explosions in, 292
  galvanized iron, 280, 281
  long, poor melting caused by, 290
  materials for, 280
  poor arrangement of, 286-288
  table of diameter and area of, 285
  table showing the necessary increase in diameter for the different lengths of, 283
  underground, 279, 280
  very best way of connecting, with tuyeres, 288-290
 positive and non-positive, 294, 295
 putting on the, 85
 taking off the, during a heat, 276-299
 time for charging the iron before putting on the, 86, 87
Blower and electric motor, 327
 Buffalo steel pressure, 301-309
 connection of tuyeres with, 5
 Connersville cycloidal, 321-328
 directions for setting up, 314, 316, 317
 efficiency of, 317, 318
 foundation for, 300
 Garden City positive blast, 328, 329

Blower, Green patented positive pressure, 314-321
 horizontal, 325
 Mackenzie, 311-314
 obtaining the best results from a, 303
 on adjustable bed, and on bed combined with countershaft, 303-305
  combined with double upright engine, 305, 306
 placed near cupola, 290
 placing, 300, 301
 prevention of the destruction of a, 300, 301
 Roots's rotary positive pressure, 329-331
 Smith's Dixie fan, 309-311
 vertical, and engine on same bed, 326
Blowers, 300-331
 cycloidal, numbers, capacities, etc., of, 326
 forced blast pressure, 311-331
 foundry, speed of, 320, 321
 standard foundry, driven by pulley, table of dimensions of, in inches, 319
 table of speed and capacities of, as applied to cupolas, 309
Blue clays for spout lining, 68
Bod bar for cutting away the, 93
 definition of, 94
 good, qualities of a, 95
 for small cupolas, 95
 horse manure as an essential of a good, 95
 material, 94, 95
  working the, 96
 mixture for, 95
 mode of making the, 96
 size and shape of, 96
 sticks, 93, 94
 wet, explosion of iron caused by a, 257
Boiling, cleaning iron by, 147, 148
Bolton Steel and Iron Co., England, use of Ireland's cupola in, 159
Bosh, taper from the, to the lining, 109
Boshing of cupola, 14
Bottom door, 4
 bolts and latches of, 4
Bottom doors, 11, 12
 devices for raising the, 133, 134

INDEX.

Bottom doors, way for reducing size and weight of, 25
Bottom, exclusively new sand in a, 340
    hard rammed, 340
    high pitch of, 66
    hollow, 66
    of cupola, height of, 3, 11
    pitch or slope of, 65, 66
    plate, 4
    plates, shape of, 26
    sand, introduction of, into the cupola, 63, 64
        preparation of, 63
        renewal of, 64
        re-use of, 63
        too wet or too hard, consequence of, 65
    tuyere, 46–49
    uneven settling and breaking of, 10
    wet sand in a, 340
Brackets, arrangement of, 23–26
Brick, curved, for lining, 22
    for casing, 111
    split, 112
Bridging, cause of, 99, 100
Buffalo blower for cupola furnaces in iron foundries, 308, 309
    electric blower built in "B" and steel pressure types, 306–308
    Forge Co., Buffalo, N. Y., banking a cupola at the, 277–279
        table of diameter of blast pipes prepared by the, 281–283
    School Furniture Co., Buffalo, N. Y., cupola of the, 54
    steel pressure blower, 301–309
Byram & Co. Iron Works, Detroit, Mich., cupola report of, 214–216

CANNON, melting of, 224
Carbon, effect of, upon cast iron, 145
    removal of, from iron, 145
Carnegie Steel Works, Homestead, Pa., cupolas in the, 275, 298, 335
Cars for removing the dump, 100
Casing, 12–14
    brick for the, 111
    cupola, construction of, 12
    lining of, 6, 7
    preventing the, from rusting off at the bottom, 25
    refractory material for lining, 21, 22

Casing, stack, 2, 5
    construction of, 12
    strain upon the, 12
    thickness of, 12
        lining, to protect the, 111
    wrought iron, 5
Casings, 4, 5
Cast, make a, 339
    iron, quantity of, that can be melted in a cupola, 224
    size and weight of a piece of, that can be melted in a cupola, 224
Casting, 339
Castings, fins on, 342
    report of, 219
Center blast cupola, Ireland's, 159–161
    tuyeres, experiments with, 298, 299
Chain blast, 294
Charcoal, experiments in softening iron with, 130–132
    use of, as fuel, 332
Charge, fuel required in each, 336
Charges, division of fuel and iron into, 212
    effect of too large a quantity of fuel in the, 128
    for experimental heats, 126, 127
    most even melting, 127
    placing the, 82–84
    table of, 200
Charging, 80–82
    bad, poor melting due to, 84
    cupola slate for, and cupola report, 220
    cupolas, different ways of, 113
    door, 6, 14
        distance of the floor of the scaffold below the, 26
        location of, 13
        wear of lining at the, 110
    doors, low, 334, 335
    flux, 84, 85
    proper way of, 84
    rule for, 333
    time for, 132, 133
Chenney tuyere, 41
Chill mould, explosion of iron in a, 259
Chipping out, 101–103
    tools for, 255
Cincinnati, O., poor melting in a cupola at, 251–253
Cinder, brittle, making a, 136
    chipping off, 106
    tendency of, in a cupola, 136, 137

INDEX.

Clam shells, 142
Clay, amount of sand in, for daubing, 104
  and sharp sand, mixtures of, 62
  effect of too much, in lining, 68
  blue, for daubing, 104
  fire, for daubing, 104
    soaking of, 104
  sands, 62
  wash, 63
  yellow, for daubing, 104
Clays for spout lining, 68
Coal and wood, bad melting caused by, 250, 251
  anthracite, amount of, required to melt iron, 90
  hard, use of, as fuel, 332
  melting with, 130
Cogniardelle, 294
Coke, Connellsville, amount of, required to melt iron, 90
  consumption of, in melting iron, 332
  picking out of, from the dump, 101
  use of, as fuel, 232
Colliau cupola, claims for the, 194–196
  history and description of, 193, 194
  patent hot-blast cupola, 192–196
  -Paxson cupola, 193, 194
    cupola, note on the, 345, 346
  tuyere, 41
Combination stick, 93, 94
Combustion, complete, 178
Connersville cycloidal blower, 321–328
Contact, point of, definition of, 323
Continuous slot tuyere, 34, 35
Copper, melting of, 223
Corry, Pa., Pevie cupola at, 186
Cost of melting, 230–232
Crandall improved cupola with Johnson patent center blast tuyere, 202–204
Crates, iron, for removing the dump, 100, 101
Crucible, experiments in a, with iron, 130
Cupola account, correctness of, 221
  accounts, 214–221
    manner of keeping, 214
  admission of air to the, 30
  and stack, weight of, 9
  banking a, 277–279
  best supports for a, 10
  Blakeney, 204, 205
  blower placed near, 290
  book, 231

Cupola, boshed, burning out of the lining of the, 109
  new lining in, 105, 106
  boshing of, 14
  bottom, height of, 11
  brackets or angle iron for the support of the lining of a, 23, 24
  brick walls for the support of a, 10
  bridged sectional view of a, 107, 108
  burning of iron in a, 88
  casing, construction of, 12
  cause of bridging and hanging up refuse in a, 99, 100
  charging a, 80–82
  chipping out the, 101–103
  Colliau patent hot blast, 192–196
  combined tuyere area of a, 49, 50
  commencement of melting in a, 86
  construction of a, 8–29
  Crandall improved, with Johnson patent center blast tuyere, 202–204
  determination of the location of the melting zone in a, 123
  does it pay to slag a? 141, 142
  dumping the refuse from the, 98
  economical melting in a, 335
  effect of limestone in a, 138
  expanding, 155–157
  experiment to learn at what point of the, iron melts, 114
  experimental, 114
  for melting tin-plate scrap, 227, 228, 229
  for tin-plate scrap, best lining material for a, 342
  foundation, 2, 9, 10
  furnace, 1–7, 332
    advantages of, 1
    chief use of, 223
    consumption of coke in, 332
    description of, 2
    fuel required in, 1
    supply of air to the, 30
  Greiner patent economical, 188–192
  hardening of iron in a, 88
  height of a, 13
    the bottom of, 3
  Herbertz, 173–182
    for melting steel, 182–184
  highest, in use, 335
  holding molten iron in the, 88
  house, novel plan of construction of a, 27, 28

INDEX. 351

Cupola, how to slag a, 140, 141
 introduction of the bottom sand into the, 63, 64
 Ireland's, 157-159
  center blast, 159-161
 iron support for a, 10
 Jumbo, 198-202, 208
 large, lighting up a, 76
 learning to manage a, 209, 210
 length of time the blast can be taken off a, 276, 277
 lining, life of, 110
  renewal of, 12
 locating the tap hole in the, 73
 location of, 8
  of slag hole in a, 74, 75
 Mackenzie, 170-173
 management, 58-112
 means of supplying air to a, 294
 melting capacity of, 14, 80
  iron in a, terms used to indicate, 339
  tin-plate scrap in a, 225-229
  zone or melting point of a, 77
 modern, casing or shell of, 12
 necessity of learning the peculiarities in the working of every, 58
  understanding the, to do good melting, 207
 newly lined, trouble in melting in a, 79
 number of hours a, will melt iron freely, 224
 old, theory of melting in the, 296
  style, construction of, 149-152
   English, 154, 155
   spark catcher in, 263, 264
 picks, 102, 103
 pit of, 3, 4
 Paxson-Colliau, 193, 194
 Pevie, 184-186
 placing charges in the, 82-84
  tuyeres in a, 20, 21
 point of melting in a, 113
 practical instructions for charging and managing a, 129
  working of a, 81
 preparation of a, for a heat, 208
 putting in two fronts and tap holes in a, 73, 74
 record, accurate, 337
 requirements of the foundryman from the, 91
 requisites for melting iron in a, 332, 333
 report, Abendroth Bros., 214, 215

Cupola report, Byram & Co's, 214, 216
 reports, misleading, 90
  unreliability of, 344
 reservoir, 152, 153
 restriction to the passage of air through the, 294
 rule for charging a, 333
 scrap, charging of, 83
 scraps, 339-345
 sectional view of a, at Cincinnati, 251, 252
 size and weight of a piece of cast iron that can be melted in a, 224
 slate for charging, and cupola report, 220
 small, lighting up a, 76
 space of melting iron in a, 77
 spout, old way of making, 67
 stationary bottom, 154, 155
 Stewart's, 186-188
 stopping in a, 88
 straight, adhesion of slag and cinder to the lining of, 106
 supports of, 2, 3
 tank or reservoir, 167-170
 tendency of slag and cinder in a, 136, 137
 tuyeres, 30-57
 two-hour, 170
 Voisin's, 161-163
 warming up a, 248-250
 waste heat from a, 274, 275
 weight of slag drawn from a, 137, 138
 what a, will melt, 223, 224
 Whiting, 196-198
 with tuyeres near the top, 129
 Woodward's steam jet, 163-167
Cupolas, amount of fuel required for the bed of, 79
 and cupola practice up to date, 332-338
 casings of, 2
 connection of blast pipes with, 284
 different styles of, 149-205
  ways of charging, 113
 fluxing of iron in, 135-148
 for heavy work, location of tuyeres in, 53
 height and size of door for, 13, 14
  of tuyeres in, above sand bottom, 19, 20
 hot blast, 271-275
 in machine and jobbing foundries, location of tuyeres in, 53
 large, tap-holes for, 16, 17
 location of a greater number of tuyeres in, 54

Cupolas, melting of lead in, 223-224
  mistake of placing small tuyeres in, 49
  modern, spark catching device for, 264-266
  number of tuyeres in, 2
  odd-shaped, shaping the lining of, 109
  of large diameter, location of tuyeres in, 52
  of very small diameter, location of tuyeres in, 52, 53
  old style, 149-152
  patent, shaping the lining of, 109
  props for, 60, 61
  shapes of, 2
  size of, 333
  sizes of, 2
  small, bod for, 95
    breaking away the bridge in, 99
    dumping of 99
    support of the stock in, 60
    tap hole for, 16
  smaller, location of tuyeres in, 52
  spark catching devices for, 263-270
  table of speed and capacities of blowers as applied to, 309
  use of limestone in, 136
  with high tuyeres, impossibility of making hot iron for light work in, 52
  with two tap holes, slope of bottom in, 67
Cylinder blower, 294
Cycloidal blowers, numbers, capacities, etc., of, 326
  curves, 322

D AUBING, 103-105
  application of, 105
  object of application of, to a lining, 105
  poor, cheap, nothing gained by using, 104, 105
  substances used for, 103
  thickness of, on a lining, 107
  wet, explosion of iron by, 262
Detroit, Mich., novel plan of construction of a scaffold and cupola house at, 27, 28
Diamond Drill and M'f'g Co., Birdsboro, Pa., cupola of the, 46
Doherty tuyere, 33, 34
Door, charging, 6, 14
  location of, 13

Door, for cupolas, height and size of, 13, 14
Doors, bottom, 11, 12
  devices for raising the, 133, 134
  devices for raising the, in place, 59, 60
  dropping the, 61, 62
  heavy, best device for raising, 134
  putting up the, 59-61
  sliding, 4
  small, device for raising, 134
  supports of, 60
Double tuyere, 42, 43
Drying the lining, 58, 59
Dump, breaking up the, 101
  chilling the, 99
  constitution of the, 100
  handling of the, 100
  picking over the, 101
  removing the, 11, 100, 101
Dumping, 98-100

E LEVATOR, 9
  Elizabethport, N. J., tests with the Herbertz cupola at, 176, 179-182
England, Ireland's cupola patented in, 157
  use of tanks in, 169
English cupola, old style, 154, 155
Expanded tuyere, 32, 33
Expanding cupola, 155-157
Experiments in melting, 113-134
Explosion of molten iron, 257-262
Explosions in blast pipes, 292
  blast gauges, blast in melting, 276, 277

F AN blower, Smith's Dixie, 309-311
  Fire clay for daubing, 104
    soaking of, 104
  Fire-proof scaffold, 26-29
  Floor, explosion of iron by falling upon the, 258
  Fluor spar, 146, 147
  Flux, charging of, 84, 85
    definition of, 135
    effect of, upon iron, 138
    fluor spar as a, 146, 147
    quantity of, required, 84
  Fluxes, action of, on lining, 139
    materials used as, 135
    mineral, effect of, on a front material, 72
    use of, 135

INDEX.

Fluxing, improper, injury to iron by, 143, 144
of iron in cupolas, 135–148
tin-plate scrap, 228
Foundation, block in the, to rest the prop upon, 60
construction of, 10
cupola, 2, 9, 10
Foundries, disturbances in melting in, 205
number of men employed in, 230
Foundry blowers, speed of, 320, 321
standard, driven by pulley, table of dimensions of, in inches, 319
department, Lebanon Stove Works, daily report of, 217
Outfitting Co., Detroit, Mich., cupola manufactured by, 202–204
result of keeping an accurate melting account in a, 231
work, furnaces employed for, 332
general, best practical results for melting for, 343
Foundryman, requirements of the, from the cupola, 91
Foundrymen, theory of melting not understood by, 212
Forced blast pressure blowers, 311–331
Front, 71–73
drying of the, 72
material, effect of mineral fluxes on, 72
for putting in the, 71
poor, effect of, 72
too wet, effect of, 72
putting in the, 71
thickness of, 72
Fuel, 90–92
air required for the combustion of, 295, 296
amount of, in each charge, 81
required for a bed, 79
in each charge, 336
and iron, division of, into charges, 212
arranging the, for lighting up, 76
consumption of too great an amount of, 336
under the tuyeres, 50, 51
depth of, in the bed, 78
distribution of the charge of, 83
effect of too large a quantity of, in a bed, 128
much, 230, 231

Fuel, guessing the weight of, 212
heated, passage of blast through, 127
heating of, in a cupola, 129
measuring of, 230
old way of placing the, in the cupola, 80
proportion of, to iron for melting, 90
quantity of, for a bed, 336
required in various furnaces, 1
too heavy charges of, effect of, 81
too light charges of, effect of, 81
under the tuyeres, 121, 122, 334
value of, wasted every year in the United States, 53
waste of, 341
weight of the charges of iron to the charges of, 82
Furnace, blast, definition of a, 136
use of limestone in the, 135, 136
cupola, 1–7, 332
kinds of, 332
pot, 332
reverberatory, 332
Furnaces, various, fuel required in, 1

GALVANIZED sheet iron scrap, 342
melting of, 227
Garden City positive blast blowers, 328, 329
Gas, preventing the passage of, into the blast pipe, 85
Gases, escaping, composition of, 177, 178
free oxygen in the, 178
Gates, charging of, 83
Gauges, blast, 292–294
Glasgow, Scotland, use of Stewart's cupola at, 186–188
Gould & Eberhardt, scaffold in the foundry of, at Newark, N. J., 28
Green patented positive pressure blower, 314–321
Greiner patent economical cupola, 188–192
tuyere, 45
Grout, composition of, 6
Grouting for lining, 22

HEARTH in the Herbertz cupola at Elizabethport, N. J., 180
movable, of the Herbertz cupola, 173
Heat, escaping, attempts to return the, to the cupola, 273, 274

23

Heat, make a, 339
 take a, 339
 taking off the blast during a, 276–299
 theory of the production of, 44, 45
 utilization of, 275
 waste, from a cupola, 274, 275
  plans for the utilization of, 271
  utilization of, 1, 2, 13
Heats, experimental, charges for, 126, 127
Height of cupola bottom, 11
 tuyeres, 50–53
Herbertz cupola, 173–182
  for melting steel, 182–184
  test-heats with the, 176, 177
Hibler, B. H., bottom tuyere patented by, 48
Horizontal and vertical slot tuyere, 36, 37
 blower, 325
 slot tuyere, 34
Horse manure as an essential of a good bed, 95
Hot blast cupolas, 271–275

IMPELLER, complete, 316
Iron, additional, charging of. 83
 affinity of limestone for, 136
 amount of anthracite coal required to melt, 90
  combined with the slag, 142
  Connellsville coke required to melt, 90
  limestone required for, 137
  placed upon the bed in the first charge, 81
 and fuel, division of, into charges, 212
 arrangement of, in the experimental cupola, 114–119
 art of melting, 211, 212
 burning of, in a cupola, 88
 carbonized, use of, as softeners, 145
 cast, effect of carbon upon, 145
  quantity of, that can be melted in a cupola, 224
  size and weight of a piece of, that can be melted in a cupola, 224

Iron, cause of the variations in the weight of the first charge of, 82
 change in the action of the, at the spout, 66
 charging, 340, 341
  large pieces of, 224
 cleaning of, by boiling, 147, 148
 consumption of coke in melting, 332
 contents of, in slag, 138
 cost of melting, 342
 cubic feet of air required to melt a ton of, 293, 294
 deception in the quality of the, at the spout, 66
 dull, cause of, 336
 effect of flux upon, 138
  silicon on, 144
  tin on, 227
  too heavy a charge of, 128
 experiments in softening with charcoal 130–132
  with, in a crucible, 130
 explosion of molten iron, by, 258, 259
 first melted, chilling and hardening of, 87
 fluxing of, in cupolas, 135–148
 furnaces employed in melting of, 332
 guessing the weight of, 212
 hard, experiments in softening, 51
 hard, softening, 130–132
  sparks from, 258
 hardening of, in a cupola, 88
 high silicon, Southern, use of, in stove foundries, 144, 145
 hot and of even temperature, melting of, 208
  for light work, impossibility of making, in cupolas with high tuyeres, 52
 impossibility of melting, under the tuyeres, 51
 improvement of, in a cupola furnace, 143
 indication of the melting by the flow of, from the tap hole, 88, 89
 injury to, by improper melting and fluxing, 143, 144
 malleable, experiments in making, 143

INDEX. 355

Iron, melted high in a cupola, cause of dullness of, 128
  melting of, correct theory of, 80
    in a cupola, no chance work in, 207
    terms used to indicate the, 339
    things to be learned in, 209
  molten, explosion of, 64, 65, 257-262
    filtering of, through slag, 84, 85
    handling of, 341
    holding of, in the cupola, 88
    poling of, 147, 148
  number of hours a cupola will melt, 224
  old way of placing the, in the cupola, 80
  over, mould for, 222
  picking out of, from the dump, 101
  placing the first charge of, on the bed, 80
    in the cupola, 83
  point of melting, in a cupola, 113
  proportion of fuel to, for melting, 90
  recovery of, from the dump, 101
  removal of carbon from, 145
  requisites for melting, in a cupola, 332, 333
  space of melting, in a cupola, 77
  theory of preventing, from running into the tuyeres, 102
  time for charging, 132, 133
    the, before the blast is put on, 86, 87
    too heavy charges of, effect of, 81
    light charges of, effect of, 81
  tuyeres to improve the quality of, 55, 56
  weight of the charges of, to the charges of fuel, 82
    first charge, 81
Ireland, Mr., bottom tuyere used by, 48
    double or two rows of tuyeres devised by, 42
Ireland's center blast cupola, 159-161
  cupola, 157-159

Isbell Porter Co., Newark, N. J., blower built by the, 312
  Mackenzie cupola, manufactured by, 170-173

JAGGER, Treadwell & Perry, cupolas constructed by, 271-273
Jamestown, N. Y., tuyeres in a cupola, 31
Jobbing foundry cupolas, location of tuyeres in, 53
Johnson, Mr., experiments of, with the center blast tuyere, 209
Johnson, John D., & Co., Hainesport, N. J., action of fluxes on lining of cupola of, 139
Jumbo cupola, 198-202, 298
  table of charges of, 200

KNOEPPEL, Mr., on banking a cupola, 277-279

LADLE, damp, iron caused to boil by a, 262
Lawrence reducing tuyere, 38, 39
Lead, melting of, in cupolas, 223
  molten, handling of, 342
Leather bellows, 294
Lebanon Stove Works, daily report of foundry department of, 217
Light-up, bad, 340
Lighting-up, 76, 77
  wood for, 340
Limestone, action of a large per cent. of, 137
  affinity of, for iron, 136
  amount required of, 137
  charging of, 140
  effect of, in a cupola, 138
  in large quantities, 136-138
  object of use of, 136
  use of, in blast furnace, 135, 136
    cupolas, 136
Lining, 21-23
  action of fluxes on, 139
  belly of, 107
  brackets or angle iron for the support of the, 23, 24
  burning away of the, at the melting zone, 110
    out of, 24, 25
  cupola, life of, 110
  destruction of, at and below the tuyeres, 110
  drying the, 58, 59

INDEX.

Lining, effect of fluor spar on the, 146
    too much clay in, 68
        sand in, 68
  false, 22, 111, 112
  filling in the, at the melting zone, 106, 107
  greatest wear of, 110
  laying up a, 22
  material, best, 342
    for spouts, 68
  new, in cupola, 105, 106
  object of applying daubing to a, 105
  of boshed cupola, burning out of a, 109
  old, false lining over the, 111, 112
  out of shape, sectional view of, 235, 237, 239, 241
  prevention of absorption of moisture by the, 25, 26
  refractory material for, 6, 7
  renewal of, 12
  selection of a, 139
  settling of, 25
  shaping the, 105–110
  split brick, 112
  spout, 67
  stack, wear of, 110
  support of, 5
  taper of, 107
    to the, from the bosh, 109
  thickness of, 22, 110
      daubing on a, 107
      to protect the casing, 111
  wear of, at the charging door, 110
Loading, old way of, 80
Loam clays for spout lining, 67
  sands, 62
Loams for bods, 94
Lobdell Car Wheel Co., Wilmington, Del., melting cannon at the foundry of, 224
Low tuyeres, 122, 123

McGILVERY, WM. & CO., accident in the foundry of, 259, 260
Machine foundry cupolas, location of tuyeres in, 53
Mackenzie blower, 311–314
  cupola, 170–173
  tuyere, 34, 35
Magee Furnace Co., Boston, Mass., triangular tuyere used by the, 39
Malleable iron, experiments in making, 143
Marble spalls, 142, 143

Marsh, James, explosion of iron in the foundry of, 260
Massachusetts, early use of bottom tuyere in, 48
Melt, run a, 339
Melter, aim of every, 255, 256
  directions by the, 210
  good, interference with a, 254, 255
  poor, 254
  practical and scientific, 254
  process of chipping out by the, 101–103
  skill of the, seen at the tap hole, 98
  whims of, 210
Melters, 254–256
  no attention paid by many, to shape the cupola, 105
  theory of some, 102
Melting, 86–89
  account, accurate, result of keeping a, 231
  art in, 205–210
  bad, cause of, 247
    caused by wood and coal, 250, 251
    examples of, 233–253
  best practical results for, 343
  blast in, 294–299
  capacity, increase in the, by two or three rows of tuyeres, 21
    of a cupola, 30
  commencement of, 86
  cost of, 230–232
    per ton, mode of figuring, 231, 232
  disturbances in, 205
  economical, 335
  experiments in, 113–134
  fast, 207
  galvanized sheet-iron scrap, 227
  good, necessity of understanding the cupola for, 207
  improper, injury to iron by, 143, 144
  indication of the, by the flow of iron from the tap hole, 88, 89
  iron, art of, 211, 212
    correct theory of, 80
    cost of, 342
    hot and of an even temperature, 208
    in a cupola, no chance work in, 207
    terms used to indicate the, 339
    things to be learned in, 209
  most even, charges for, 127
  point, 77

INDEX. 357

Melting point, discovery of the, 77
   in a cupola, 113
   to find the, 78
   poor, caused by long blast pipes, 290
      due to the bed being burned too much, 249, 250
      in a Cincinnati cupola, 251–253
   preparation of tin-plate scrap for, 225
   reduction of, to a system, 214, 337
   scrap sheet iron, 227
   sheet of Syracuse Stove Works, 218
   slow, 207, 208
      and irregular, 337
      cause of, 337
   study of the materials used in, 209
   theory of, in the old cupola, 296
      not understood by foundrymen, 212
   tin-plate scrap, cost of, 342
      experiments in, 226, 227
      in a cupola, 225–229
   with coal, 130
   zone, 77, 123–129
      burning away of the lining at the, 110
      depth of, 125
      determination of the location of the, 123, 335
      top of, 335, 336
      development of the, 129
      filling in the lining at the, 106, 107
      location of the, 77
      raising and lowering the, 120, 123
Mercury gauges, 292
Metal from tin plate scrap, doctoring of, 227
      quality of, 225, 226
   gray, from tin-plate scrap, 226
Moisture, prevention of the absorption of, by the lining, 25, 26
Molding sand for spout lining, 67
Molten iron, explosion of, 257–262
Mould for over iron, 222
Moulder, aim of every, 210
Moulding, 339
   floors, cleanings from, 63
   sand, use of, for daubing, 103
   sands for bods, 94

M. Steel Co., Springfield, O., Blakeney cupola manufactured by the, 204, 205
   tuyere used in the cupola, constructed by the, 35, 36
Mud, explosion of iron when poured into, 260

NAU, J. B., on the Herbertz cupola, 173–182
North Bros., explosion of iron in the foundry of, 260, 261

O'KEEFE, J., spark catcher, designed by, 266–268
Oval tuyere, 32
Over iron, mould for, 222
Oyster shells, 142
Oxygen, free, in the escaping gases, 178

PARIS, D. E., & CO., West Troy, N. Y., bad melting at the foundry of, 242–248
Paxson-Colliau cupola, 193, 194
   note on the, 345, 346
Perry & Co., cause of having to dump a cupola at the foundry of, 253
   examples of bad melting at the foundry of, 233–242
Pevie cupola, 184–186
Picks, cupola, 102, 103
Pig iron, placing the, in the cupola, 82, 83
Pipe, main, perfect manner of connecting the, with an air chamber, 290
Pipes, blast, 279–281
   branch, area of, 284
   friction of air in, 281–282
Piston blower, 294
Pit, cupola, 3, 4
Pit lining, 3
Platform scales, 211
Point of contact, definition of, 323
   melting, discovery of the, 77
Poking the tuyeres, 89
Poling molten iron, 147, 148
Poor melting in a Cincinnati cupola, 251–253
Pot furnace, 332
   consumption of coke in, 332
   fuel required in, 1

Power, rule for estimating the amount of, to displace a given amount of air at a given pressure, 318
Pratt & Whitney, charging large pieces of iron at the foundry of, 224
Props, 60, 61
  removing the, 61, 62, 99
Providence Locomotive Works, visit to the plant of, 248-250

RANSOM & CO., Albany, N. Y., cupola at the stove foundry of, 273
Records, blanks for, 221
  value of, 214
Reducing tuyere, Lawrence, 38, 39
  Truesdale, 37, 38
Refractory material for lining, 6, 7
Relining and repairing, 110-112
Repairing and relining, 110-112
Report, cupola, Abendroth Bros.', 214, 215
  Byram & Co.'s, 214, 216
  daily, of Foundry department Lebanon Stove works, 217
  of castings, 219
Reports, blanks for, 221
  false, 221
  keeping of, 221
Reservoir cupola, 152, 153
  or tank cupola, 167-170
Return flue cupola spark catcher, 266-268
Reverberatory furnace, 332
  consumption of coke in, 332
  fuel required in, 1
Reversed T tuyere, 37
Richmond, Ind., sectional view of a bridged cupola at, 107, 108
Riddles, increase in size of, 212
Roots's rotary positive pressure blower, 329-331
Round tuyere, 31, 32

ST. LOUIS, MO., large cupola with two tuyeres in, 54
Sand, amount of, in clay for daubing, 104
  bottom, 62-67
    destruction of the, 64
    elements to contend with in the, 65
    height of tuyeres above, 19, 20, 333, 334
    leakage of, 64
    perfect joint between the, and the spout lining, 68, 69

Sand bottom, pitch or slope of, 64
    riddling out of the, 101
    slushing the, 64
    effect of too much, in lining, 68
    for bottom, working of, 65
    for sand-bottom, 62
    mould, explosion of iron in a, 259
    moulding, use of, for daubing, 103
    sharp, for daubing, 104
Sash weights, 226
Scaffold, 7, 8, 9
    construction of, 9
    distance of the floor of, below the charging door, 26
    exposure of, to fire, 26, 27
    fire proof, 9
    location of, 26
    novel plan of construction of a, 27, 28
    old worn-out scales upon the, 213
    scales in the floor of, 211
Scaffolds, best and safest, 28, 29
    devices for rendering fire-proof, 27, 28
Scale, size of, 211
Scales and their use, 211-213
    old worn-out, 213
Scrap, charging of, 83
    galvanized sheet iron, melting of, 227
    heavy government melting of, 224
    rusted, explosion of iron when brought in contact with, 260, 262
    sheet iron, melting of, 227
    tin-plate, cupola for melting, 227, 228, 229
    experiments in melting, 226, 227
    fluxing of, 228
    melting of, in a cupola, 225-229
    preparation of, for melting, 225
Shaping the lining, 105-110
Shavings for lighting up, 76
Sheet blast tuyere, 34
    iron scrap, galvanized, melting of, 227
    melting of, 227
Shells, 142
    crackling of, 142
Silicon, effect of, on iron, 144
Size of tuyeres, 49, 50
Skinner Engine Co., explosion in the the cupola of, 261, 262
Slag, amount of iron combined with the, 142

Slag, chilling of, 75
   chipping off, 106
   closing up of the tap hole with, 339
   contents of, 138
   filtering molten iron through, 84, 85
   formation of, in a spout, 339
   hole, 74, 75
     front, 75
       location of the, 17, 18, 140, 141
   impurities in the, 141
   in cupola, breaking down the, 102
   position of, in the cupola, 18
   removal of, from the spout, 70
   tapping of, 17, 51, 53
   tendency of, in a cupola, 136, 137
   time for drawing of, 141
   weight of, drawn from a cupola, 137, 138
Slagging a cupola, 140, 141
   cost of, 141, 142
   saving effected by, 141
   trouble in, 140
Slate, cupola, for charging, and cupola report, 220
Sledging, bars for, 92
Sliding doors, 4
Smith's Dixie fan blower, 309-311
Smithfield, N. J., Pevie cupola at, 186
Soapstone for daubing, 104
Softening hard iron, 130-132
Spark catcher, return flue cupola, 266-268
   catching device, best, 269, 270
         for modern cupolas, 264-266
         oldest and most efficient, 263, 264
      devices for cupolas, 263-270
         various, 268, 269
Sparks, 258, 342
   objections to, 5
Speed of foundry blowers, 320-321
   ordinary, definition of, 326
Split-brick, 112
Spout, 17, 18, 67-70
   building the sides of the lining of, 69
   change in the action of the iron at the, 66
   choking up of the, 69
   coating of the, 70
   damp, boiling of iron in a, 257
   deception in the quality of the iron at the, 66
   formation of slag in a, 339

Spout, lining, 67
   cutting out of the, in holes, 339
   drying of the, 72
   greatest strain upon the, 69, 70
   making up of the, 68
   perfect joint between the, and the sand bottom, 68, 69
   old way of making, 67
   removal of slag from the, 70
   shaping the lining of the, 70
   size of, 17
   wet, explosion of iron in a, 257
   with a broad flat bottom, 339, 340
Spouts, lining material for, 68
   modern, 67
   short, common practice in, 69
Stack and cupola, weight of, 9
   casing, 2. 5
   construction of, 12
   contracted, 5
   contraction of, 12
   enlarged, 5, 269
   enlarging of, 12, 13
   height of, 5
   lining, renewal of, 12
     thickness of, 22, 23
     wear of, 110
   size of, 5
Standard foundry blowers driven by pulley; dimensions in inches, table of, 319
Stationary bottom cupola, 154, 155
Steam jet, advantages of the, 175
   cupola, Woodward's, 163-167
Steel, Herbertz cupola for melting, 182-184
   spring gauges, 292
Stewart's cupola, 186-188
Stocking, modern way of, 80
   old way of, 80
Stopping-in and tapping, 96-98
   devices to assist in, 97, 98
   difficulties in, 97
   knack in, 97
   mode of, 97
Stove foundries, breakage in, 145
   height of tuyeres in, 20
   sand for bottom used in, 62, 63
   use of high silicon Southern iron in, 144, 145
Straight Line Engine Co., Syracuse, N. Y., scaffold in the foundry of, 28, 29
Straw for lighting up, 76
Syracuse Stove Works, melting sheet of, 218

TABLE of diameter and area of blast pipes, 285
— speed and capacities of blowers as applied to cupolas, 309
— standard foundry blowers driven by pulley; dimensions in inches, 319
— showing the necessary increase in diameter for the different lengths of blast pipes, 283
Taking off the blast during a heat, 276-299
Tank or reservoir cupola, 167-170
Tanks, use of, in England, 169
Tap-hole, 16, 17
— chilling of slag in the, 75
— closing up of the, with slag, 339
— indication of the melting by the flow of iron from the, 88, 89
— skill of the melter seen at the, 98
— mode of forming, 71
— preventing the cutting of, 73
— reducing the size of the, 87, 88
— sizes of, 73
— too long, effect of, 72
Tap-holes, locating the, 17, 73, 74
Tap, making a, when iron is handled in large ladles, 87
— mode of making the, 96, 97
Tapping and stopping in, 96-98
— bar, explosion of iron caused by, 258
Tapping bars, 92, 93
— burning up the, 341
Tin deposited upon iron, recovery of, 225
— effect of, on iron, 227
— -plate scrap, 342
— — cost of melting, 342
— — cupola for melting, 227, 228
— — doctoring metal from, 227
— — experiments in melting, 226, 227
— — fluxing of, 228
— — gray metal from, 226
— — loss of metal in melting, 342
— — melting of, in a cupola, 225-229
— — preparation of, for melting, 225
— — quality of metal from, 225, 226
Treat, C. A., remarks of, 344
Triangular tuyere, 39
Trompe, 294

Trucks for removing the dump, 100
Truesdale reducing tuyere, 37, 38
Tuyere area, combined, of a cupola, 49, 50
— Blakeney, 35, 36
— bottom, 46-49
— boxes, 19, 56, 57
— Chenney, 42
— Colliau, 41
— continuous slot, 34, 35
— Doherty, 33, 34
— double, 42, 43
— expanded, 32, 33
— expanding, 297
— Greiner, 45
— horizontal and vertical slot, 36, 37
— — slot, 34
— invention, epidemics of, 30, 31
— Lawrence reducing, 38, 39
— Mackenzie, 34, 35
— outlet area of a, 297
— oval, 32
— reversed T, 37
— round, 31, 32
— sheet blast, 34
— triangular, 26, 39
— Truesdale reducing, 37, 38
— vertical slot, 37
— water, 40, 41
— Whiting, 41
Tuyeres, 18-20
— adjustable, 45, 46
— air chamber for supplying the, with blast, 14
— center blast, experiments with, 298, 299
— connecting blast pipes direct with, from a belt air chamber, 286-290
— connection of, with the blower, 5
— consumption of fuel under the, 50, 51
— cupola, 30-57
— destruction of the lining, at and below the, 110
— direct delivery of blast to, 284
— fuel under the, 121, 122, 334
— general improvement made in, 341
— height of, 50-53
— — above sand bottom, 19, 20, 333, 334
— high, 341
— — reason in favor of, 51
— increase in the melting capacity by two or three rows of, 21
— liability of, to be closed, 50
— location of, 18, 19
— low, 122, 123

Tuyeres, number of, in a cupola, 2, 18, 19, 54, 55
  placing of, in a cupola, 20, 21
  poking the, 89
  projection or hump on the lining over the, 102
  shape of, 19, 55
  size of, 49, 50
  small, objection to, 49
  theory of preventing iron from running into the, 102
  three rows of, 43-45
  to improve the quality of the iron, 55, 56
  triangular, 297
  two rows of, 42, 43
  or more rows of, 20, 21
  vertical slot, 297
  very best way of connecting blast pipes with, 288-290

VERTICAL blower and engine on same bed plate, 326
  slot tuyere, 37
Voisin, bottom tuyere used by, 48
Voisin's cupola, 161-163
  double tuyere used in, 42

WARMING up a cupola, 248-250
  Waste heat from a cupola, 274, 275
  utilization of, 12, 13
Water blast, 294
  cylinder blast, 294
  gauges, 292
  tuyere, 40, 41

West, Thomas D., experiments with the bottom tuyere by, 48, 49
  with the center blast tuyere, 299
West Troy Stove Works, bad melting at a, 242-248
Whiting cupola, 196-198
  Foundry Equipment Co., Chicago, tuyere manufactured by the, 41
Wilbraham-Baker Blower Co., blower built by the, 314-321
Wood and coal, bad melting caused by, 250, 251
  arranging the, for lighting up, 76
  for lighting up, 340
Woodward's steam-jet cupola, 163-167
  double tuyere used in, 42

ZINC, effect of, on the fire in the cupola, 227
Zone, melting, 77, 123-129
  burning away of the lining at the, 110
  depth of, 125
  determination of the location of the, 123, 335
  of the top of, 335, 336
  development of the, 129
  filling in the lining at the, 106, 107
  location of the, 77
  raising and lowering the, 120, 123

The Largest and Most Reliable Foundry Supply House in the World.

The S. Obermayer Co.

Manufactures "Everything you need in your Foundry."

Cincinnati, O., U. S. A., and Chicago, Ill., U. S. A.

Cable address "Esso." Use A. B. C. code.

Estimates given on complete Brass, Iron or Steel Foundries.

IMPORTERS AND REFINERS

EAST INDIA PLUMBAGO, SILVER, LEAD, GRAPHITE and TALC.

General and special Catalogues sent on application.

Special Attention given to Export Orders.

Shippers; Cupola Blocks, Fire Brick Moulding Sands, Ganister and Silica.

SPECIAL ARTICLES.

Cupolas,	Crucibles,	Chaplets,
Cranes,	Brushes,	Core Wash,
Tumbling Mills,	Riddles,	Heavy Machinery Facing,
Moulding Machines,	Shovels,	Stove Plate Facings,
Ladles,	Bituminous Facing,	Sand Sifting Machines,
Brass Melting Furnaces,	Core Compound,	Pattern Lumber, Etc., Etc.
Crucible Tongs,	Snap Flasks,	

J. W. PAXSON CO.,

Quaker City Facing Mills,

PHILADELPHIA, PA.

A Train of Nine Exhaust Tumblers,

ELECTRIC DRIVEN.

(Two Egg-shaped of Cast Iron, Seven with Steel Plate Centers, ribbed.)

BUILT BY US.

WE EQUIP

Iron, Brass or Steel Foundries

WITH ALMOST EVERYTHING WANTED FROM THE FURNACE DOWN TO A MOLDER'S TOOL FROM OUR OWN WORKS.

J. W. PAXSON CO.

J. W. PAXSON CO.,
Manufacturers, Philadelphia, Pa.

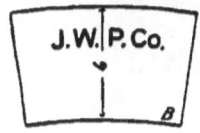

Cupolas.
Spark Arresters.
Linings.
Blast Pipe.
Gates and Gauges.
Mica Schist.
Cupola Breast and Runner.
Screen Charging Doors.
Mica for Tuyeres.
Cupola Picks.
Cupola Fluxes.
Ladles.
Trucks and Turntables.
Core Ovens.
Sand Sifters.
Sand Grinders.
Sand Dryers.
Rumblers for Iron and Brass.
Wheelbarrows.
Pig Iron Barrows.
Core Box Machines.
Crucible Tongs.
Coke and Coal Baskets.

Cramp's Pneumatic Rammer.
Chain and Rope Slings.
Brass Furnaces.
Brass Foundry Drying Stoves.
Emery Grinders.
Brass Founder's Flasks.
Hay Rope Machines.
Foundry Lamps.
Magnetic Separators.
Power Cleaning Brushes.
Sprue Cutters.
Skimming Gates.
Universal Trimmers.
Molding Machines.
Sand Blast Machines.
Molder's Tools.
Foundry Facings,
Molding Sand.
Wire and Bristle Brushes.
Tinned Chaplets,
Snap Flasks.
Fire Clay.
Shovels and Riddles, etc.

→ SEND FOR CATALOGUE, ←

IT CONTAINS
VALUABLE INFORMATION FOR

Iron, Brass or Steel Foundry,

STURTEVANT
PRESSURE BLOWERS

Made with **MOTOR DIRECT** Connected.

Send for Illustrated Catalogue No. 99.

B. F. STURTEVANT CO.,
BOSTON, MASS.

131 Liberty Street, *NEW YORK.*   135 N. Third St., *PHILADELPHIA.*
16 S. Canal St., *CHICAGO.*   75 Queen Victoria St., *LONDON.*

STOCK CARRIED AT BRANCHES.

The Sturtevant Pressure Blowers

FOR

CUPOLA FURNACES

AND

FORGE FIRES.

Blower on Adjustable Bed.

MADE IN ALL SIZES AND TYPES.

Blower on Adjustable Bed with Double Enclosed Engine.

WE ALSO MANUFACTURE

Heating and Ventilating Apparatus for
 Foundries and other Industrial Establishments,
Steam Engines, Generating Sets,
 Piping for Blast and Exhaust Systems.

Whitehead Brothers Company,

DEALERS IN

Molding Sand

FIRE SAND,
PHILADELPHIA SAND
FRENCH SAND,
FIRE CLAY, KAOLIN,
FOUNDRY FACINGS
and SUPPLIES
OF ALL KINDS.
LEAD FACINGS
A SPECIALTY.

Albany Sand for Brass and Stove Plate Castings.

537-39 West 27th St., NEW YORK.

Providence Office,
42 S. Water Street.

Buffalo Office,
70 and 72 Columbia St.

WORKS AT

CHEESEQUAKE CREEK, N. J.
SOUTH RIVER, N. J.
RARITAN RIVER, N. J.
SOUTH AMBOY, N. J.
WAREHAM, MASS.
CENTRE ISLAND, L. I.
WATERFORD, N. Y.
ALBANY, N. Y.

CEDAR HILL, N. Y.
COXSACKIE, N. Y.
CLINTON POINT, N. Y.
ATHENS, N. Y.
COEYMANS, N. Y.
WILLOW SPRINGS, N. Y.
POUGHKEEPSIE, N. Y.
CRESCENT, N. Y.

The Green Rotary Pressure Blower.

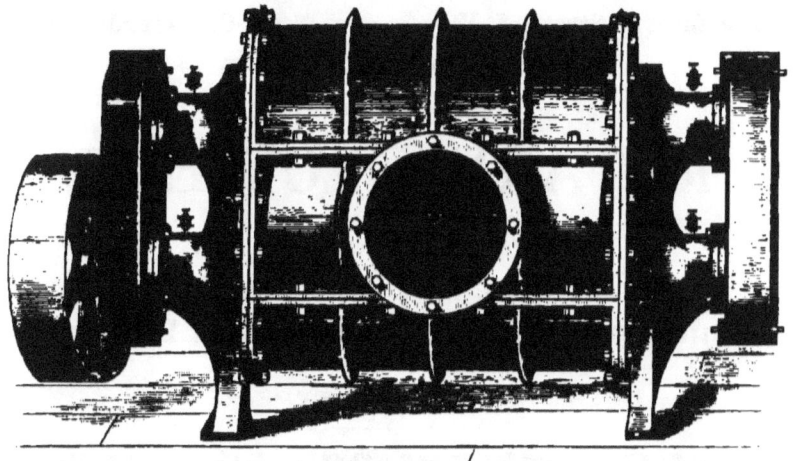

Working parts are two perfectly balanced impellers, mounted on the best steel shafts, running in bronze bushes, detachable journal bearings, cut gearing at both ends of the Blower enclosed in oil-tight casing.

Weighs less, occupies less space, costs less for freight, than any other POSITIVE BLOWER in existence.

FOUNDRY BLOWERS.....
High Pressure Blowers.

GAS EXHAUSTERS.....
Exhausters for Hot Gases.

The GREEN gives the Most Value for a $.

Wilbraham Baker Blower Co.,
2518 Frankford Ave., Philadelphia, Pa.

Sole Agents for Mining and Smelting Industries,
FRASER & CHALMERS, Inc., CHICAGO, ILL., and Branch Houses.

P. H. & F. M. ROOTS CO,

109 Liberty Street, N. Y.   Connersville, Ind.
Vandergrift Building, Pittsburg, Pa.

MANUFACTURERS OF

ROTARY BLOWERS,

Unrivaled. The Standard.
Steam Driven. Electrically Driven.
Minimum of Friction. Maximum of Efficiency.

OVER FORTY YEARS' EXPERIENCE.

MORE THAN 30,000 SOLD.

SEND FOR CATALOGUE.

IN THE PLUMBAGO LINE WE HAVE FOUR WINNERS.

X Plumbago-Pure Ceylon
Corona Plumbago
C. F. M. Founders Perfect Wash
Corona Wash.

Send for Free Sample.

Manufactured only by

The Cleveland Facing Mill Co.,

Foundry Facings and Supplies,

CLEVELAND, O.

CATALOGUE
OF
Practical and Scientific Books
PUBLISHED BY
HENRY CAREY BAIRD & CO.
INDUSTRIAL PUBLISHERS, BOOKSELLERS AND IMPORTERS.
810 Walnut Street, Philadelphia.

☞ Any of the Books comprised in this Catalogue will be sent by mail, free of postage, to any address in the world, at the publication prices.

☞ A Descriptive Catalogue, 90 pages, 8vo., will be sent free and free of postage, to any one in any part of the world, who will furnish his address.

☞ Where not otherwise stated, all of the Books in this Catalogue are bound in muslin.

AMATEUR MECHANICS' WORKSHOP:
A treatise containing plain and concise directions for the manipulation of Wood and Metals, including Casting, Forging, Brazing, Soldering and Carpentry. By the author of the "Lathe and Its Uses." Seventh edition. Illustrated. 8vo. . . . $2.50

ANDES.—Animal Fats and Oils:
Their Practical Production, Purification and Uses; their Properties, Falsification and Examination. 62 illustrations. 8vo. . $4.00

ANDES.—Vegetable Fats and Oils:
Their Practical Preparation, Purification and Employment; their Properties, Adulteration and Examination. 94 illustrations. 8vo.
$4.00

ARLOT.—A Complete Guide for Coach Painters:
Translated from the French of M. ARLOT, Coach Painter, for eleven years Foreman of Painting to M. Eherler, Coach Maker, Paris. By A. A. FESQUET, Chemist and Engineer. To which is added an Appendix, containing Information respecting the Materials and the Practice of Coach and Car Painting and Varnishing in the United States and Great Britain. 12mo. . . . $1.25

(1)

ARMENGAUD, AMOROUX, AND JOHNSON.—The Practical Draughtsman's Book of Industrial Design, and Machinist's and Engineer's Drawing Companion:
Forming a Complete Course of Mechanical Engineering and Architectural Drawing. From the French of M. Armengaud the elder, Prof. of Design in the Conservatoire of Arts and Industry, Paris, and MM. Armengaud the younger, and Amoroux, Civil Engineers. Rewritten and arranged with additional matter and plates, selections from and examples of the most useful and generally employed mechanism of the day. By WILLIAM JOHNSON, Assoc. Inst. C. E. Illustrated by fifty folio steel plates, and fifty wood-cuts. A new edition, 4to., cloth. $6.00

ARMSTRONG.—The Construction and Management of Steam Boilers:
By R. ARMSTRONG, C. E. With an Appendix by ROBERT MALLET, C. E., F. R. S. Seventh Edition. Illustrated. 1 vol. 12mo. .60

ARROWSMITH.—Paper-Hanger's Companion:
A Treatise in which the Practical Operations of the Trade are Systematically laid down: with Copious Directions Preparatory to Papering; Preventives against the Effect of Damp on Walls; the various Cements and Pastes Adapted to the Several Purposes of the Trade; Observations and Directions for the Panelling and Ornamenting of Rooms, etc. By JAMES ARROWSMITH. 12mo., cloth $1.00

ASHTON.—The Theory and Practice of the Art of Designing Fancy Cotton and Woollen Cloths from Sample:
Giving full instructions for reducing drafts, as well as the methods of spooling and making out harness for cross drafts and finding any required reed; with calculations and tables of yarn. By FREDERIC T. ASHTON, Designer, West Pittsfield, Mass. With fifty-two illustrations. One vol. folio $5.00

ASKINSON.—Perfumes and their Preparation:
A Comprehensive Treatise on Perfumery, containing Complete Directions for Making Handkerchief Perfumes, Smelling-Salts, Sachets, Fumigating Pastils; Preparations for the Care of the Skin, the Mouth, the Hair; Cosmetics, Hair Dyes, and other Toilet Articles. By G.W. ASKINSON. Translated from the German by ISIDOR FURST. Revised by CHARLES RICE. 32 Illustrations. 8vo. $3.00

BRONGNIART.—Coloring and Decoration of Ceramic Ware.
8vo. $2.00

BAIRD.—The American Cotton Spinner, and Manager's and Carder's Guide:
A Practical Treatise on Cotton Spinning; giving the Dimensions and Speed of Machinery, Draught and Twist Calculations, etc.; with notices of recent Improvements: together with Rules and Examples for making changes in the sizes and numbers of Roving and Yarn. Compiled from the papers of the late ROBERT H. BAIRD. 12mo.
$1.50

BAIRD.—Standard Wages Computing Tables:
An Improvement in all former Methods of Computation, so arranged that wages for days, hours, or fractions of hours, at a specified rate per day or hour, may be ascertained at a glance. By T. SPANGLER BAIRD. Oblong folio $5.00

BAKER.—Long-Span Railway Bridges:
Comprising Investigations of the Comparative Theoretical and Practical Advantages of the various Adopted or Proposed Type Systems of Construction; with numerous Formulæ and Tables. By B. BAKER. 12mo. $1.00

BAKER.—The Mathematical Theory of the Steam-Engine:
With Rules at length, and Examples worked out for the use of Practical Men. By T. BAKER, C. E., with numerous Diagrams. Sixth Edition, Revised by Prof. J. R. YOUNG. 12mo. . 75

BARLOW.—The History and Principles of Weaving, by Hand and by Power:
Reprinted, with Considerable Additions, from "Engineering," with a chapter on Lace-making Machinery, reprinted from the Journal of the "Society of Arts." By ALFRED BARLOW. With several hundred illustrations. 8vo., 443 pages $10.00

BARR.—A Practical Treatise on the Combustion of Coal:
Including descriptions of various mechanical devices for the Economic Generation of Heat by the Combustion of Fuel, whether solid, liquid or gaseous. 8vo. $2.50

BARR.—A Practical Treatise on High Pressure Steam Boilers:
Including Results of Recent Experimental Tests of Boiler Materials, together with a Description of Approved Safety Apparatus, Steam Pumps, Injectors and Economizers in actual use. By WM. M. BARR. 204 Illustrations. 8vo. $3.60

BAUERMAN.—A Treatise on the Metallurgy of Iron:
Containing Outlines of the History of Iron Manufacture, Methods of Assay, and Analysis of Iron Ores, Processes of Manufacture of Iron and Steel, etc., etc. By H. BAUERMAN, F. G. S., Associate of the Royal School of Mines. Fifth Edition, Revised and Enlarged. Illustrated with numerous Wood Engravings from Drawings by J. B. JORDAN. 12mo. $2.00

BRANNT.—The Metallic Alloys: A Practical Guide
For the Manufacture of all kinds of Alloys, Amalgams, and Solders, used by Metal-Workers; together with their Chemical and Physical Properties and their Application in the Arts and the Industries; with an Appendix on the Coloring of Alloys and the Recovery of Waste Metals. By WILLIAM T. BRANNT. 34 Engravings. A New, Revised, and Enlarged Edition. 554 pages. 8vo. . . $4.50

BEANS.—A Treatise on Railway Curves and Location of Railroads:
By E. W. BEANS, C. E. Illustrated. 12mo. Tucks . $1.50

BECKETT.—A Rudimentary Treatise on Clocks, and Watches and Bells:
By Sir EDMUND BECKETT, Bart., LL. D., Q. C. F. R. A. S. With numerous illustrations. Seventh Edition, Revised and Enlarged. 12mo. $1.80

BELL.—Carpentry Made Easy:
Or, The Science and Art of Framing on a New and Improved System. With Specific Instructions for Building Balloon Frames, Barn Frames, Mill Frames, Warehouses, Church Spires, etc. Comprising also a System of Bridge Building, with Bills, Estimates of Cost, and valuable Tables. Illustrated by forty-four plates, comprising nearly 200 figures. By WILLIAM E. BELL, Architect and Practical Builder. 8vo. $5.00

BEMROSE.—Fret-Cutting and Perforated Carving:
With fifty-three practical illustrations. By W. BEMROSE, JR. 1 vol. quarto $2.50

BEMROSE.—Manual of Buhl-work and Marquetry:
With Practical Instructions for Learners, and ninety colored designs. By W. BEMROSE, JR. 1 vol. quarto . . . $3.00

BEMROSE.—Manual of Wood Carving:
With Practical Illustrations for Learners of the Art, and Original and Selected Designs. By WILLIAM BEMROSE, JR. With an Introduction by LLEWELLYN JEWITT, F. S. A., etc. With 128 illustrations, 4to. $2.50

BILLINGS.—Tobacco:
Its History, Variety, Culture, Manufacture, Commerce, and Various Modes of Use. By E. R. BILLINGS. Illustrated by nearly 200 engravings. 8vo. $3.00

BIRD.—The American Practical Dyers' Companion:
Comprising a Description of the Principal Dye-Stuffs and Chemicals used in Dyeing, their Natures and Uses; Mordants, and How Made; with the best American, English, French and German processes for Bleaching and Dyeing Silk, Wool, Cotton, Linen, Flannel, Felt, Dress Goods, Mixed and Hosiery Yarns, Feathers, Grass, Felt, Fur, Wool, and Straw Hats, Jute Yarn, Vegetable Ivory, Mats, Skins, Furs, Leather, etc., etc. By Wood, Aniline, and other Processes, together with Remarks on Finishing Agents, and Instructions in the Finishing of Fabrics, Substitutes for Indigo, Water-Proofing of Materials, Tests and Purification of Water, Manufacture of Aniline and other New Dye Wares, Harmonizing Colors, etc., etc.; embracing in all over 800 Receipts for Colors and Shades, *accompanied by 170 Dyed Samples of Raw Materials and Fabrics.* By F. J. BIRD, Practical Dyer, Author of "The Dyers' Hand-Book." 8vo. $7.50

BLINN.—A Practical Workshop Companion for Tin, Sheet-Iron, and Copper-plate Workers:
Containing Rules for describing various kinds of Patterns used by Tin, Sheet-Iron and Copper-plate Workers; Practical Geometry; Mensuration of Surfaces and Solids; Tables of the Weights of Metals, Lead-pipe, etc.; Tables of Areas and Circumferences of Circles; Japan, Varnishes, Lackers, Cements, Compositions, etc., etc. By LEROY J. BLINN, Master Mechanic. With One Hundred and Seventy Illustrations. 12mo. $2.50

BOOTH.—Marble Worker's Manual:
Containing Practical Information respecting Marbles in general, their Cutting, Working and Polishing; Veneering of Marble; Mosaics; Composition and Use of Artificial Marble, Stuccos, Cements, Receipts, Secrets, etc., etc. Translated from the French by M. L. BOOTH. With an Appendix concerning American Marbles. 12mo., cloth **$1.50**

BOOTH and MORFIT.—The Encyclopædia of Chemistry, Practical and Theoretical:
Embracing its application to the Arts, Metallurgy, Mineralogy, Geology, Medicine and Pharmacy. By JAMES C. BOOTH, Melter and Refiner in the United States Mint, Professor of Applied Chemistry in the Franklin Institute, etc., assisted by CAMPBELL MORFIT, author of "Chemical Manipulations," etc. Seventh Edition. Complete in one volume, royal 8vo., 978 pages, with numerous wood-cuts and other illustrations **$3.50**

BRAMWELL.—The Wool Carder's Vade-Mecum.
A Complete Manual of the Art of Carding Textile Fabrics. By W. C. BRAMWELL. Third Edition, revised and enlarged. Illustrated. Pp. 400. 12mo. **$2.50**

BRANNT.—A Practical Treatise on Animal and Vegetable Fats and Oils:
Comprising both Fixed and Volatile Oils, their Physical and Chemical Properties and Uses, the Manner of Extracting and Refining them, and Practical Rules for Testing them; as well as the Manufacture of Artificial Butter and Lubricants, etc., with lists of American Patents relating to the Extraction, Rendering, Refining, Decomposing, and Bleaching of Fats and Oils. By WILLIAM T. BRANNT, Editor of the "Techno-Chemical Receipt Book." Second Edition, Revised and in a great part Rewritten. Illustrated by 302 Engravings. In Two Volumes. 1304 pp. 8vo. **$10.00**

BRANNT.—A Practical Treatise on the Manufacture of Soap and Candles:
Based upon the most Recent Experiences in the Practice and Science; comprising the Chemistry, Raw Materials, Machinery, and Utensils and Various Processes of Manufacture, including a great variety of formulas. Edited chiefly from the German of Dr. C. Deite, A. Engelhardt, Dr. C. Schaedler and others; with additions and lists of American Patents relating to these subjects. By WM. T. BRANNT. Illustrated by 163 engravings. 677 pages. 8vo. . . **$7.50**

BRANNT.—India Rubber, Gutta Percha and Balata:
Occurrence, Geographical Distribution, and Cultivation, Obtaining and Preparing the Raw Materials, Modes of Working and Utilizing them, Including Washing, Maceration, Mixing, Vulcanizing, Rubber and Gutta-Percha Compounds, Utilization of Waste, etc. By WILLIAM T. BRANNT. Illustrated. 12mo. (1900.) . . **$3.00**

BRANNT—WAHL.—The Techno-Chemical Receipt Book:
Containing several thousand Receipts covering the latest, most important, and most useful discoveries in Chemical Technology, and their Practical Application in the Arts and the Industries. Edited chiefly from the German of Drs. Winckler, Elsner, Heintze, Mierzinski, Jacobsen, Koller, and Heinzerling, with additions by WM. T. BRANNT and WM. H. WAHL, PH. D. Illustrated by 78 engravings. 12mo. 495 pages $2.00

ROWN.—Five Hundred and Seven Mechanical Movements:
Embracing all those which are most important in Dynamics, Hydraulics, Hydrostatics, Pneumatics, Steam-Engines, Mill and other Gearing, Presses, Horology and Miscellaneous Machinery; and including many movements never before published, and several of which have only recently come into use. By HENRY T. BROWN. 12mo. $1.00

BUCKMASTER.—The Elements of Mechanical Physics:
By J. C. BUCKMASTER. Illustrated with numerous engravings. 12mo. $1.00

BULLOCK.—The American Cottage Builder:
A Series of Designs, Plans and Specifications, from $200 to $20,000, for Homes for the People; together with Warming, Ventilation, Drainage, Painting and Landscape Gardening. By JOHN BULLOCK, Architect and Editor of "The Rudiments of Architecture and Building," etc., etc. Illustrated by 75 engravings. 8vo. $2.50

BULLOCK.—The Rudiments of Architecture and Building:
For the use of Architects, Builders, Draughtsmen, Machinists, Engineers and Mechanics. Edited by JOHN BULLOCK, author of "The American Cottage Builder." Illustrated by 250 Engravings. 8vo. $2.50

BURGH.—Practical Rules for the Proportions of Modern Engines and Boilers for Land and Marine Purposes.
By N. P. BURGH, Engineer. 12mo. $1.50

BYLES.—Sophisms of Free Trade and Popular Political Economy Examined.
By a BARRISTER (SIR JOHN BARNARD BYLES, Judge of Common Pleas). From the Ninth English Edition, as published by the Manchester Reciprocity Association. 12mo. . . . $1.25

BOWMAN.—The Structure of the Wool Fibre in its Relation to the Use of Wool for Technical Purposes:
Being the substance, with additions, of Five Lectures, delivered at the request of the Council, to the members of the Bradford Technical College, and the Society of Dyers and Colorists. By F. H. BOWMAN, D. Sc., F. R. S. E., F. L. S. Illustrated by 32 engravings. 8vo. . . . . $5.00.

YRNE.—Hand-Book for the Artisan, Mechanic, and Engineer:
Comprising the Grinding and Sharpening of Cutting Tools, Abrasive Processes, Lapidary Work, Gem and Glass Engraving, Varnishing and Lackering, Apparatus, Materials and Processes for Grinding and

Polishing, etc. By OLIVER BYRNE. Illustrated by 185 wood engravings. 8vo. $5.00
BYRNE.—Pocket-Book for Railroad and Civil Engineers:
Containing New, Exact and Concise Methods for Laying out Railroad Curves, Switches, Frog Angles and Crossings; the Staking out of work; Levelling; the Calculation of Cuttings; Embankments; Earthwork, etc. By OLIVER BYRNE. 18mo., full bound, pocket-book form $1.50
BYRNE.—The Practical Metal-Worker's Assistant:
Comprising Metallurgic Chemistry; the Arts of Working all Metals and Alloys; Forging of Iron and Steel; Hardening and Tempering; Melting and Mixing; Casting and Founding; Works in Sheet Metal; the Processes Dependent on the Ductility of the Metals; Soldering; and the most Improved Processes and Tools employed by Metal-Workers. With the Application of the Art of Electro-Metallurgy to Manufacturing Processes; collected from Original Sources, and from the works of Holtzapffel, Bergeron, Leupold, Piumier, Napier, Scoffern, Clay, Fairbairn and others. By OLIVER BYRNE. A new, revised and improved edition, to which is added an Appendix, containing The Manufacture of Russian Sheet-Iron. By JOHN PERCY, M. D., F. R. S. The Manufacture of Malleable Iron Castings, and Improvements in Bessemer Steel. By A. A. FESQUET, Chemist and Engineer. With over Six Hundred Engravings, Illustrating every Branch of the Subject. 8vo. $5.00
BYRNE.—The Practical Model Calculator:
For the Engineer, Mechanic, Manufacturer of Engine Work, Naval Architect, Miner and Millwright. By OLIVER BYRNE. 8vo., nearly 600 pages $3.00
CABINET MAKER'S ALBUM OF FURNITURE:
Comprising a Collection of Designs for various Styles of Furniture. Illustrated by Forty-eight Large and Beautifully Engraved Plates. Oblong, 8vo. $1.50
CALLINGHAM.—Sign Writing and Glass Embossing:
A Complete Practical Illustrated Manual of the Art. By JAMES CALLINGHAM. 12mo. $1.50
CAMPIN.—A Practical Treatise on Mechanical Engineering:
Comprising Metallurgy, Moulding, Casting, Forging, Tools, Workshop Machinery, Mechanical Manipulation, Manufacture of Steam-Engines, etc. With an Appendix on the Analysis of Iron and Iron Ores. By FRANCIS CAMPIN, C. E. To which are added, Observations on the Construction of Steam Boilers, and Remarks upon Furnaces used for Smoke Prevention; with a Chapter on Explosions. By R. ARMSTRONG, C. E., and JOHN BOURNE. Rules for Calculating the Change Wheels for Screws on a Turning Lathe, and for a Wheel-cutting Machine. By J. LA NICCA. Management of Steel, Including Forging, Hardening, Tempering, Annealing, Shrinking and Expansion; and the Case-hardening of Iron. By G. EDE. 8vo. Illustrated with twenty-nine plates and 100 wood engravings $5.00

CAREY.—A Memoir of Henry C. Carey.
By Dr. Wm. Elder. With a portrait. 8vo., cloth . . 75

CAREY.—The Works of Henry C. Carey:
Harmony of Interests: Agricultural, Manufacturing and Commercial. 8vo. $1.25
Manual of Social Science. Condensed from Carey's "Principles of Social Science." By Kate McKean. 1 vol. 12mo. . $2.00
Miscellaneous Works. With a Portrait. 2 vols. 8vo. $10.00
Past, Present and Future. 8vo. $2.50
Principles of Social Science. 3 volumes, 8vo. . . $7.50
The Slave-Trade, Domestic and Foreign; Why it Exists, and How it may be Extinguished (1853). 8vo. . . $2.00
The Unity of Law: As Exhibited in the Relations of Physical, Social, Mental and Moral Science (1872). 8vo. . . $2.50

CLARK.—Tramways, their Construction and Working:
Embracing a Comprehensive History of the System. With an exhaustive analysis of the various modes of traction, including horse-power, steam, heated water and compressed air; a description of the varieties of Rolling stock, and ample details of cost and working expenses. By D. Kinnear Clark. Illustrated by over 200 wood engravings, and thirteen folding plates. 1 vol. 8vo. . $7.50

COLBURN.—The Locomotive Engine:
Including a Description of its Structure, Rules for Estimating its Capabilities, and Practical Observations on its Construction and Management. By Zerah Colburn. Illustrated. 12mo. . $1.00

COLLENS.—The Eden of Labor; or, the Christian Utopia.
By T. Wharton Collens, author of "Humanics," "The History of Charity," etc. 12mo. Paper cover, $1.00; Cloth . $1.2¡

COOLEY.—A Complete Practical Treatise on Perfumery:
Being a Hand-book of Perfumes, Cosmetics and other Toilet Articles With a Comprehensive Collection of Formulæ. By Arnold Cooley. 12mo. . . . . $1.5.

COOPER.—A Treatise on the use of Belting for the Transmission of Power.
With numerous illustrations of approved and actual methods of arranging Main Driving and Quarter Twist Belts, and of Belt Fastenings. Examples and Rules in great number for exhibiting and calculating the size and driving power of Belts. Plain, Particular and Practical Directions for the Treatment, Care and Management of Belts. Descriptions of many varieties of Beltings, together with chapters on the Transmission of Power by Ropes; by Iron and Wood Frictional Gearing; on the Strength of Belting Leather; and on the Experimental Investigations of Morin, Briggs, and others. By John H. Cooper, M. E. 8vo. $3.50

CRAIK.—The Practical American Millwright and Miller.
By David Craik, Millwright. Illustrated by numerous wood engravings and two folding plates. 8vo. $3.50

CROSS.—The Cotton Yarn Spinner:
Showing how the Preparation should be arranged for Different Counts of Yarns by a System more uniform than has hitherto been practiced; by having a Standard Schedule from which we make all our Changes. By RICHARD CROSS. 122 pp. 12mo. . 75

CRISTIANI.—A Technical Treatise on Soap and Candles:
With a Glance at the Industry of Fats and Oils. By R. S. CRISTIANI, Chemist. Author of "Perfumery and Kindred Arts." Illustrated by 176 engravings. 581 pages, 8vo.

COURTNEY.—The Boiler Maker's Assistant in Drawing, Templating, and Calculating Boiler Work and Tank Work, etc.
Revised by D. K. CLARK. 102 ills. Fifth edition. . . 80

COURTNEY.—The Boiler Maker's Ready Reckoner:
With Examples of Practical Geometry and Templating. Revised by D. K. CLARK, C. E. 37 illustrations. Fifth edition. . $1.60

DAVIDSON.—A Practical Manual of House Painting, Graining, Marbling, and Sign-Writing:
Containing full information on the processes of House Painting in Oil and Distemper, the Formation of Letters and Practice of Sign-Writing, the Principles of Decorative Art, a Course of Elementary Drawing for House Painters, Writers, etc., and a Collection of Useful Receipts. With nine colored illustrations of Woods and Marbles, and numerous wood engravings. By ELLIS A DAVIDSON. 12mo.
$2.00

DAVIES.—A Treatise on Earthy and Other Minerals and Mining:
By D. C. DAVIES, F. G. S., Mining Engineer, etc. Illustrated by 76 Engravings. 12mo. $5 00

DAVIES.—A Treatise on Metalliferous Minerals and Mining:
By D. C. DAVIES, F. G. S, Mining Engineer, Examiner of Mines, Quarries and Collieries. Illustrated by 148 engravings of Geological Formations, Mining Operations and Machinery, drawn from the practice of all parts of the world. Fifth Edition, thoroughly Revised and much Enlarged by his son, E. Henry Davies. 12mo., 524 pages $5.00

DAVIES.—A Treatise on Slate and Slate Quarrying:
Scientific, Practical and Commercial. By D C. DAVIES, F. G. S., Mining Engineer, etc. With numerous illustrations and folding plates. 12mo $2.00

DAVIS.—A Practical Treatise on the Manufacture of Brick, Tiles and Terra-Cotta:
Including Stiff Clay, Dry Clay, Hand Made, Pressed or Front, and Roadway Paving Brick, Enamelled Brick, with Glazes and Colors, Fire Brick and Blocks, Silica Brick, Carbon Brick, Glass Pots, Re-

torts, Architectural Terra-Cotta, Sewer Pipe, Drain Tile, Glazed and Unglazed Roofing Tile, Art Tile, Mosaics, and Imitation of Intarsia or Inlaid Surfaces. Comprising every product of Clay employed in Architecture, Engineering, and the Blast Furnace. With a Detailed Description of the Different Clays employed, the Most Modern Machinery, Tools, and Kilns used, and the Processes for Handling, Disintegrating, Tempering, and Moulding the Clay into Shape, Drying, Setting, and Burning. By Charles Thomas Davis. Third Edition. Revised and in great part rewritten. Illustrated by 261 engravings. 662 pages $5.00

DAVIS.—A Treatise on Steam-Boiler Incrustation and Methods for Preventing Corrosion and the Formation of Scale:
By CHARLES T. DAVIS. Illustrated by 65 engravings. 8vo. $2.00

DAVIS.—The Manufacture of Paper:
Being a Description of the various Processes for the Fabrication, Coloring and Finishing of every kind of Paper, Including the Different Raw Materials and the Methods for Determining their Values, the Tools, Machines and Practical Details connected with an intelligent and a profitable prosecution of the art, with special reference to the best American Practice. To which are added a History of Paper, complete Lists of Paper-Making Materials, List of American Machines, Tools and Processes used in treating the Raw Materials, and in Making, Coloring and Finishing Paper. By CHARLES T. DAVIS. Illustrated by 156 engravings. 608 pages, 8vo. $6.00

DAVIS.—The Manufacture of Leather:
Being a Description of all the Processes for the Tanning and Tawing with Bark, Extracts, Chrome and all Modern Tannages in General Use, and the Currying, Finishing and Dyeing of Every Kind of Leather; Including the Various Raw Materials, the Tools, Machines, and all Details of Importance Connected with an Intelligent and Profitable Prosecution of the Art, with Special Reference to the Best American Practice. To which are added Lists of American Patents (1884-1897) for Materials, Processes, Tools and Machines for Tanning, Currying, etc. By CHARLES THOMAS DAVIS. Second Edition, Revised, and in great part Rewritten. Illustrated by 147 engravings and 14 Samples of Quebracho Tanned and Aniline Dyed Leathers. 8vo, cloth, 712 pages. Price $7.50

DAWIDOWSKY—BRANNT.—A Practical Treatise on the Raw Materials and Fabrication of Glue, Gelatine, Gelatine Veneers and Foils, Isinglass, Cements, Pastes, Mucilages, etc.:
Based upon Actual Experience. By F. DAWIDOWSKY, Technical Chemist. Translated from the German, with extensive additions, including a description of the most Recent American Processes, by WILLIAM T. BRANNT, Graduate of the Royal Agricultural College of Eldena, Prussia. 35 Engravings. 12mo. . . . $2.50

DE GRAFF.—The Geometrical Stair-Builders' Guide:
Being a Plain Practical System of Hand-Railing, embracing all its necessary Details, and Geometrically Illustrated by twenty-two Steel Engravings; together with the use of the most approved principles of Practical Geometry. By SIMON DE GRAFF, Architect. 4to.
$2.00

DE KONINCK—DIETZ.—A Practical Manual of Chemical Analysis and Assaying:
As applied to the Manufacture of Iron from its Ores, and to Cast Iron, Wrought Iron, and Steel, as found in Commerce. By L. L. DE KONINCK, Dr. Sc., and E. DIETZ, Engineer. Edited with Notes, by ROBERT MALLET, F. R. S., F. S. G., M. I. C. E., etc. American Edition, Edited with Notes and an Appendix on Iron Ores, by A. A. FESQUET, Chemist and Engineer. 12mo. . . . $1.50

DUNCAN.—Practical Surveyor's Guide:
Containing the necessary information to make any person of common capacity, a finished land surveyor without the aid of a teacher By ANDREW DUNCAN. Revised. 72 engravings, 214 pp. 12mo. $1.50

DUPLAIS.—A Treatise on the Manufacture and Distillation of Alcoholic Liquors:
Comprising Accurate and Complete Details in Regard to Alcohol from Wine, Molasses, Beets, Grain, Rice, Potatoes, Sorghum, Asphodel, Fruits, etc.; with the Distillation and Rectification of Brandy, Whiskey, Rum, Gin, Swiss Absinthe, etc., the Preparation of Aromatic Waters, Volatile Oils or Essences, Sugars, Syrups, Aromatic Tinctures, Liqueurs, Cordial Wines, Effervescing Wines, etc., the Ageing of Brandy and the improvement of Spirits, with Copious Directions and Tables for Testing and Reducing Spirituous Liquors, etc., etc. Translated and Edited from the French of MM. DUPLAIS, By M. MCKENNIE, M. D. Illustrated. 743 pp. 8vo. $15.00

DYER AND COLOR-MAKER'S COMPANION:
Containing upwards of two hundred Receipts for making Colors, on the most approved principles, for all the various styles and fabrics now in existence; with the Scouring Process, and plain Directions for Preparing, Washing-off, and Finishing the Goods. 12mo. $1 00

EIDHERR.—The Techno-Chemical Guide to Distillation:
A Hand-Book for the Manufacture of Alcohol and Alcoholic Liquors, including the Preparation of Malt and Compressed Yeast. Edited from the German of Ed. Eidherr. Fully illustrated. (In preparation.)

EDWARDS.—A Catechism of the Marine Steam-Engine,
For the use of Engineers, Firemen, and Mechanics. A Practical Work for Practical Men. By EMORY EDWARDS, Mechanical Engineer. Illustrated by sixty-three Engravings, including examples of the most modern Engines. Third edition, thoroughly revised, with much additional matter. 12mo. 414 pages . . . $2 00

EDWARDS.—Modern American Locomotive Engines,
Their Design, Construction and Management. By EMORY EDWARDS. Illustrated 12mo. $2.00

EDWARDS.—The American Steam Engineer:
Theoretical and Practical, with examples of the latest and most approved American practice in the design and construction of Steam Engines and Boilers. For the use of engineers, machinists, boilermakers, and engineering students. By EMORY EDWARDS. Fully illustrated, 419 pages. 12mo. . . . $2.50

EDWARDS.—Modern American Marine Engines, Boilers, and Screw Propellers,
Their Design and Construction. Showing the Present Practice of the most Eminent Engineers and Marine Engine Builders in the United States. Illustrated by 30 large and elaborate plates. 4to. $5.00

EDWARDS.—The Practical Steam Engineer's Guide
In the Design, Construction, and Management of American Stationary, Portable, and Steam Fire-Engines, Steam Pumps, Boilers, Injectors, Governors, Indicators, Pistons and Rings, Safety Valves and Steam Gauges. For the use of Engineers, Firemen, and Steam Users. By EMORY EDWARDS. Illustrated by 119 engravings. 420 pages. 12mo. $2.50

EISSLER.—The Metallurgy of Gold:
A Practical Treatise on the Metallurgical Treatment of Gold-Bearing Ores, including the Processes of Concentration and Chlorination, and the Assaying, Melting, and Refining of Gold. By M. EISSLER. With 132 Illustrations. 12mo. $5.00

EISSLER.—The Metallurgy of Silver:
A Practical Treatise on the Amalgamation, Roasting, and Lixiviation of Silver Ores, including the Assaying, Melting, and Refining of Silver Bullion. By M. EISSLER. 124 Illustrations. 336 pp. 12mo. $4.25

ELDER.—Conversations on the Principal Subjects of Political Economy.
By DR. WILLIAM ELDER. 8vo. . . $2.50

ELDER.—Questions of the Day,
Economic and Social. By DR. WILLIAM ELDER. 8vo. . $3.00

ERNI.—Mineralogy Simplified.
Easy Methods of Determining and Classifying Minerals, including Ores, by means of the Blowpipe, and by Humid Chemical Analysis, based on Professor von Kobell's Tables for the Determination of Minerals, with an Introduction to Modern Chemistry. By HENRY ERNI, A.M., M.D., Professor of Chemistry. Second Edition, rewritten, enlarged and improved. 12mo. . . . .

FAIRBAIRN.—The Principles of Mechanism and Machinery of Transmission
Comprising the Principles of Mechanism, Wheels, and Pulleys, Strength and Proportions of Shafts, Coupling of Shafts, and Engaging and Disengaging Gear. By SIR WILLIAM FAIRBAIRN, Bart. C. E. Beautifully Illustrated by over 150 wood-cuts. In one volume, 12mo $2.00

FLEMING.—Narrow Gauge Railways in America.
A Sketch of their Rise, Progress, and Success. Valuable Statistics as to Grades, Curves, Weight of Rail, Locomotives, Cars, etc. By HOWARD FLEMING. Illustrated, 8vo. $1.00

FORSYTH.—Book of Designs for Headstones, Mural, and other Monuments:
Containing 78 Designs. By JAMES FORSYTH. With an Introduction by CHARLES BOUTELL, M. A. 4to., cloth . . $3.50

FRANKEL—HUTTER.—A Practical Treatise on the Manufacture of Starch, Glucose, Starch-Sugar, and Dextrine:
Based on the German of LADISLAUS VON WAGNER, Professor in the Royal Technical High School, Buda-Pest, Hungary, and other authorities. By JULIUS FRANKEL, Graduate of the Polytechnic School of Hanover. Edited by ROBERT HUTTER, Chemist, Practical Manufacturer of Starch-Sugar. Illustrated by 58 engravings, covering every branch of the subject, including examples of the most Recent and Best American Machinery. 8vo., 344 pp. . $3.50

GARDNER.—The Painter's Encyclopædia:
Containing Definitions of all Important Words in the Art of Plain and Artistic Painting, with Details of Practice in Coach, Carriage, Railway Car, House, Sign, and Ornamental Painting, including Graining, Marbling, Staining, Varnishing, Polishing, Lettering, Stenciling, Gilding, Bronzing, etc. By FRANKLIN B. GARDNER. 158 Illustrations. 12mo. 427 pp. $2.00

GARDNER.—Everybody's Paint Book:
A Complete Guide to the Art of Outdoor and Indoor Painting. 38 illustrations. 12mo, 183 pp. $1.00

GEE.—The Jeweller's Assistant in the Art of Working in Gold:
A Practical Treatise for Masters and Workmen. 12mo. . $3.00

GEE.—The Goldsmith's Handbook:
Containing full instructions for the Alloying and Working of Gold, including the Art of Alloying, Melting, Reducing, Coloring, Collecting, and Refining; the Processes of Manipulation, Recovery of Waste; Chemical and Physical Properties of Gold; with a New System of Mixing its Alloys; Solders, Enamels, and other Useful Rules and Recipes. By GEORGE E. GEE. 12mo. . . $1.25

GEE.—The Silversmith's Handbook:
Containing full instructions for the Alloying and Working of Silver, including the different modes of Refining and Melting the Metal; its Solders; the Preparation of Imitation Alloys; Methods of Manipulation; Prevention of Waste; Instructions for Improving and Finishing the Surface of the Work; together with other Useful Information and Memoranda. By GEORGE E. GEE. Illustrated. 12mo. $1.25

GOTHIC ALBUM FOR CABINET-MAKERS:
Designs for Gothic Furniture. Twenty-three plates. Oblong $1.50

GRANT.—A Handbook on the Teeth of Gears:
Their Curves, Properties, and Practical Construction. By GEORGE B. GRANT. Illustrated. Third Edition, enlarged. 8vo. $1.00

GREENWOOD.—Steel and Iron:
Comprising the Practice and Theory of the Several Methods Pursued in their Manufacture, and of their Treatment in the Rolling-Mills, the Forge, and the Foundry. By WILLIAM HENRY GREENWOOD, F. C. S. With 97 Diagrams, 536 pages. 12mo. $1.75

GREGORY.—Mathematics for Practical Men:
Adapted to the Pursuits of Surveyors, Architects, Mechanics, and Civil Engineers. By OLINTHUS GREGORY. 8vo., plates . . $3.00

GRISWOLD.—Railroad Engineer's Pocket Companion for the Field:
Comprising Rules for Calculating Deflection Distances and Angles, Tangential Distances and Angles, and all Necessary Tables for Engineers; also the Art of Levelling from Preliminary Survey to the Construction of Railroads, intended Expressly for the Young Engineer, together with Numerous Valuable Rules and Examples. By W. GRISWOLD. 12mo., tucks $1.50

GRUNER.—Studies of Blast Furnace Phenomena:
By M. L. GRUNER, President of the General Council of Mines of France, and lately Professor of Metallurgy at the Ecole des Mines. Translated, with the author's sanction, with an Appendix, by L. D. B. GORDON, F. R. S. E., F. G. S. 8vo. . . . $2.50

Hand-Book of Useful Tables for the Lumberman, Farmer and Mechanic:
Containing Accurate Tables of Logs Reduced to Inch Board Measure, Plank, Scantling and Timber Measure; Wages and Rent, by Week or Month; Capacity of Granaries, Bins and Cisterns; Land Measure, Interest Tables, with Directions for Finding the Interest on any sum at 4, 5, 6, 7 and 8 per cent., and many other Useful Tables. 32 mo., boards. 186 pages 25

HASERICK.—The Secrets of the Art of Dyeing Wool, Cotton, and Linen,
Including Bleaching and Coloring Wool and Cotton Hosiery and Random Yarns. A Treatise based on Economy and Practice. By E. C. HASERICK. *Illustrated by* 323 *Dyed Patterns of the Yarns or Fabrics.* 8vo. $7.50

HATS AND FELTING:
A Practical Treatise on their Manufacture. By a Practical Hatter. Illustrated by Drawings of Machinery, etc. 8vo. . . $1.25

HERMANN.—Painting on Glass and Porcelain, and Enamel Painting:
A Complete Introduction to the Preparation of all the Colors and Fluxes Used for Painting on Glass, Porcelain, Enamel, Faience and Stoneware, the Color Pastes and Colored Glasses, together with a Minute Description of the Firing of Colors and Enamels, on the Basis of Personal Practical Experience of the Art up to Date. 18 illustrations. Second edition. $4.00

HAUPT.—Street Railway Motors:
With Descriptions and Cost of Plants and Operation of the Various Systems now in Use. 12mo. $1.75

HERZFELD.—The Technical Testing of Yarns and Textile Fabrics.
69 illustrations. 8vo. $4.00

HURST.—Lubricating Oils, Fats and Greases:
Their Origin, Preparation, Properties, Uses and Analysis. 65 illustrations. 8vo. $4.00

HURST.—Soaps:
A Practical Manual of the Manufacture of Domestic, Toilet and Other Soaps. 66 illustrations. 8vo. $5.00

HUGHES.—American Miller and Millwright's Assistant:
By WILLIAM CARTER HUGHES. 12mo. $1.50

HULME.—Worked Examination Questions in Plane Geometrical Drawing:
For the Use of Candidates for the Royal Military Academy, Woolwich; the Royal Military College, Sandhurst; the Indian Civil Engineering College, Cooper's Hill; Indian Public Works and Telegraph Departments; Royal Marine Light Infantry; the Oxford and Cambridge Local Examinations, etc. By F. EDWARD HULME, F. L. S., F. S. A., Art-Master Marlborough College. Illustrated by 300 examples. Small quarto $2.50

JERVIS.—Railroad Property:
A Treatise on the Construction and Management of Railways; designed to afford useful knowledge, in the popular style, to the holders of this class of property; as well as Railway Managers, Officers, and Agents. By JOHN B. JERVIS, late Civil Engineer of the Hudson River Railroad, Croton Aqueduct, etc. 12mo., cloth $2.00

KEENE.—A Hand-Book of Practical Gauging:
For the Use of Beginners, to which is added a Chapter on Distillation, describing the process in operation at the Custom-House for ascertaining the Strength of Wines. By JAMES B. KEENE, of H. M. Customs. 8vo. $1.00

KELLEY.—Speeches, Addresses, and Letters on Industrial and Financial Questions:
By HON. WILLIAM D. KELLEY, M. C. 544 pages, 8vo. . $2.50

KELLOGG.—A New Monetary System:
The only means of Securing the respective Rights of Labor and Property, and of Protecting the Public from Financial Revulsions. By EDWARD KELLOGG. 12mo. Paper cover, $1.00. Bound in cloth. $1.25

KEMLO.—Watch-Repairer's Hand-Book:
Being a Complete Guide to the Young Beginner, in Taking Apart, Putting Together, and Thoroughly Cleaning the English Lever and other Foreign Watches, and all American Watches. By F. KEMLO, Practical Watchmaker. With Illustrations. 12mo. . $1.25

KENTISH.—A Treatise on a Box of Instruments,
And the Slide Rule; with the Theory of Trigonometry and Logarithms, including Practical Geometry, Surveying, Measuring of Timber, Cask and Malt Gauging, Heights, and Distances. By THOMAS KENTISH. In one volume. 12mo. $1.00

KERL.—The Assayer's Manual:
An Abridged Treatise on the Docimastic Examination of Ores, and Furnace and other Artificial Products. By BRUNO KERL, Professor in the Royal School of Mines. Translated from the German by WILLIAM T. BRANNT. Second American edition, edited with Extensive Additions by F. LYNWOOD GARRISON, Member of the American Institute of Mining Engineers, etc. Illustrated by 87 engravings. 8vo. $3.00

KICK.—Flour Manufacture.
A Treatise on Milling Science and Practice. By FREDERICK KICK Imperial Regierungsrath, Professor of Mechanical Technology in the Imperial German Polytechnic Institute, Prague. Translated from the second enlarged and revised edition with supplement by H. H. P. POWLES, Assoc. Memb. Institution of Civil Engineers. Illustrated with 28 Plates, and 167 Wood-cuts. 367 pages. 8vo. . $10.00

KINGZETT.—The History, Products, and Processes of the Alkali Trade:
Including the most Recent Improvements. By CHARLES THOMAS KINGZETT, Consulting Chemist. With 23 illustrations. 8vo. $2.50

KIRK.—The Cupola Furnace:
A Practical Treatise on the Construction and Management of Foundry Cupolas. By EDWARD KIRK, Practical Moulder and Melter, Consulting Expert in Melting. Author of "The Founding of Metals." Illustrated by 78 engravings. 8vo. 379 pages. . . $3.50

LANDRIN.—A Treatise on Steel:
Comprising its Theory, Metallurgy, Properties, Practical Working, and Use. By M. H. C. LANDRIN, JR. From the French, by A. A. FESQUET. 12mo. $2.50

LANGBEIN.—A Complete Treatise on the Electro-Deposition of Metals:
Comprising Electro-Plating and Galvanoplastic Operations, the Deposition of Metals by the Contact and Immersion Processes, the Coloring of Metals, the Methods of Grinding and Polishing, as well as Descriptions of the Electric Elements Dynamo-Electric Machines, Thermo-Piles and of the Materials and Processes used in Every Department of the Art. From the German of DR. GEORGE LANGBEIN, with additions by WM. T. BRANNT. Third Edition, thoroughly revised and much enlarged. 150 Engravings. 528 pages. 8vo. $4.00

LARDNER.—The Steam-Engine:
For the Use of Beginners. Illustrated. 12mo. 60

LEHNER.—The Manufacture of Ink:
Comprising the Raw Materials, and the Preparation of Writing, Copying and Hektograph Inks, Safety Inks, Ink Extracts and Powders, etc. Translated from the German of SIGMUND LEHNER, with additions by WILLIAM T. BRANNT. Illustrated. 12mo. $2.00

LARKIN.—The Practical Brass and Iron Founder's Guide:
A Concise Treatise on Brass Founding, Moulding, the Metals and their Alloys, etc.; to which are added Recent Improvements in the Manufacture of Iron, Steel by the Bessemer Process, etc., etc. By JAMES LARKIN, late Conductor of the Brass Foundry Department in Reany, Neafie & Co.'s Penn Works, Philadelphia. New edition, revised, with extensive additions. 12mo. . . . $2.50

LEROUX.—A Practical Treatise on the Manufacture of Worsteds and Carded Yarns:
Comprising Practical Mechanics, with Rules and Calculations applied to Spinning; Sorting, Cleaning, and Scouring Wools; the English and French Methods of Combing, Drawing, and Spinning Worsteds, and Manufacturing Carded Yarns. Translated from the French of CHARLES LEROUX, Mechanical Engineer and Superintendent of a Spinning-Mill, by HORATIO PAINE, M. D., and A. A. FESQUET, Chemist and Engineer. Illustrated by twelve large Plates. To which is added an Appendix, containing Extracts from the Reports of the International Jury, and of the Artisans selected by the Committee appointed by the Council of the Society of Arts, London, on Woolen and Worsted Machinery and Fabrics, as exhibited in the Paris Universal Exposition, 1867. 8vo. $5.00

LEFFEL.—The Construction of Mill-Dams:
Comprising also the Building of Race and Reservoir Embankments and Head-Gates, the Measurement of Streams, Gauging of Water Supply, etc. By JAMES LEFFEL & Co. Illustrated by 58 engravings. 8vo. $2.50

LESLIE.—Complete Cookery:
Directions for Cookery in its Various Branches. By MISS LESLIE. Sixtieth thousand. Thoroughly revised, with the addition of New Receipts. 12mo. $1.50

LE VAN.—The Steam Engine and the Indicator:
Their Origin and Progressive Development; including the Most Recent Examples of Steam and Gas Motors, together with the Indicator, its Principles, its Utility, and its Application. By WILLIAM BARNET LE VAN. Illustrated by 205 Engravings, chiefly of Indicator-Cards. 469 pp. 8vo. $4.00

LIEBER.—Assayer's Guide:
Or, Practical Directions to Assayers, Miners, and Smelters, for the Tests and Assays, by Heat and by Wet Processes, for the Ores of all the principal Metals, of Gold and Silver Coins and Alloys, and of Coal, etc. By OSCAR M. LIEBER. Revised. 283 pp. 12mo. $1.50

Lockwood's Dictionary of Terms:
Used in the Practice of Mechanical Engineering, embracing those Current in the Drawing Office, Pattern Shop, Foundry, Fitting, Turning, Smith's and Boiler Shops, etc., etc., comprising upwards of Six Thousand Definitions. Edited by a Foreman Pattern Maker, author of "Pattern Making." 417 pp. 12mo. . . . $3.00

LUKIN.—The Lathe and Its Uses:
Or Instruction in the Art of Turning Wood and Metal. Including a Description of the Most Modern Appliances for the Ornamentation of Plane and Curved Surfaces, an Entirely Novel Form of Lathe for Eccentric and Rose-Engine Turning; A Lathe and Planing Machine Combined; and Other Valuable Matter Relating to the Art. Illustrated by 462 engravings. Seventh edition. 315 pages. 8vo. $4.25

MAIN and BROWN.—Questions on Subjects Connected with the Marine Steam-Engine:
And Examination Papers; with Hints for their Solution. By THOMAS J. MAIN, Professor of Mathematics, Royal Naval College, and THOMAS BROWN, Chief Engineer, R. N. 12mo., cloth . $1.00

MAIN and BROWN.—The Indicator and Dynamometer:
With their Practical Applications to the Steam-Engine. By THOMAS J. MAIN, M. A. F. R., Ass't S. Professor Royal Naval College, Portsmouth, and THOMAS BROWN, Assoc. Inst. C. E., Chief Engineer R. N., attached to the R. N. College. Illustrated. 8vo. .

MAIN and BROWN.—The Marine Steam-Engine.
By THOMAS J. MAIN, F. R. Ass't S. Mathematical Professor at the Royal Naval College, Portsmouth, and THOMAS BROWN, Assoc. Inst. C. E., Chief Engineer R. N. Attached to the Royal Naval College. With numerous illustrations. 8vo. . .

MAKINS.—A Manual of Metallurgy:
By GEORGE HOGARTH MAKINS. 100 engravings. Second edition rewritten and much enlarged. 12mo., 592 pages . . $3.00

MARTIN.—Screw-Cutting Tables, for the Use of Mechanical Engineers:
Showing the Proper Arrangement of Wheels for Cutting the Threads of Screws of any Required Pitch; with a Table for Making the Universal Gas-Pipe Thread and Taps. By W. A. MARTIN, Engineer. 8vo. 50

MICHELL.—Mine Drainage:
Being a Complete and Practical Treatise on Direct-Acting Underground Steam Pumping Machinery. With a Description of a large number of the best known Engines, their General Utility and the Special Sphere of their Action, the Mode of their Application, and their Merits compared with other Pumping Machinery. By STEPHEN MICHELL. Illustrated by 137 engravings. 8vo., 277 pages . $6.00

MOLESWORTH.—Pocket-Book of Useful Formulæ and Memoranda for Civil and Mechanical Engineers.
By GUILFORD L. MOLESWORTH, Member of the Institution of Civil Engineers, Chief Resident Engineer of the Ceylon Railway. Full-bound in Pocket-book form $1.00

MOORE.—The Universal Assistant and the Complete Mechanic:
Containing over one million Industrial Facts, Calculations, Receipts, Processes, Trades Secrets, Rules, Business Forms, Legal Items, Etc., in every occupation, from the Household to the Manufactory. By R. MOORE. Illustrated by 500 Engravings. 12mo. . **$2.50**

MORRIS.—Easy Rules for the Measurement of Earthworks:
By means of the Prismoidal Formula. Illustrated with Numerous Wood-Cuts, Problems, and Examples, and concluded by an Extensive Table for finding the Solidity in cubic yards from Mean Areas. The whole being adapted for convenient use by Engineers, Surveyors, Contractors, and others needing Correct Measurements of Earthwork. By ELWOOD MORRIS, C. E. 8vo. **$1.50**

MAUCHLINE.—The Mine Foreman's Hand-Book
Of Practical and Theoretical Information on the Opening, Ventilating, and Working of Collieries. Questions and Answers on Practical and Theoretical Coal Mining. Designed to Assist Students and Others in Passing Examinations for Mine Foremanships. By ROBERT MAUCHLINE, Ex-Inspector of Mines. A New, Revised and Enlarged Edition. Illustrated by 114 engravings. 8vo. 337 pages **$3.75**

NAPIER.—A System of Chemistry Applied to Dyeing.
By JAMES NAPIER, F. C. S. A New and Thoroughly Revised Edition. Completely brought up to the present state of the Science, including the Chemistry of Coal Tar Colors, by A. A. FESQUET, Chemist and Engineer. With an Appendix on Dyeing and Calico Printing, as shown at the Universal Exposition, Paris, 1867. Illustrated. 8vo. 422 pages **$3.00**

NEVILLE.—Hydraulic Tables, Coefficients, and Formulæ, for finding the Discharge of Water from Orifices, Notches, Weirs, Pipes, and Rivers:
Third Edition, with Additions, consisting of New Formulæ for the Discharge from Tidal and Flood Sluices and Siphons; general information on Rainfall, Catchment-Basins, Drainage, Sewerage, Water Supply for Towns and Mill Power. By JOHN NEVILLE. C. E. M R I. A.; Fellow of the Royal Geological Society of Ireland. Thick 12mo. **$5.50**

NEWBERY.—Gleanings from Ornamental Art of every style:
Drawn from Examples in the British, South Kensington, Indian, Crystal Palace, and other Museums, the Exhibitions of 1851 and 1862, and the best English and Foreign works. In a series of 100 exquisitely drawn Plates, containing many hundred examples. By ROBERT NEWBERY. 4to. . . . . .

NICHOLLS.—The Theoretical and Practical Boiler-Maker and Engineer's Reference Book:
Containing a variety of Useful Information for Employers of Labor Foremen and Working Boiler-Makers. Iron, Copper, and Tinsmiths

Draughtsmen, Engineers, the General Steam-using Public, and for the Use of Science Schools and Classes. By SAMUEL NICHOLLS. Illustrated by sixteen plates, 12mo. $2.50

NICHOLSON.—A Manual of the Art of Bookbinding:
Containing full instructions in the different Branches of Forwarding, Gilding, and Finishing. Also, the Art of Marbling Book-edges and Paper. By JAMES B. NICHOLSON. Illustrated. 12mo., cloth $2.25

NICOLLS.—The Railway Builder:
A Hand-Book for Estimating the Probable Cost of American Railway Construction and Equipment. By WILLIAM J. NICOLLS, Civil Engineer. Illustrated, full bound, pocket-book form .

NORMANDY.—The Commercial Handbook of Chemical Analysis;
Or Practical Instructions for the Determination of the Intrinsic or Commercial Value of Substances used in Manufactures, in Trades, and in the Arts. By A. NORMANDY. New Edition, Enlarged, and to a great extent rewritten. By HENRY M. NOAD, Ph.D., F.R.S., thick 12mo. $5.00

NORRIS.—A Handbook for Locomotive Engineers and Machinists:
Comprising the Proportions and Calculations for Constructing Locomotives; Manner of Setting Valves; Tables of Squares, Cubes, Areas, etc., etc. By SEPTIMUS NORRIS, M. E. New edition. Illustrated, 12mo. $1.50

NYSTROM.—A New Treatise on Elements of Mechanics:
Establishing Strict Precision in the Meaning of Dynamical Terms; accompanied with an Appendix on Duodenal Arithmetic and Metrology. By JOHN W. NYSTROM, C. E. Illustrated. 8vo. $3.00

NYSTROM.—On Technological Education and the Construction of Ships and Screw Propellers:
For Naval and Marine Engineers. By JOHN W. NYSTROM, late Acting Chief Engineer, U. S. N. Second edition, revised, with additional matter. Illustrated by seven engravings. 12mo. . $1.2

O'NEILL.—A Dictionary of Dyeing and Calico Printing:
Containing a brief account of all the Substances and Processes in use in the Art of Dyeing and Printing Textile Fabrics; with Practical Receipts and Scientific Information. By CHARLES O'NEILL, Analytical Chemist. To which is added an Essay on Coal Tar Colors and their application to Dyeing and Calico Printing. By A. A. FESQUET, Chemist and Engineer. With an appendix on Dyeing and Calico Printing, as shown at the Universal Exposition, Paris, 1867. 8vo., 491 pages $3.00

ORTON.—Underground Treasures:
How and Where to Find Them. A Key for the Ready Determination of all the Useful Minerals within the United States. By JAMES ORTON, A.M., Late Professor of Natural History in Vassar College, N. Y.; Cor. Mem. of the Academy of Natural Sciences, Philadelphia, and of the Lyceum of Natural History, New York; author of the "Andes and the Amazon," etc. A New Edition, with Additions. Illustrated $1.50

OSBORN.—The Prospector's Field Book and Guide.
In the Search For and the Easy Determination of Ores and Other Useful Minerals. By Prof. H. S. OSBORN, LL. D. Illustrated by 58 Engravings. 12mo. Fourth Edition. Revised and Enlarged (1899) $1.50

OSBORN—A Practical Manual of Minerals, Mines and Mining:
Comprising the Physical Properties, Geologic Positions, Local Occurrence and Associations of the Useful Minerals; their Methods of Chemical Analysis and Assay; together with Various Systems of Excavating and Timbering, Brick and Masonry Work, during Driving, Lining, Bracing and other Operations, etc. By Prof. H. S. OSBORN, LL. D., Author of "The Prospector's Field-Book and Guide." 171 engravings. Second Edition, revised. 8vo. . . . $4.50

OVERMAN.—The Manufacture of Steel:
Containing the Practice and Principles of Working and Making Steel. A Handbook for Blacksmiths and Workers in Steel and Iron, Wagon Makers, Die Sinkers, Cutlers, and Manufacturers of Files and Hardware, of Steel and Iron, and for Men of Science and Art. By FREDERICK OVERMAN, Mining Engineer, Author of the "Manufacture of Iron," etc. A new, enlarged, and revised Edition. By A. A. FESQUET, Chemist and Engineer. 12mo. . . $1.50

OVERMAN.—The Moulder's and Founder's Pocket Guide:
A Treatise on Moulding and Founding in Green-sand, Dry-sand, Loam, and Cement; the Moulding of Machine Frames, Mill-gear, Hollowware, Ornaments, Trinkets, Bells, and Statues; Description of Moulds for Iron, Bronze, Brass, and other Metals; Plaster of Paris, Sulphur, Wax, etc.; the Construction of Melting Furnaces, the Melting and Founding of Metals; the Composition of Alloys and their Nature, etc., etc. By FREDERICK OVERMAN, M. E. A new Edition, to which is added a Supplement on Statuary and Ornamental Moulding, Ordnance, Malleable Iron Castings, etc. By A. A. FESQUET, Chemist and Engineer. Illustrated by 44 engravings. 12mo. . $2.00

PAINTER, GILDER, AND VARNISHER'S COMPANION.
Comprising the Manufacture and Test of Pigments, the Arts of Painting, Graining, Marbling, Staining, Sign-writing, Varnishing, Glass-staining, and Gilding on Glass; together with Coach Painting and Varnishing, and the Principles of the Harmony and Contrast of Colors. Twenty-seventh Edition. Revised, Enlarged, and in great part Rewritten. By WILLIAM T. BRANNT, Editor of "Varnishes, Lacquers, Printing Inks and Sealing Waxes." Illustrated. 395 pp. 12mo. $1.50

PALLETT.—The Miller's, Millwright's, and Engineer's Guide.
By HENRY PALLETT. Illustrated. 12mo. . . . $2.00

PERCY.—The Manufacture of Russian Sheet-Iron.
By JOHN PERCY, M. D., F. R. S. Paper. . . . 25 cts.

PERKINS.—Gas and Ventilation:
Practical Treatise on Gas and Ventilation. Illustrated. 12mo. $1.25

PERKINS AND STOWE.—A New Guide to the Sheet-iron and Boiler Plate Roller:
Containing a Series of Tables showing the Weight of Slabs and Piles to Produce Boiler Plates, and of the Weight of Piles and the Sizes of Bars to produce Sheet-iron; the Thickness of the Bar Gauge in decimals; the Weight per foot, and the Thickness on the Bar or Wire Gauge of the fractional parts of an inch; the Weight per sheet, and the Thickness on the Wire Gauge of Sheet-iron of various dimensions to weigh 112 lbs. per bundle; and the conversion of Short Weight into Long Weight, and Long Weight into Short.
$1.50

POSSELT.—Recent Improvements in Textile Machinery Relating to Weaving:
Giving the Most Modern Points on the Construction of all Kinds of Looms, Warpers, Beamers, Slashers, Winders, Spoolers, Reeds, Temples, Shuttles, Bobbins, Heddles, Heddle Frames, Pickers, Jacquards, Card Stampers, etc., etc. 600 illus. . . $3 00

POSSELT.—Technology of Textile Design:
The Most Complete Treatise on the Construction and Application of Weaves for all Textile Fabrics and the Analysis of Cloth. By E. A. Posselt. 1,500 illustrations. 4to. $5.00

POSSELT.—Textile Calculations:
A Guide to Calculations Relating to the Manufacture of all Kinds of Yarns and Fabrics, the Analysis of Cloth, Speed, Power and Belt Calculations. By E. A. POSSELT. Illustrated. 4to. . $2.00

REGNAULT.—Elements of Chemistry:
By M. V. REGNAULT. Translated from the French by T. FORREST BETTON, M. D., and edited, with Notes, by JAMES C. BOOTH, Melter and Refiner U. S. Mint, and WILLIAM L. FABER, Metallurgist and Mining Engineer. Illustrated by nearly 700 wood-engravings. Comprising nearly 1,500 pages. In two volumes, 8vo., cloth . $6.00

RICHARDS.—Aluminium:
Its History, Occurrence, Properties, Metallurgy and Applications, including its Alloys. By JOSEPH W. RICHARDS, A. C., Chemist and Practical Metallurgist, Member of the Deutsche Chemische Gesellschaft. Illust. Third edition, enlarged and revised (1895) . $6.00

RIFFAULT, VERGNAUD, and TOUSSAINT.—A Practical Treatise on the Manufacture of Colors for Painting:
Comprising the Origin, Definition, and Classification of Colors; the Treatment of the Raw Materials; the best Formulæ and the Newest Processes for the Preparation of every description of Pigment, and the Necessary Apparatus and Directions for its Use; Dryers; the Testing, Application, and Qualities of Paints, etc., etc. By MM. RIFFAULT, VERGNAUD, and TOUSSAINT. Revised and Edited by M.

F. MALEPEYRE. Translated from the French, by A. A. FESQUET, Chemist and Engineer. Illustrated by Eighty engravings. In one vol., 8vo., 659 pages $5.00

ROPER.—A Catechism of High-Pressure, or Non-Condensing Steam-Engines:
Including the Modelling, Constructing, and Management of Steam-Engines and Steam Boilers. With valuable illustrations. By STEPHEN ROPER. Engineer. Sixteenth edition, revised and enlarged. 18mo., tucks, gilt edge $2.00

ROPER.—Engineer's Handy-Book:
Containing a full Explanation of the Steam-Engine Indicator, and its Use and Advantages to Engineers and Steam Users. With Formulæ for Estimating the Power of all Classes of Steam-Engines; also, Facts, Figures, Questions, and Tables for Engineers who wish to qualify themselves for the United States Navy, the Revenue Service, the Mercantile Marine, or to take charge of the Better Class of Stationary Steam-Engines. Tenth edition. 16mo., 690 pages, tucks, gilt edge $3.50

ROPER.—Hand-Book of Land and Marine Engines:
Including the Modelling, Construction, Running, and Management of Land and Marine Engines and Boilers. With illustrations. By STEPHEN ROPER. Engineer. Sixth edition. 12mo., tucks, gilt edge. $3.50

ROPER.—Hand-Book of the Locomotive:
Including the Construction of Engines and Boilers, and the Construction, Management, and Running of Locomotives. By STEPHEN ROPER. Eleventh edition. 18mo., tucks, gilt edge . $2.50

ROPER.—Hand-Book of Modern Steam Fire-Engines.
With illustrations. By STEPHEN ROPER, Engineer. Fourth edition, 12mo., tucks, gilt edge $3.50

ROPER.—Questions and Answers for Engineers.
This little book contains all the Questions that Engineers will be asked when undergoing an Examination for the purpose of procuring Licenses, and they are so plain that any Engineer or Fireman of ordinary intelligence may commit them to memory in a short time. By STEPHEN ROPER, Engineer. Third edition . . . $2.00

ROPER.—Use and Abuse of the Steam Boiler.
By STEPHEN ROPER, Engineer. Eighth edition, with illustrations. 18mo., tucks, gilt edge $2.00

ROSE.—The Complete Practical Machinist:
Embracing Lathe Work, Vise Work, Drills and Drilling, Taps and Dies, Hardening and Tempering, the Making and Use of Tools. Tool Grinding, Marking out Work, Machine Tools, etc. By JOSHUA ROSE. 395 Engravings. Nineteenth Edition, greatly Enlarged with New and Valuable Matter. 12mo., 504 pages. . . $2.50

ROSE.—Mechanical Drawing Self-Taught:
Comprising Instructions in the Selection and Preparation of Drawing Instruments, Elementary Instruction in Practical Mechanical Draw-

ing, together with Examples in Simple Geometry and Elementary Mechanism, including Screw Threads, Gear Wheels, Mechanical Motions, Engines and Boilers. By JOSHUA ROSE, M. E. Illustrated by 330 engravings. 8vo , 313 pages $4.00

ROSE.—The Slide-Valve Practically Explained:
Embracing simple and complete Practical Demonstrations of the operation of each element in a Slide-valve Movement, and illustrating the effects of Variations in their Proportions by examples carefully selected from the most recent and successful practice. By JOSHUA ROSE, M. E. Illustrated by 35 engravings . $1.00

ROSS.—The Blowpipe in Chemistry, Mineralogy and Geology:
Containing all Known Methods of Anhydrous Analysis, many Working Examples, and Instructions for Making Apparatus. By LIEUT.- COLONEL W. A. ROSS, R. A., F. G. S. With 120 Illustrations. 12mo. $2.00

SHAW.—Civil Architecture:
Being a Complete Theoretical and Practical System of Building, containing the Fundamental Principles of the Art. By EDWARD SHAW, Architect. To which is added a Treatise on Gothic Architecture, etc. By THOMAS W. SILLOWAY and GEORGE M. HARDING, Architects. The whole illustrated by 102 quarto plates finely engraved on copper. Eleventh edition. 4to. $6.00

SHUNK.—A Practical Treatise on Railway Curves and Location, for Young Engineers.
By W. F. SHUNK, C. E. 12mo. Full bound pocket-book form $2.00

SLATER.—The Manual of Colors and Dye Wares.
By J. W. SLATER. 12mo. $3.00

SLOAN.—American Houses:
A variety of Original Designs for Rural Buildings. Illustrated by 26 colored engravings, with descriptive references. By SAMUEL SLOAN, Architect. 8vo. 75

SLOAN.—Homestead Architecture:
Containing Forty Designs for Villas, Cottages, and Farm-houses, with Essays on Style, Construction, Landscape Gardening, Furniture, etc., etc. Illustrated by upwards of 200 engravings. By SAMUEL SLOAN, Architect. 8vo. $2.50

SLOANE.—Home Experiments in Science.
By T. O'CONOR SLOANE, E. M., A. M., Ph. D. Illustrated by 91 engravings. 12mo. $1.00

SMEATON.—Builder's Pocket-Companion:
Containing the Elements of Building, Surveying, and Architecture; with Practical Rules and Instructions connected with the subject. By A. C. SMEATON, Civil Engineer, etc. 12mo. . . 75 cts.

SMITH.—A Manual of Political Economy.
By E. PESHINE SMITH. A New Edition, to which is added a full Index. 12mo. $1 25

SMITH.—Parks and Pleasure-Grounds:
Or Practical Notes on Country Residences, Villas, Public Parks, and Gardens. By CHARLES H. J. SMITH, Landscape Gardener and Garden Architect, etc., etc. 12mo. $2.00

SMITH.—The Dyer's Instructor:
Comprising Practical Instructions in the Art of Dyeing Silk, Cotton, Wool, and Worsted, and Woolen Goods; containing nearly 800 Receipts. To which is added a Treatise on the Art of Padding; and the Printing of Silk Warps, Skeins, and Handkerchiefs, and the various Mordants and Colors for the different styles of such work. By DAVID SMITH, Pattern Dyer. 12mo. . . . $1.50

SMYTH.—A Rudimentary Treatise on Coal and Coal-Mining.
By WARRINGTON W. SMYTH, M. A., F. R. G., President R. G. S. of Cornwall. Fifth edition, revised and corrected. With numerous illustrations. 12mo. $1.75

SNIVELY.—Tables for Systematic Qualitative Chemical Analysis.
By JOHN H. SNIVELY, Phr. D. 8vo. $1.00

SNIVELY.—The Elements of Systematic Qualitative Chemical Analysis:
A Hand-book for Beginners. By JOHN H. SNIVELY, Phr. D. 16mo. $2.00

STOKES.—The Cabinet-Maker and Upholsterer's Companion:
Comprising the Art of Drawing, as applicable to Cabinet Work; Veneering, Inlaying, and Buhl-Work; the Art of Dyeing and Staining Wood, Ivory, Bone, Tortoise-Shell, etc. Directions for Lackering, Japanning, and Varnishing; to make French Polish, Glues. Cements, and Compositions; with numerous Receipts, useful to workmen generally. By STOKES. Illustrated. A New Edition, with an Appendix upon Ench Polishing, Staining, Imitating, Varnishing, etc., etc. 12mo $1.25

STRENGTH AND OTHER PROPERTIES OF METALS:
Reports of Experiments on the Strength and other Properties of Metals for Cannon. With a Description of the Machines for Testing Metals, and of the Classification of Cannon in service. By Officers of the Ordnance Department, U. S. Army. By authority of the Secretary of War. Illustrated by 25 large steel plates. Quarto . $5.00

SULLIVAN.—Protection to Native Industry.
By Sir EDWARD SULLIVAN, Baronet, author of "Ten Chapters on Social Reforms." 8vo. $1.00

SHERRATT.—The Elements of Hand-Railing:
Simplified and Explained in Concise Problems that are Easily Understood. The whole illustrated with Thirty-eight Accurate and Original Plates, Founded on Geometrical Principles, and Showing how to Make Rail Without Centre Joints, Making Better Rail of the Same Material, with Half the Labor, and Showing How to Lay Out Stairs of all Kinds. By R. J. SHERRATT. Folio. . . . $2.50

SYME.—Outlines of an Industrial Science.
By DAVID SYME. 12mo. $2.00
TABLES SHOWING THE WEIGHT OF ROUND, SQUARE, AND FLAT BAR IRON, STEEL, ETC.,
By Measurement. Cloth 63
TAYLOR.—Statistics of Coal:
Including Mineral Bituminous Substances employed in Arts and Manufactures; with their Geographical, Geological, and Commercial Distribution and Amount of Production and Consumption on the American Continent. With Incidental Statistics of the Iron Manufacture. By R. C. TAYLOR. Second edition, revised by S. S. HALDEMAN. Illustrated by five Maps and many wood engravings. 8vo., cloth $6.00
TEMPLETON.—The Practical Examiner on Steam and the Steam-Engine:
With Instructive References relative thereto, arranged for the Use of Engineers, Students, and others. By WILLIAM TEMPLETON, Engineer. 12mo. $1.00
THAUSING.—The Theory and Practice of the Preparation of Malt and the Fabrication of Beer:
With especial reference to the Vienna Process of Brewing. Elaborated from personal experience by JULIUS E. THAUSING, Professor at the School for Brewers, and at the Agricultural Institute, Mödling, near Vienna. Translated from the German by WILLIAM T. BRANNT, Thoroughly and elaborately edited, with much American matter, and according to the latest and most Scientific Practice, by A. SCHWARZ and DR. A. H. BAUER. Illustrated by 140 Engravings. 8vo., 815 pages $10.00
THOMAS.—The Modern Practice of Photography:
By R. W. THOMAS, F. C. S. 8vo. 25
THOMPSON.—Political Economy. With Especial Reference to the Industrial History of Nations:
By ROBERT E. THOMPSON, M. A., Professor of Social Science in the University of Pennsylvania. 12mo. $1.50
THOMSON.—Freight Charges Calculator:
By ANDREW THOMSON, Freight Agent. 24mo. . . $1.25
TURNER'S (THE) COMPANION:
Containing Instructions in Concentric, Elliptic, and Eccentric Turning; also various Plates of Chucks, Tools, and Instruments; and Directions for using the Eccentric Cutter, Drill, Vertical Cutter, and Circular Rest; with Patterns and Instructions for working them 12mo. $1.00
TURNING: Specimens of Fancy Turning Executed on the Hand or Foot-Lathe:
With Geometric, Oval, and Eccentric Chucks, and Elliptical Cutting Frame. By an Amateur. Illustrated by 30 exquisite Photographs. 4to. $2.50

VAILE.—Galvanized-Iron Cornice-Worker's Manual:
Containing Instructions in Laying out the Different Mitres, and Making Patterns for all kinds of Plain and Circular Work. Also, Tables of Weights, Areas and Circumferences of Circles, and other Matter calculated to Benefit the Trade. By CHARLES A. VAILE. Illustrated by twenty-one plates. 4to. $5.00

VILLE.—On Artificial Manures:
Their Chemical Selection and Scientific Application to Agriculture. A series of Lectures given at the Experimental Farm at Vincennes, during 1867 and 1874-75. By M. GEORGES VILLE. Translated and Edited by WILLIAM CROOKES, F. R. S. Illustrated by thirty-one engravings. 8vo., 450 pages $6.00

VILLE.—The School of Chemical Manures:
Or, Elementary Principles in the Use of Fertilizing Agents. From the French of M. GEO. VILLE, by A. A. FESQUET, Chemist and Engineer. With Illustrations. 12mo. $1.25

VOGDES.—The Architect's and Builder's Pocket-Companion and Price-Book:
Consisting of a Short but Comprehensive Epitome of Decimals, Duodecimals, Geometry and Mensuration; with Tables of United States Measures, Sizes, Weights, Strengths, etc., of Iron, Wood, Stone, Brick, Cement and Concretes, Quantities of Materials in given Sizes and Dimensions of Wood, Brick and Stone; and full and complete Bills of Prices for Carpenter's Work and Painting; also, Rules for Computing and Valuing Brick and Brick Work, Stone Work, Painting, Plastering, with a Vocabulary of Technical Terms, etc. By FRANK W. VOGDES, Architect, Indianapolis, Ind. Enlarged, revised, and corrected. In one volume, 368 pages, full-bound, pocket-book form, gilt edges $2.00
Cloth 1.50

VAN CLEVE.—The English and American Mechanic:
Comprising a Collection of Over Three Thousand Receipts, Rules, and Tables, designed for the Use of every Mechanic and Manufacturer. By B. FRANK VAN CLEVE. Illustrated. 500 pp. 12mo. $2.00

WAHNSCHAFFE.—A Guide to the Scientific Examination of Soils:
Comprising Select Methods of Mechanical and Chemical Analysis and Physical Investigation. Translated from the German of Dr. F. WAHNSCHAFFE. With additions by WILLIAM T. BRANNT. Illustrated by 25 engravings. 12mo. 177 pages . . . $1.50

WALL.—Practical Graining:
With Descriptions of Colors Employed and Tools Used. Illustrated by 47 Colored Plates, Representing the Various Woods Used in Interior Finishing. By WILLIAM E. WALL. 8vo. . $2.50

WALTON.—Coal-Mining Described and Illustrated:
By THOMAS H. WALTON, Mining Engineer. Illustrated by 24 large and elaborate Plates, after Actual Workings and Apparatus. $5.00

WARE.—The Sugar Beet.
Including a History of the Beet Sugar Industry in Europe, Varieties of the Sugar Beet, Examination, Soils, Tillage, Seeds and Sowing, Yield and Cost of Cultivation, Harvesting, Transportation, Conservation, Feeding Qualities of the Beet and of the Pulp, etc. By LEWIS S. WARE, C. E., M. E. Illustrated by ninety engravings. 8vo.
$4.00

WARN.—The Sheet-Metal Worker's Instructor:
For Zinc, Sheet-Iron, Copper, and Tin-Plate Workers, etc. Containing a selection of Geometrical Problems; also, Practical and Simple Rules for Describing the various Patterns required in the different branches of the above Trades. By REUBEN H. WARN, Practical Tin-Plate Worker. To which is added an Appendix, containing Instructions for Boiler-Making, Mensuration of Surfaces and Solids, Rules for Calculating the Weights of different Figures of Iron and Steel, Tables of the Weights of Iron, Steel, etc. Illustrated by thirty-two Plates and thirty-seven Wood Engravings. 8vo. .. $3.00

WARNER.—New Theorems, Tables, and Diagrams, for the Computation of Earth-work:
Designed for the use of Engineers in Preliminary and Final Estimates, of Students in Engineering, and of Contractors and other non-professional Computers. In two parts, with an Appendix. Part I. A Practical Treatise; Part II. A Theoretical Treatise, and the Appendix. Containing Notes to the Rules and Examples of Part I.; Explanations of the Construction of Scales, Tables, and Diagrams, and a Treatise upon Equivalent Square Bases and Equivalent Level Heights. By JOHN WARNER, A. M., Mining and Mechanical Engineer. Illustrated by 14 Plates. 8vo. $4.00

WILSON.—Carpentry and Joinery:
By JOHN WILSON, Lecturer on Building Construction, Carpentry and Joinery, etc., in the Manchester Technical School. Third Edition, with 65 full page plates, in flexible cover, oblong . . .80

WATSON.—A Manual of the Hand-Lathe:
Comprising Concise Directions for Working Metals of all kinds, Ivory, Bone and Precious Woods; Dyeing, Coloring, and French Polishing; Inlaying by Veneers, and various methods practised to produce Elaborate work with Dispatch, and at Small Expense. By EGBERT P. WATSON, Author of "The Modern Practice of American Machinists and Engineers." Illustrated by 78 engravings. $1.50

WATSON.—The Modern Practice of American Machinists and Engineers
Including the Construction, Application, and Use of Drills, Lathe Tools, Cutters for Boring Cylinders, and Hollow-work generally, with the most Economical Speed for the same; the Results verified by Actual Practice at the Lathe, the Vise, and on the Floor. Together

with Workshop Management, Economy of Manufacture, the Steam Engine, Boilers, Gears, Belting, etc., etc. By EGBERT P. WATSON. Illustrated by eighty-six engravings. 12mo. . . . $2.50

WATT.—The Art of Soap Making:
A Practical Hand-Book of the Manufacture of Hard and Soft Soaps, Toilet Soaps, etc. Fifth Edition, Revised, to which is added an Appendix on Modern Candle Making. By ALEXANDER WATT. Ill. 12mo. $3.00

WEATHERLY.—Treatise on the Art of Boiling Sugar, Crystallizing, Lozenge-making, Comfits, Gum Goods,
And other processes for Confectionery, etc., in which are explained, in an easy and familiar manner, the various Methods of Manufacturing every Description of Raw and Refined Sugar Goods, as sold by Confectioners and others. 12mo. $1.50

WILL.—Tables of Qualitative Chemical Analysis:
With an Introductory Chapter on the Course of Analysis. By Professor HEINRICH WILL, of Giessen, Germany. Third American, from the eleventh German edition. Edited by CHARLES F. HIMES, Ph. D., Professor of Natural Science, Dickinson College, Carlisle, Pa. 8vo. $1.50

WILLIAMS.—On Heat and Steam:
Embracing New Views of Vaporization, Condensation and Explosion. By CHARLES WYE WILLIAMS, A. I. C. E. Illustrated. 8vo.
$2.50

WILSON.—First Principles of Political Economy:
With Reference to Statesmanship and the Progress of Civilization. By Professor W. D. WILSON, of the Cornell University. A new and revised edition. 12mo. $1.50

WILSON.—The Practical Tool-Maker and Designer:
A Treatise upon the Designing of Tools and Fixtures for Machine Tools and Metal Working Machinery, Comprising Modern Examples of Machines with Fundamental Designs for Tools for the Actual Production of the work; Together with Special Reference to a Set of Tools for Machining the Various Parts of a Bicycle. Illustrated by 189 engravings. 1898. . . . . . . $2.50

CONTENTS: Introductory. Chapter I. Modern Tool Room and Equipment. II. Files, Their Use and Abuse. III. Steel and Tempering. IV. Making Jigs. V. Milling Machine Fixtures. VI. Tools and Fixtures for Screw Machines VII. Broaching. VIII. Punches and Dies for Cutting and Drop Press. IX. Tools for Hollow-Ware. X. Embossing: Metal, Coin, and Stamped Sheet-Metal Ornaments. XI. Drop Forging. XII. Solid Drawn Shells or Ferrules; Cupping or Cutting, and Drawing; Breaking Down Shells. XIII. Annealing, Pickling and Cleaning. XIV. Tools for Draw Bench. XV. Cutting and Assembling Pieces by Means of Ratchet Dial Plates at One Operation. XVI. The Header. XVII. Tools for Fox Lathe. XVIII Suggestions for a Set of Tools for Machining the Various Parts of a Bicycle. XIX. The Plater's Dynamo. XX. Conclusion— With a Few Random Ideas. Appendix. Index.

WOODS.—Compound Locomotives:
By ARTHUR TANNATT WOODS. Second edition, revised and enlarged by DAVID LEONARD BARNES, A M., C. E. 8vo. 330 pp. $3.00

WOHLER.—A Hand-Book of Mineral Analysis:
By F. WÖHLER, Professor of Chemistry in the University of Göttingen. Edited by HENRY B. NASON, Professor of Chemistry in the Renssalaer Polytechnic Institute, Troy, New York. Illustrated. 12mo. $2.50

WORSSAM.—On Mechanical Saws:
From the Transactions of the Society of Engineers, 1869. By S. W. WORSSAM, JR. Illustrated by eighteen large plates. 8vo. $1.50

RECENT ADDITIONS.

BRANNT.—Varnishes, Lacquers, Printing Inks and Sealing Waxes:
Their Raw Materials and their Manufacture, to which is added the Art of Varnishing and Lacquering, including the Preparation of Putties and of Stains for Wood, Ivory, Bone, Horn, and Leather. By WILLIAM T. BRANNT. Illustrated by 39 Engravings, 338 pages. 12mo. $3.00

BRANNT—The Practical Scourer and Garment Dyer:
Comprising Dry or Chemical Cleaning; the Art of Removing Stains; Fine Washing; Bleaching and Dyeing of Straw Hats, Gloves, and Feathers of all kinds; Dyeing of Worn Clothes of all fabrics, including Mixed Goods, by One Dip; and the Manufacture of Soaps and Fluids for Cleansing Purposes. Edited by WILLIAM T. BRANNT, Editor of "The Techno-Chemical Receipt Book." Illustrated. 203 pages. 12mo. $2.00

BRANNT.—Petroleum.
Its History, Origin, Occurrence, Production, Physical and Chemical Constitution, Technology, Examination and Uses; Together with the Occurrence and Uses of Natural Gas. Edited chiefly from the German of Prof. Hans Hoefer and Dr. Alexander Veith, by WM. T. BRANNT. Illustrated by 3 Plates and 284 Engravings. 743 pp. 8vo. $7.50

BRANNT.—A Practical Treatise on the Manufacture of Vinegar and Acetates, Cider, and Fruit-Wines:
Preservation of Fruits and Vegetables by Canning and Evaporation; Preparation of Fruit-Butters, Jellies, Marmalades, Catchups, Pickles, Mustards, etc. Edited from various sources. By WILLIAM T. BRANNT. Illustrated by 79 Engravings. 479 pp. 8vo. $6.00

BRANNT.—The Metal Worker's Handy-Book of Receipts and Processes:
Being a Collection of Chemical Formulas and Practical Manipulations for the working of all Metals; including the Decoration and Beautifying of Articles Manufactured therefrom, as well as their Preservation. Edited from various sources. By WILLIAM T. BRANNT. Illustrated. 12mo. $2.50

DEITE.—A Practical Treatise on the Manufacture of Perfumery.
Comprising directions for making all kinds of Perfumes, Sachet Powders, Fumigating Materials, Dentifrices, Cosmetics, etc., with a full account of the Volatile Oils, Balsams, Resins, and other Natural and Artificial Perfume-substances, including the Manufacture of Fruit Ethers, and tests of their purity. By Dr. C. DEITE, assisted by L. BORCHERT, F. EICHBAUM, E. KUGLER, H. TOEFFNER, and other experts. From the German, by WM. T. BRANNT. 28 Engravings. 358 pages. 8vo. $3.00

EDWARDS.—American Marine Engineer, Theoretical and Practical:
With Examples of the latest and most approved American Practice. By EMORY EDWARDS. 85 illustrations. 12mo. . . $2.50

EDWARDS.—900 Examination Questions and Answers:
For Engineers and Firemen (Land and Marine) who desire to obtain a United States Government or State License. Pocket-book form, gilt edge $1.50

KIRK.—The Cupola Furnace:
A Practical Treatise on the Construction and Management of Foundry Cupolas. By EDWARD KIRK, Practical Moulder and Melter, author of "The Founding of Metals." Illustrated by 80 Engravings. 8vo. (1899) $3.50

POSSELT.—The Jacquard Machine Analysed and Explained:
With an Appendix on the Preparation of Jacquard Cards, and Practical Hints to Learners of Jacquard Designing. By E. A. POSSELT. With 230 illustrations and numerous diagrams. 127 pp. 4to. $3.00

POSSELT.—The Structure of Fibres, Yarns and Fabrics:
Being a Practical Treatise for the Use of all Persons Employed in the Manufacture of Textile Fabrics, containing a Description of the Growth and Manipulation of Cotton, Wool, Worsted, Silk, Flax, Jute, Ramie, China Grass and Hemp, and Dealing with all Manufacturers' Calculations for Every Class of Material, also Giving Minute Details for the Structure of all kinds of Textile Fabrics, and an Appendix of Arithmetic, specially adapted for Textile Purposes. By E. A. POSSELT. Over 400 Illustrations. quarto. . $5.00

RICH.—Artistic Horse-Shoeing:
A Practical and Scientific Treatise, giving Improved Methods of Shoeing, with Special Directions for Shaping Shoes to Cure Different Diseases of the Foot, and for the Correction of Faulty Action in Trotters. By GEORGE E. RICH. 62 Illustrations. 153 pages 12mo $1.00

RICHARDSON.—Practical Blacksmithing:
A Collection of Articles Contributed at Different Times by Skilled Workmen to the columns of "The Blacksmith and Wheelwright," and Covering nearly the Whole Range of Blacksmithing, from the Simplest Job of Work to some of the Most Complex Forgings. Compiled and Edited by M. T. RICHARDSON.
Vol. I.   210 Illustrations.   224 pages.   12mo.   .   . $1.00
Vol. II. 230 Illustrations.   262 pages.   12mo.   .   . $1.00
Vol. III. 390 Illustrations.   307 pages.   12mo.   .   . $1.00
Vol. IV. 226 Illustrations.   276 pages.   12mo.   .   . $1.00

RICHARDSON.—The Practical Horseshoer:
Being a Collection of Articles on Horseshoeing in all its Branches which have appeared from time to time in the columns of "The Blacksmith and Wheelwright," etc. Compiled and edited by M. T. RICHARDSON. 174 illustrations. $1.00

ROPER.—Instructions and Suggestions for Engineers and Firemen:
By STEPHEN ROPER, Engineer. 18mo. Morocco . $2.00

ROPER.—The Steam Boiler: Its Care and Management:
By STEPHEN ROPER, Engineer. 12mo., tuck, gilt edges. $2.00

ROPER.—The Young Engineer's Own Book:
Containing an Explanation of the Principle and Theories on which the Steam Engine as a Prime Mover is Based. By STEPHEN ROPER, Engineer. 160 illustrations, 363 pages. 18mo., tuck . $2.50

ROSE.—Modern Steam-Engines:
An Elementary Treatise upon the Steam-Engine, written in Plain language; for Use in the Workshop as well as in the Drawing Office. Giving Full Explanations of the Construction of Modern Steam Engines: Including Diagrams showing their Actual operation. Together with Complete but Simple Explanations of the operations of Various Kinds of Valves, Valve Motions, and Link Motions, etc., thereby Enabling the Ordinary Engineer to clearly Understand the Principles Involved in their Construction and Use, and to Plot out their Movements upon the Drawing Board. By JOSHUA ROSE. M. E. Illustrated by 422 engravings. Revised. 358 pp. . . $6.00

ROSE.—Steam Boilers:
A Practical Treatise on Boiler Construction and Examination, for the Use of Practical Boiler Makers, Boiler Users, and Inspectors; and embracing in plain figures all the calculations necessary in Designing or Classifying Steam Boilers. By JOSHUA ROSE, M. E. Illustrated by 73 engravings. 250 pages. 8vo. $2.50

SCHRIBER.—The Complete Carriage and Wagon Painter:
A Concise Compendium of the Art of Painting Carriages, Wagons, and Sleighs, embracing Full Directions in all the Various Branches, including Lettering, Scrolling, Ornamenting, Striping, Varnishing, and Coloring, with numerous Recipes for Mixing Colors. 73 Illustrations. 177 pp. 12mo. $1.00

www.ingramcontent.com/pod-product-compliance
Lightning Source LLC
Chambersburg PA
CBHW030552300426
44111CB00009B/953